ADVANCES IN
ORGAN BIOLOGY

Volume 3 • 1997

RETINOIDS: THEIR PHYSIOLOGICAL FUNCTION AND THERAPEUTIC POTENTIAL

ADVANCES IN ORGAN BIOLOGY

RETINOIDS: THEIR PHYSIOLOGICAL FUNCTION AND THERAPEUTIC POTENTIAL

Series Editor: E. EDWARD BITTAR
Department of Physiology
University of Wisconsin
Madison, Wisconsin

Guest Editor: G.V. SHERBET
Cancer Research Unit
The Medical School
University of Newcastle upon Tyne
Newcastle upon Tyne, England

VOLUME 3 • 1997

JAI PRESS INC

Greenwich, Connecticut *London, England*

55 Old Post Road, No. 2
Greenwich, Connecticut 06836

JAI PRESS LTD.
38 Tavistock Street
Covent Garden
London WC2E 7PB
England

ISBN: 0-7623-0285-2

Manufactured in the United States of America

CONTENTS

LIST OF CONTRIBUTORS vii

PREFACE
G.V. Sherbet ix

RETINIOID STRUCTURE, CHEMISTRY, AND BIOLOGICALLY ACTIVE DERIVATIVES
Robert W. Curley and Michael J. Robarge 1

MOLECULAR MECHANISMS OF RETINOID FUNCTION
Christopher P.F. Redfern 35

RETINOIDS IN MAMMALIAN EMBRYONIC DEVELOPMENT
Gillian M. Morriss-Kay 79

THE ROLE OF RETINOIDS IN PATTERNING FISH, AMPHIBIAN, AND CHICK EMBRYOS
Malcolm Maden and John Pizzey 93

RETINOID AND GROWTH FACTOR SIGNAL TRANSDUCTION
G.V. Sherbet and M.S. Lakshmi 141

RETINOIDS AND APOPTOSIS
Lin-Xin Zhang and Anton M. Jetten 161

RETINOIDS IN TUMOR CELL ADHESION, INVASION, AND METASTASIS
Michael Edward 197

RETINOID RECEPTORS AND CANCER
Joseph A. Fontana and Arun K. Rishi 219

RETINOIDS IN THE MANAGEMENT OF CENTRAL NERVOUS SYSTEM (CNS) TUMORS
M.E. Westarp 231

RETINOIDS AND LUNG CANCER
Andrew M. Arnold and Richard G. Tozer 261

INDEX 289

LIST OF CONTRIBUTORS

Andrew M. Arnold	Hamilton Regional Cancer Centre Department of Medicine McMaster University Hamilton, Ontario, Canada
Robert W. Curley	Division of Medicinal Chemistry and Pharmacognosy College of Pharmacy The Ohio State University Columbus, Ohio
Michael Edward	Department of Dermatology University of Glasgow Glasgow, Scotland
Joseph A. Fontana	Department of Medicine and Cancer Center University of Maryland at Baltimore Baltimore, Maryland
Anton M. Jetten	Laboratory of Pulmonary Pathobiology National Institute of Environmental Health Sciences National Institutes of Health Research Triangle Park, North Carolina
M.S. Lakshmi	Cancer Research Unit University of Newcastle upon Tyne Newcastle upon Tyne, England
Malcolm Maden	Developmental Biology Research Centre Division of Biomedical Sciences King's College London London, England

Gillian M. Morriss-Kay	Department of Human Anatomy University of Oxford Oxford, England
John Pizzey	Developmental Biology Research Centre Division of Biomedical Sciences King's College London London, England
Christopher P.F. Redfern	Medical Molecular Biology Group Department of Medicine University of Newcastle upon Tyne Newcastle upon Tyne, England
Arun K. Rishi	Department of Medicine and Cancer Center University of Maryland at Baltimore Baltimore, Maryland
Michael J. Robarge	Division of Medicinal Chemistry and Pharmacognosy College of Pharmacy The Ohio State University Colombus, Ohio
G.V. Sherbet	Cancer Research Unit University of Newcastle upon Tyne Newcastle upon Tyne, England
Richard G. Tozer	Hamilton Regional Cancer Centre Department of Medicine McMaster University Hamilton, Ontario, Canada
M.E. Westarp	Department of Neurology Ulm University Bad Orb, Germany
Lin-Xin Zhang	Laboratory of Pulmonary Pathobiology National Institute of Environmental Health Sciences National Institutes of Health Research Triangle Park, North Carolina

PREFACE

விரைந்து தொழில் கேட்கும் ஞாலம் நிரந்தினிது
சொல்லுதல் வல்லார்ப் பெறின்

The world will rush to listen
when knowledgeable people convey their thoughts sweetly and accurately

Thiru Valluvar
(Tamil Poet, second century, India)
Thirukkural, Chapter 65, Verse 648

Retinoids, which are vitamin A derivatives, influence a wide variety of biological processes such as cell differentiation, morphogenesis, cell adhesion, and motility. Some years ago I edited a symposium for *Pathobiology* on *Retinoids and Their Biological Effects* (Sherbet, 1992). The response of the scientific community to this short series of articles has been so great that when I was approached by the series editor of *Advances in Organ Biology* with an invitation to edit a volume on the role of retinoids in biology and medicine, I gladly accepted it, especially on account of the wide spectrum of new developments and discoveries in the field which, in my view, had a powerful influence not only on scientific thought relating to the basic functions subserved by retinoids but also on the exciting possibility of the application of retinoids in the control and management of disease processes.

Retinoids have received considerable attention in recent years and due cognizance has been given to their versatility as biological response modifiers, as evidenced by the virtually explosive growth of literature in this field in the past few years. This volume has been designed to give a current state-of-the-art picture of retinoids. The perceived potential of retinoids in the treatment of certain disease states has initiated attempts at identifying and synthesizing new retinoid derivatives with definable and selective effects on aberrant biological phenomena. Appropriately, therefore, we begin with the chemistry of retinoids and their derivatives together with discussions of their biological activity. Major advances have been made in understanding the mechanisms by which retinoids modulate physiological and phenotypic traits of cells. The transduction of retinoid signaling by the mediation of nuclear receptors of the steroid/thyroid receptor superfamily has now been studied extensively and the cloning and defining the characteristics of these receptors has been a focus of discussion in this volume. Retinoids also markedly modulate the transduction of extracellular signals such as those imparted by growth factors and hormones, and thus actively influence and control cellular proliferative patterns. Retinoids can alter epidermal growth factor receptor expression (Kawaguchi et al., 1994), responsiveness to thyroid hormone (Esfandiari et al., 1994; Pallet et al., 1994), inhibit the proliferative responses of hematopoietic progenitor cells to granulocyte colony stimulating factor (Smeland et al., 1994), and modulate the secretion of interleukins by leukaemic cells (Balitrand et al., 1994), among other things. This has obvious implications for pharmacological manipulation of deregulated growth (Dickens and Colletta, 1993; Mulshine et al., 1993). Apoptosis is another component in the regulation of growth control. Apoptotic cell death is influenced by several agents and retinoids may function by interfering with apoptotic pathways of regulation of growth control and quite legitimately, therefore, the importance of this aspect of retinoid function has been duly recognized here.

The panoply of biological processes modulated by retinoids has, over a long period of time, provided the rationale for investigating the potential of retinoids as therapeutic tools. The highly significant role which retinoids play in cell differentiation and morphogenesis and much recent work aimed at understanding the basis of their involvement in these fundamental biological processes has most persuasively emphasized their relevance to cancer development, growth, and progression. We have here two excellent reviews of the field, which not only provide up to date accounts but clearly also knit the processes of neoplasia with those occurring in developmental biology.

Several years ago, it was demonstrated that retinoic acid (RA) inhibited cellular transformation by oncogenes (Datto et al., 1985; Roberts et al., 1985), as well as chemically induced carcinogenesis (Shklar et al., 1980; Hill and Grubs, 1982; Morn, et al., 1983). The most insidious ability of the cancer cell is its ability to invade neighboring tissue, and also gain entry into the vascular system preparatory to achieving distant dissemination leading to the development of metastatic disease. These processes involve a large number of genetic determinants (Sherbet and Lakshmi, 1997), some of which have been found to be targets of RA action. These targets include genes that regulate vertebrate development, hormone and growth factor encoding genes, genes involved in the remodeling of the extracellular matrix, as well as those whose expression is strongly associated with metastatic spread of cancer (Parker and Sherbet, 1992). RA is known to modulate, with some degree of specificity, the expression of some immediate early response genes which code for transcription factors (Simonson, 1994).

The inhibitory effects of retinoic acid on cancer cell invasion are well documented and the biochemical bases of these effects are discussed here in some detail; these discussions provide keen insights into the possible modes of the invasion inhibitory action exerted by retinoic acid. Thus one can envisage retinoids modulating the expression of genes which code for adhesion-mediating plasma membrane proteins and inhibiting the invasive ability of neoplastic cells. A link-up between retinoid function and invasive behavior is also suggested by the demonstration that RA can upregulate the murine *18A2/mts1* gene (Parker et al., 1991) which codes for a Ca^{2+}-binding protein belonging to the S-100 family, and interestingly the 18A2/mts1 protein promotes depolymerization of cytoskeletal structures (Lakshmi et al., 1993). This fits in with the finding that RA produces marked changes in the morphology of human melanocytes, presumably an effect of RA action on cytoskeletal proteins (Ortonne, 1992).

The avowed objective of the scientific community naturally is to see if the accumulated knowledge of retinoid function in biological systems can be harnessed for designing new modalities of treatment of human diseases. Retinoids have been known for a long time to regulate the growth and differentiation of epidermal cells and they have been employed in the treatment of nonmalignant skin conditions (Biro and Shalita, 1993; Orfanos et al., 1997), and much effort has gone into the evaluation of retinoids with minimal topical and systemic toxicity.

In recent years, retinoids have been used in the management of several forms of human cancer, e.g. cancers of head and neck, breast, lung, uterine cervix, prostate, basal cell and squamous cell, carcinomas of the skin and tumors of the central nervous system and hemopoietic malignancies. Several extensive reviews relating to retinoid use in the clinical context are available in the literature (Bollag and Holdener, 1992; Tallman and Wiernik, 1992; Fountzillas, 1994; Warrell, 1994; Pienta, 1997; Trump, 1997). Retinoid therapy has met with qualified success in the treatment of some forms of leukemia and several problems have been encountered which relate to retinoid pharmacology and to side effects of treatment (Parkinson and Smith, 1992). The potential teratogenic effects of retinoids has to be carefully considered before administering the drugs as a mode of cancer management to women of child-bearing age. The incidence of ophthalmological side effects of retinoids has also been described (Mariani et al., 1996). Some clinical trials have made notable progress and on these attention has been focused here. The somewhat emotive word chemoprevention has often been used in the context of human cancer, but in the present state of our knowledge, retinoids may be viewed more appropriately as tools in cancer treatment and patient management rather than as preventive agents. This issue has been expertly addressed in the last chapter. Admittedly the preventive effects of retinoids have been demonstrated using animal tumor models. For example, 9-cis RA can effectively counteract the carcinogenicity of methyl nitrosourea, and some retinoid derivatives are more effective than others (Moon et al., 1992; Anzano et al., 1994; Shealy et al., 1997). Such studies can aid in determining the adverse effects of the retinoids derivatives and can also provide valuable information about the differential accumulation of the drugs upon sytemic administration. The potential of retinoids as chemopreventive agents can be tested in familial cancer such as those carrying germ line mutations of breast and ovarian cancer susceptibility BRCA1 tumor suppressor gene (Easton et al., 1993; Sherbet and Lakshmi, 1997). The BRCA1-encoded protein is known to function as a suppressor of cell proliferation and growth, besides being strongly implicated in cell differentiation and embryonic development. It is needless, however, to emphasize that there is much groundwork to be done before such epidemiological studies can be fruitfully undertaken. Among other options which should be considered in the context of evaluating the chemopreventive action of retinoids could be the prevention of the development of second primary tumors in patients treated for head and neck tumors. It has been reported that 4-7% of patients treated for early stage head and neck cancer may develop second primary tumors, and these could possibly provide a suitable epidemiological area for investigation (Khuri et al., 1997).

This volume has obviously focused much attention on the basic biology of retinoid function and the therapeutic potential of retinoids, and I am confident that it will attract wide readership. I thank my esteemed colleagues for so readily agreeing to contribute to this volume. Its success is essentially the success of the contributors. As the Tamil quotation above states, these scientists, acknowledged experts in their respective fields, have written their reviews in a highly readable and pleasing style, and I am sure that the readership at large would receive their writings with approbation. I would like to thank the series editor of *Advances in Organ Biology* for inviting me to edit this volume and it has been a great privilege to work with him and with the staff of JAI Press Inc. I also thank Dr. M.S. Lakshmi for providing the Kural and its translation. Finally, it would be remiss of me not to acknowledge the unflinching support that I have received over the past several years from the North of England Cancer Research Campaign.

G.V. Sherbet
Guest Editor

REFERENCES

Anzano, M.A., Byers, S.W., Smith, J.M., Peer, C.W., Mullen, L.T., Brown, C.C., Roberts, A.B., & Sporn, M.B. (1994). Prevention of breast cancer in the rat with 9-cis retinoic acid as a single agent in combination with tamoxifen. Cancer Res. 54, 4614-4617.

Balitrand, N., Schlageter, M.H.L., De Gentile, A., Toubert, M.E., Krawice, I., Fenaux, P., Castaigne, S., Najean,Y., Degos, L., & Chomienne, C. (1994). Modulation of IL-8, IL-1β, and G-CSF by all-*trans* retinoic acid in acute promyelocytic leukaemia. Leukemia 8, 1750-1757.

Biro, D.E., & Shalita, A.R. (1993). Clinical aspects of topical retinoids. Skin Pharmacol. 6, 53-60.

Bollag, W., & Holdener, E.E. (1992). Retinoids in cancer prevention and therapy. Annals Oncol. 3, 512-526.

Datto, G.P., Parada, L.F., & Weinberg, R.A. (1985). Specific growth response of ras transformed embryo fibroblasts to tumour promoters. Nature 318, 472-475.

Dickens, T.A., & Colletta, A.A. (1993). The pharmacological manipulation of members of the transforming growth factor beta family in the chemo-prevention of breast cancer. BioEssays 15, 71-74.

Easton, D.F., Bishop, D.T., Ford, D., & Crockford, G.P. (1993). Genetic linkage analysis in familial breast and ovarian cancer. Results from 214 families. Amer. J. Hum. Genet. 52, 678-701.

Esfandiari, A., Gagelin, C., Gavaret, J.M., Pavlka, S., Lennon, A.M., Pierre, M., & Courtin. F. (1994). Induction of type III de-iodinase activity in astroglial cells by retinoids. Glia 11, 255-261.

Fountzillas, G. (1994). Retinoids in the management of head and neck cancer. An update. J. Chemotherapy 6, 127-138.

Hill, D.L., & Grubbs, C.J. (1982). Retinoids as chemopreventive anticancer agents. Anticaner Res. 2, 111-124.

Kawaguchi, M., Kawashima, F., Ohshima, K., Kawaguchi, S., & Wada, H. (1994). Retinoic acid enhances the number of epidermal growth factor receptors in rat glomerular epithelial cells *in vitro*. Nephron 68, 97-103

Khuri, F.R., Lippman, S.M., Spitz, M.R., Lotan, R., & Hong, W.K. (1997). Molecular epidemiology and retinoid chemoprevention of head and neck cancer. J. Natl. Cancer Inst. 89, 199-211.

Lakshmi, M.S., Parker, C., & Sherbet, G.V. (1993). Metastasis associated *mts1* and *nm23* genes affect tubulin polymerisation in B16 melanomas. A possible mechanism of their regulation of metastatic behaviour of tumours. Anticancer Res. 13, 299-304.

Mariani, L., Formelli, F., DePalo, G., Manzari, A., Camerini, T., & Campa, T. (1996). Chemoprevention of breast cancer with fenretinide (4-HPR): Study of long term visual and ophthalmologic tolerability. Tumori 82, 444-449.

Moon, R.C., McCormick, D.L., & Mehta, R.G. (1983). Inhibition of carcinogenesis by retinoids. Cancer Res 43 (suppl), 2469-2475.

Moon, R.C., Mehta, R.B., & Detrisac, C.J. (1992). Retinoids as chemopreventive agents for breast cancer. Cancer Detection Prevention 16, 73-79.

Mulshine, J.L., Jett, M., Cuttitta, F., Treston, A.M., Quinn, K., Scott, K., Scott, F., Iwai, N., Avis, I., Linnoila, R.I., & Shaw, G.L. (1993). Scientific basis for cancer prevention. Intermediate cancer markers. Cancer 72, 978-983.

Orfanos, C.E., Zouboulis, C.C., Almond-Rosler, B., & Geilen, C.C. (1997). Current use and future potential role of retinoids in dermatology. Drugs 53, 358-388.

Ortonne, J.P. (1992). Retinoic acid and pigment cells. A review of *in vitro* and *in vivo* studies. Br. J. Dermatol. 127, 43-47.

Pallet, V., Audouin-Chevallier, I., Verret, C., Garcin, H., & Higuret, P. (1994). Retinoic acid differentially modulates triiodothyronine and retinoic acid receptors in rat liver according to thyroid status. Eur. J. Endocrinol. 131, 377-384.

Parker, C., Whittaker, P.A., Weeks, R.J., Thody, A.J., & Sherbet, G.V. (1991). Modulation of metastatic behaviour alters the expression of metastasis associated genes *mts1* and *nm23* in metastatic variants of the B16 murine melanoma. Clin. Biotechnol. 3, 217-222.

Parker, C., & Sherbet, G.V. (1992). Modulation of gene function by retinoic acid. Pathobiology 60, 278-283.

Parkinson, D.R., & Smith, M.A. (1992). Retinoid therapy for acute promyelocytic leukaemia. A coming of age for the differentiative therapy of malignancy. Annals Int. Med. 117, 338-340.

Pienta, J.K., Esper, P.S., Zwas, F., Krzeminski, R., & Flaherty, L.E. (1997). Phase II chemoprevention trial of oral fenretinide in patients at risk for adenocarcinoma of the prostate. Am. J. Clin. Oncol. 20, 36-39.

Roberts, A.B., Roche, N.S., & Sporn, M.B. (1985). Selective inhibition of the anchorage-independent growth of *myc*-transfected fibroblasts by retinoic acid. Nature 315, 237-239.

Shealy, Y.F., Frye, J.L.K., Riordn, J.M., Hill, D.L., McPhillips, M., Wille, J.J., Sani, B.P., Kalin, J.R., Eto, I., & Grubbs, C.J. (1997). Retinyl ethers as cancer chemopreventive agents. Suppression of mammary cancer. Anticancer Drug Design 12, 15-33.

Sherbet, G.V. (Ed.) (1992). Symposium on retinoids and their biological effects. Pathobiology 60, 253-283.

Sherbet, G.V., & Lakshmi, M.S. (1997). *The Genetics of Cancer. Genes Associated with Cancer Invasion, Metastasis and Cell Proliferation.* Academic Press, London & New York.

Shklar, G., Schwartz, J., Brau, D., Trickler, D., & Wallace, K.D. (1980). Inhibition of hamster cheek pouch carcinogenesis by 13-*cis*-retinoic acid. Oral Surg 50, 45-53.

Simonson, M.S. (1994). Anti-AP1 activity of all-trans-retinoic acid in glomerular mesangial cells. Amer. J. Physiol. 36, F805-F815.

Smeland, E.B., Rusten, L., Jacobsen, S.E.W., Skrede, B., Blomhoff, R., Wang, M.Y., Funderud, S., Kvalheim, G., & Blomhoff, H.K. (1994). All-*trans* retinoic acid directly inhibits granulocyte colony stimulating factor-induced proliferation of CD34(+) human haematopoietic progenitor cells. Blood 84, 2940-2945.

Tallman, M.S., & Wiernik, P.H. (1992). Retinoids in cancer treatment. J. Clin. Pharmacol. 32, 868-888.

Trump, D.L., Smith, D.C., Stiff, D., Adedoyin, A., Day, R., Bahnson, R.R., Hofacker, J., & Branch, R.A. (1997). A phase II trial of all-trans-retinoic acid in hormone refractory prostate cancer: A clinical trial with detailed pharmacokinetic analysis. Cancer Chemother. Pharmacol. 39, 349-356.

Warrell, R.P. (1994). Applications for retinoids in cancer therapy. Introduction. Sem. Haematol. 31, 1-13.

RETINOID STRUCTURE, CHEMISTRY, AND BIOLOGICALLY ACTIVE DERIVATIVES

Robert W. Curley and Michael J. Robarge

Abstract 1
I. Introduction 2
A. Retinoic Acid, Its Structure and Formation 2
B. Overview of Retinoic Acid Metabolism 5
II. Actions of Retinoic Acid Metabolites 6
A. Isomerization 8
B. Oxidation 9
C. Conjugation 12
III. Retinoic Acid Analogues 13
A. Modifications of Retinoic Acid 13
B. Synthetic Retinoids 21
References 26

ABSTRACT

After a description of the structure and important physical properties of retinoic acid is completed, a brief overview of the metabolic processes leading to the production

Advances in Organ Biology
Volume 3, pages 1-34.

ISBN: 0-7623-0285-2

of retinoic acid *in vivo* is provided. A reasonably detailed discussion of the quantitatively and/or biologically important metabolites of retinoic acid then follows. The chapter is concluded with a description of efforts in the development of synthetic retinoids from earlier developments of simpler retinoic acid analogues to more recent efforts on the development of more heavily structurally modified analogues, particularly those with retinoid receptor or receptor subtype specificity. Because of the expanding efforts in this area, emphasis is focussed on those molecules that appear to be the most important or successful.

I. INTRODUCTION

A. Retinoic Acid, Its Structure and Formation

Retinoic acid (**1**) is a polyene carboxylic acid derived from metabolic oxidation of vitamin A (retinol). Systematically, this yellow crystalline material is named as 3,7-dimethyl-9-(2,6,6-trimethyl-1-cyclohexen-1-yl)-2,4,6,8-nonatetraenoic acid. The more common numbering scheme for this molecule is shown in Figure 1. The resemblance of this compound to a long-chain polyunsaturated fatty acid results in its limited aqueous solubility while the presence of the conjugated network of double bonds confers a long wavelength UV absorption maxima of about 350 nm and a high molar extinction coefficient (ε~40,000) when dissolved in most organic solvents. Also, because of this conjugated polyene component of its structure, retinoic acid is relatively susceptible to acid-catalyzed, thermal, and light-induced isomerization, which can have profound effects on its biological activity and site of action.

Physiologically, retinoic acid (RA) is produced via irreversible oxidation of retinal (**2**), the visual pigment chromophore which is itself biosynthesized by reversible oxidation of retinol (**3**) (Blaner and Olson, 1994). With the exception of its visual function as well as in supporting some aspects of reproduction in mammals, it is generally believed that RA is the active form of retinol in controlling most of the other functions of vitamin A. It now appears that many of these functions result from regulation of transcription by noncovalent binding of RA to a closely related family of nuclear retinoic acid receptors (RARs) (Giguere, 1994; Mangelsdorf et al., 1994).

The main sources of vitamin A are dietary and are derived from provitamin A carotenoids from vegetables or retinyl esters from animal sources or dietary supplements. A number of the carotenoids in nature can be converted in the intestinal mucosa to vitamin A. Symmetric, central cleavage of carotenoids such as β-carotene by a β-carotenoid-15,15′-dioxygenase yields two molecules of retinal while other carotenoids suffer eccentric cleavage to eventually generate retinol (Olson, 1989; Ganguly, 1989). It now has been shown that a cellular retinol-binding protein type II (CRBP-II) binds this retinal and permits its ready reduction to retinol

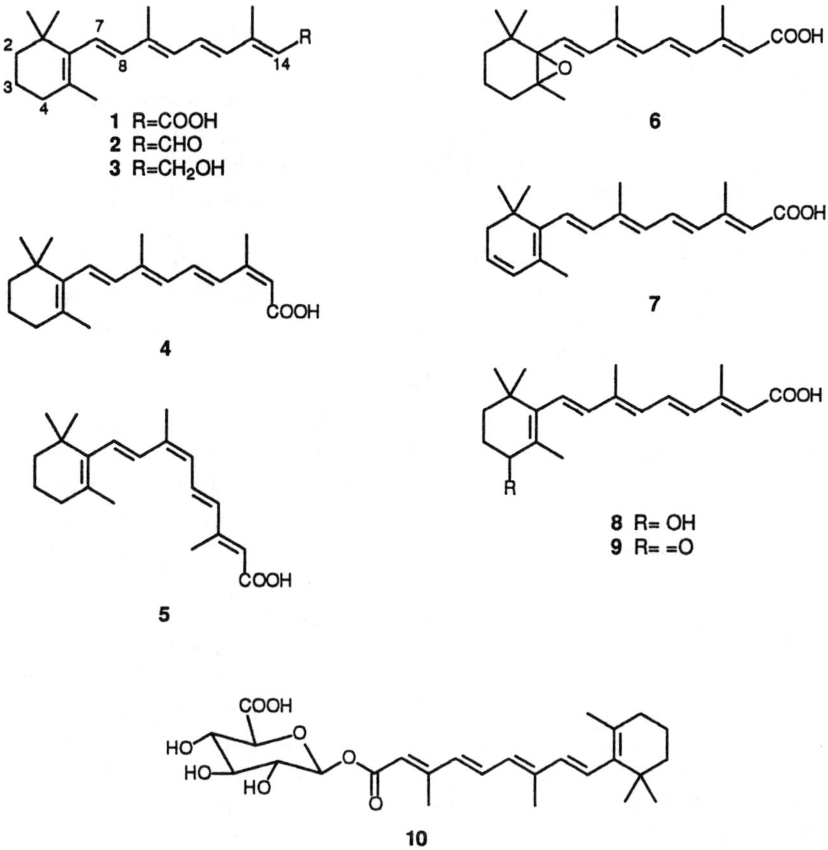

Figure 1. Structures of retinoids.

by a mucosal retinal reductase (Kakkad and Ong, 1988). Retinyl esters, on the other hand, are hydrolyzed by lipases or esterases in the intestinal lumen prior to absorption. When present at physiological concentrations, the resulting free retinol is absorbed into the enterocytes by facilitated diffusion, which is a receptor or carrier protein-mediated process, or by passive diffusion when at higher, pharmacological levels. Regardless of the source of retinol appearing in the enterocytes, the vitamin is reesterified by lecithin:retinol acyltransferase or acyl-CoA:retinol acyltransferase to form primarily retinyl palmitate and stearate. These resulting hydrophobic esters are subsequently packaged in chylomicrons and secreted into the lymphatic system. While in this system, the chylomicrons are converted to chylomicron remnants containing the retinyl esters by the actions of lipoprotein lipase (Blomhoff et al., 1991; Norum and Blomhoff, 1992). While the primary mode of removal of these remnants from the circulation is via uptake by specialized liver cells, recent studies suggest some dietary vitamin A may be removed via chylomicron uptake by the bone marrow (Hussain et al., 1989).

In liver, the primary site for uptake, two cell types play important roles in retinoid storage and metabolism. Initially, retinyl esters are taken up by the parenchymal cells (hepatocytes) where a retinyl ester hydrolase cleaves the esters to free retinol. The resulting retinol is bound to the retinol-binding protein (RBP), which serves as the carrier protein for the transport of the insoluble ligand in plasma. The 1:1 RBP-retinol complex is secreted from the liver into the plasma when the vitamin A status of the individual requires it (Soprano and Blaner, 1994). This complex circulates in the plasma bound to a second protein, the thyroid hormone binding protein transthyretin, formerly known as prealbumin.

When vitamin A status is sufficient, most of the retinol is transferred to the second specialized cell, the stellate or fat-storing cells, perhaps while bound to RBP. Here, as well as to some extent in the parenchymal cells, retinol is reesterified as primarily palmitate and stearate esters by both lecithin:retinol acyltransferase and acyl-CoA:retinol acyltransferase. As is the case for CRBP-II bound retinol in the intestine, this reesterification of retinol in the liver uses CRBP-bound retinol as substrate (Ong et al., 1988).

Retinol bound to RBP in the plasma is transported into many peripheral tissues, particularly those that are vitamin A-dependent target tissues. It is not clear whether there is a single general mechanism for the uptake of retinol from RBP by tissues, although the presence of an RBP receptor on the cell surface has been an attractive hypothesis. However, only in the case of the retinal pigmented epithelium has any substantial evidence for an RBP receptor been gathered (Bok and Heller, 1976; Bavik et al., 1991). Research in other tissues has led to the suggestion that uptake of retinol into cells is driven by the intracellular apo-CRBP concentration (Noy and Blaner, 1991), although more research is needed in this area to clarify the mechanism(s) involved in uptake of retinol by target tissues.

While RA appears to be the important metabolite of retinol in maintaining most vitamin A functions, the biosynthesis in, and delivery to target tissues remains relatively poorly understood. While a trace of RA may be derived from dietary sources, it is not clear whether this is sufficient to account for the serum RA that circulates bound to serum albumin at about 0.5 percent the concentration of plasma retinol levels (De Leenheer et al., 1982). In tissues, it is likely that RA is produced by oxidation of retinol, although which enzyme system(s) is involved remains unclear. In some experiments, forms of aldehyde dehydrogenase have been shown capable of oxidizing retinol to RA (Frolik, 1984). Doubts about whether this relatively nonspecific family of enzymes can regulate the small quantities of RA required led to the discovery that RA can be produced in porcine kidney cells in the presence of inhibitors of alcohol and aldehyde metabolism, suggesting a more specific machinery for oxidation exists (Napoli, 1986). Many researchers have now demonstrated that CRBP-bound retinol is the preferred substrate for oxidation to retinal by an $NADP^+$-dependent dehydrogenase, while an NAD^+-requiring dehydrogenase may be used to produce RA from retinol (Connor and Smit, 1987). The details of this transformation(s) needs further study but no doubt utilizes a variety of enzymes that may differ from tissue to tissue.

Recently, it has been demonstrated that in a number of tissues, β-carotene may be converted directly to RA without the presence of any detectable retinal intermediate (Napoli and Race, 1988). While the mechanistic details of this putative pathway to RA remain to be elucidated, the fact that human tissues can accumulate carotenoids raises the possibility that this mechanism may contribute in part to the concentration of intracellular RA.

B. Overview of Retinoic Acid Metabolism

Retinoic acid has long been known to have a relatively short biological half-life. One of the primary reasons for this is the ready susceptibility of the parent molecule to undergo metabolism to products that, for the most part, appear to be catabolites with reduced biological activity. While details of the biochemistry, synthetic chemistry, and bioactivity of important metabolites will be presented in the next section of this chapter, a brief overview of important features will be presented here.

Early studies of urinary and fecal metabolites from supraphysiological doses of RA led to the identification of a number of structures; many of which resulted from both extensive changes in the polyene side chain as well as oxidations of the trimethylcyclohexenyl ring (Rietz et al., 1974; Hänni et al., 1976). Few of these metabolites showed any useful biological activity and efforts shifted to study of the more prominent metabolites found at physiological concentrations.

As mentioned earlier, retinoid isomerization is a facile process chemically and physically; this is also the case metabolically. While isomerization of the 11,12-double bond of retinal provides the important visual pigment chromophore 11-*cis* retinal, isomerization of this bond does not appear to be an important pathway for RA metabolism. However, despite the fact that no specific retinoid isomerases have been identified, conversion of RA to its 13-*cis* (**4**) and 9-*cis* (**5**) isomers clearly has significant impact in terms of biological activity.

One of the prominent general pathways for RA metabolism are oxidation processes, many of which appear to be mediated by cytochrome P_{450} monooxygenases (Roberts et al., 1979; Leo et al., 1984). In addition to epoxidation of the tetrasubstituted ring double bond to give 5,6-epoxyretinoic acid (6) and dehydrogenation to give 3,4-didehydroretinoic acid (**7**), cytochrome P_{450}-mediated allylic oxidations to give 4-hydroxy-(**8**) and 4-oxoretinoic acid (**9**) are important reactions. That these may be inactivation processes is suggested by the ability of imidazole antifungal agents such as liarozole to enhance the antiproliferative actions of retinoic acid by inhibiting a retinoic acid 4-hydroxylase (Wouters et al., 1992).

In addition, conjugation reactions have been observed for RA. One of the earliest observed products was the O-acyl glucuronide of RA, retinoyl-β-glucuronide (**10**). This may be a biologically active metabolite and is a prominent component of retinoid metabolites in the bile (Lippel and Olson, 1968; Zile et al., 1987). A taurine conjugate of RA has also been discovered after extensive metabolic "degradation" of the parent structure (Skare et al., 1982a). Finally, the existence of a coenzyme A ester of RA has been hypothesized and synthesized and this retinoyl-CoA may be the activated form of RA involved in the recently observed retinoylation of proteins (Kutner et al., 1986; Renstrom and DeLuca, 1989; Takahashi and Breitman, 1989a).

Finally, as might be expected, a number of known retinoid metabolites are the result of multiple processes occurring to the parent molecule, some of which may generate biologically significant metabolites.

Much as for retinol, there are small specific cytosolic binding proteins for RA: the cellular retinoic acid-binding proteins (CRABP) types I and II. While less clear in the case of RA, it has been suggested that it is CRABP-bound RA that is an important substrate for metabolism, particularly oxidative processes (Fiorella and Napoli, 1991). Alternatively, or additionally, it has been suggested that binding to these proteins serves to sequester RA for: (1) protection of ligand from certain forms of metabolism, (2) protection of the cell from toxic concentrations of ligand, and (3) delivery of ligand to nuclear binding sites (Blaner and Olson, 1994). Much work remains to establish with certainty the existence and importance of these suggestions.

II. ACTIONS OF RETINOIC ACID METABOLITES

Biotransformation often results in inactivation of a substance by oxidative or conjugation processes. In some cases, the biotransformed substance retains its original function. Alternatively, the transformation may result in a substance with altered biological function. Vitamin A (retinol) may have little biological function on its own. However, biotransformation of Vitamin A results in formation of a host of active forms, which control or regulate many biological functions, as well as inactive forms, which are normally excreted. As mentioned earlier, the main sources of retinoids are dietary and are derived from provitamin A carotenoids or retinyl esters. These precursors are metabolized into Vitamin A (Scheme 1).

Discussion of retinoid metabolism will concentrate on the major metabolites of RA, which is a critical metabolite of retinol, since it is generally believed RA is the active form controlling most of the functions of Vitamin A. The exception here being 11-*cis* retinal, which serves as the visual pigment chromophore. For simplicity, RA metabolism can be divided into three major areas: (1) isomerization, (2) oxidation, and (3) conjugation. An overview of RA metabolism is shown in Scheme 2.

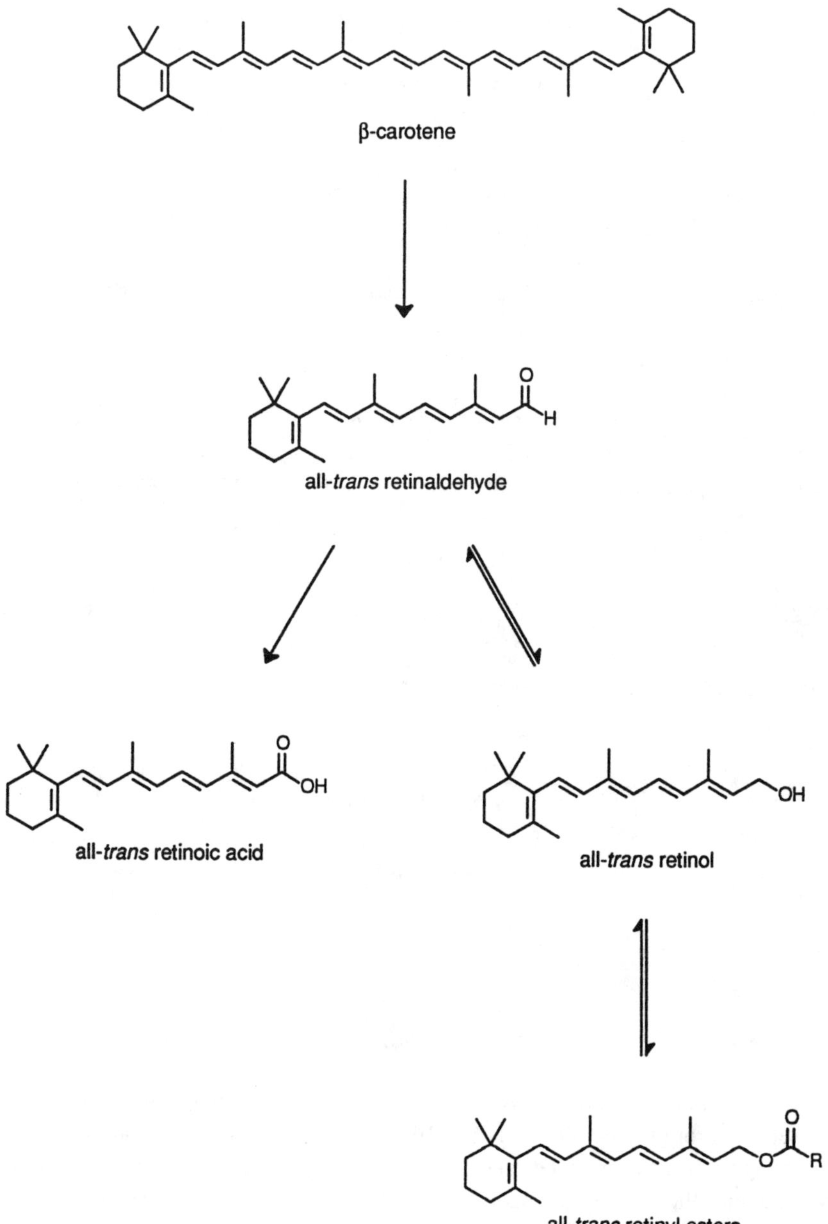

Scheme 1. Metabolism of β-Carotene.

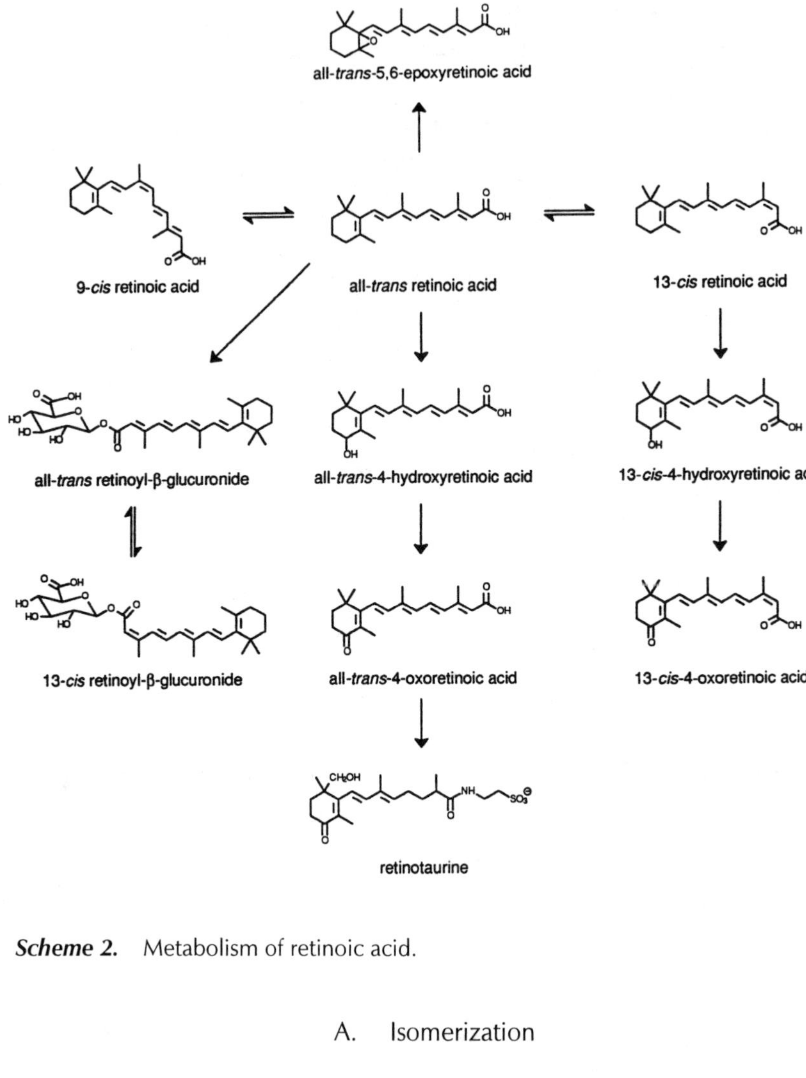

Scheme 2. Metabolism of retinoic acid.

A. Isomerization

The isolation of 13-*cis*-RA (**4**) from tissue extracts of rats given all-*trans*-RA was reported for the first time in 1967 (Zile et al.,1967). It was later found that the isolation procedures employed caused extensive isomerization of RA to 13-*cis*-RA, consequently the validity of this observation was in question. Other groups had similar problems. The presence of 13-*cis*-RA in rats was not confirmed until the use of high performance liquid chromatography (Zile et al., 1982). Since these initial uncertainties, the 13-*cis*-RA concentration has been assessed in human serum (Eckhoff et al., 1991; Tang and Russell, 1991). It is not known if this isomerization is enzymatically controlled as it may be with 11-*cis* retinaldehyde (Canada et al., 1990; Rando, 1990; Rando et al., 1991). The importance of isomerization of RA to 13- *cis*-RA is not yet known. 13-*cis*-RA, which is sometimes called isotretinoin, is a naturally occurring metabolite that is equivalent in biological activity to RA, both *in vivo* (Zile and DeLuca, 1968) and *in vitro* (Newton et al., 1980). The fasting serum levels of all-*trans*- and 13-*cis*-RA were determined in 26 volunteers and ranged from 3.7 to 6.3 nmol/L and from 3.7-7.2 nmol/L, respectively (Tang and Russell, 1990). Clinically, isotretinoin is currently used for the treatment of severe cystic acne under the trade name Accutane®.

As mentioned earlier, 11-*cis* retinal is the only retinoid shown conclusively to have biological activity at an oxidation state lower than RA. Early extraction studies showed many forms of retinaldehydes (then called retinene) from the retina (Wald, 1934-1935; 1935-1936). The photocycle was then shown to involve isomerization of 11-*cis* retinal to the all-*trans* form (Hubbard and Wald, 1952; Yoshizawa and Wald, 1963; Wald, 1968). Formation of 11-*cis*-retinal is thought to occur from the metabolism of all-*trans* retinyl esters, which are the storage form of vitamin A. It is believed that the energy released from all-*trans* retinyl ester hydrolysis is used to drive the isomerization to 11-*cis* retinol in one concerted reaction step perhaps catalyzed by an enzyme termed an isomerohydrolase (Canada et al., 1990). Oxidation of 11-*cis*-retinol produces the visual pigment chromophore 11-*cis* retinal.

Recent work has revealed a further very important isomeric metabolite of RA. The discovery of 9-*cis*-RA (**5**) RA has encouraged more research in the identification of bioactive retinoids. The existence of 9-*cis*-RA was first reported by Levin and colleagues (1992) and by Heyman and coworkers (1992). These authors also demonstrated that 9-*cis*-RA is a ligand for RXRα. The RXR receptors are a relatively new receptor family closely related to the nuclear RAR receptors (Giguere, 1994; Mangelsdorf et al., 1994). Further work by Mangelsdorf and colleagues (1992) showed 9-*cis*-RA to be 40-times more potent at activating the RXR nuclear receptor family (α,β, and γ) in rats than any other retinoid known including RA. The metabolic pathway for the generation of 9-*cis*-RA is not known. However, recent synthetic efforts toward the stereoselective synthesis of 9-*cis*-RA for biological studies have been successful (Boehm et al, 1994).

Other isomers have been found, including isomers of further oxidized metabolites. Many of these metabolites are inactive and the importance of the role they play in RA activity/degradation is yet to be determined.

B. Oxidation

Further oxidation of RA often results in inactivation and excretion of the more polar metabolite. In 1978, McCormick and colleagues isolated 5,6-epoxyretinoic acid (**6**) from the intestinal mucosa of vitamin A deficient rats given [^{3}H]-retinoic acid. Epoxide **6** was present in significant concentrations in the liver, small intestine, and intestinal contents, but not in the kidneys of vitamin A-deficient rats (Napoli et al., 1982). Vitamin A-deficient rats given physiologic doses of [^{3}H]-retinol also synthesized **6** in the kidneys (McCormick and Napoli, 1982). It was concluded that **6** is an endogenous metabolite of retinol in the rat kidney. It is believed that this epoxidation is not necessary for RA function. When *in vivo* epoxidation in vitamin A-deficient rats is blocked by N,N′-diphenyl-*p*-phenylenediamine, a free radical scavenger, RA is still able to promote growth and differentiation of vaginal epithelium (Sietsema and DeLuca, 1982). In addition **6** was shown to be only 0.5% as active as RA in promoting growth in vitamin A deficient rats (Zile et al., 1980).

Biotransformation of RA to 4-hydroxy (**8**) and subsequently to 4-oxo-RA (**9**) has been demonstrated both *in vitro* and *in vivo* in rats as well as hamsters (Hänni and Bigler, 1977; Frolik et al., 1978; Frolik et al., 1979; Skare et al., 1982a). The chemical synthesis of **9** helped in the identification of this metabolite (Barua and Ghosh, 1972). Roberts and coworkers (Roberts et al., 1980) studied the formation of **8** and **9** from RA in hamster liver microsomes. The first step involves oxidation of RA at the C-4 position of the dimethylcyclohexenyl ring by an enzyme whose properties are consistent with those of a cytochrome P_{450} monoxygenase, in that it required NADPH and molecular oxygen, and is inhibited by carbon monoxide. The second step, oxidation of **8** to **9**, involves oxidation by an enzyme whose properties are consistent with a dehydrogenase because it required NAD^+ but did not require oxygen. Leo and colleagues (Leo et al., 1984; 1989) and Roberts and coworkers (Roberts et al., 1992) further showed that cytochrome P_{450} isoforms in rat and in human liver preparations induce the conversion of RA to its 4-hydroxy form. Further metabolism of 4-oxo-RAs to more polar metabolites seems to be dependent on the cytochrome P_{450} system (Roberts et al., 1979; Roberts et al., 1980; Frolik, 1984).

The biological activity of **8** and **9** is one tenth that of retinoic acid in causing epithelial differentiation (Frolik et al., 1979) and in promoting growth in rats (Surekha Rao et al., 1972). Several urinary, biliary, and fecal metabolites have been identified that contain the 4-oxo moiety as a common feature (Hänni et al., 1976; Hänni and Bigler, 1977; Skare et al., 1982a). Consequently, oxidation of the C-4 position of RA is probably an early step in a series of chemical transformations that lead to the deactivation and excretion of RA from the body. Even though **9** seems to have no useful biological function, it still retains its teratogenic effects (Kraft et al., 1989).

The last oxidative metabolite of RA to be mentioned is 3,4-didehydroretinoic acid (**7**). The isolation, formation, and properties of **7** were reported by Thaller and Eichele (1990). It was shown that **7** was able to induce digit pattern duplication, like RA, in the developing chick limb bud. The metabolite was found to be derived from retinol *in situ* through a 3,4-didehydroretinol intermediate. Retinoic acid and **7** are the only known endogenous retinoids that have morphogenetic properties in the chick wing bud.

Although not an oxidative metabolite of RA, it is worth mentioning an oxidative metabolite of retinol that has lead to the belief that there exists another signal transduction pathway for retinoids other than the RA pathway. Buck and coworkers (1991a) demonstrated that B lymphocytes formed several metabolites of retinol that did not include RA or **7**. Some of these metabolites were able to sustain growth of B-cells in the absence of external vitamin A. One of these metabolites was later identified as optically active 14-hydroxy-4,14-*retro*-retinol (**11**) (Buck et al., 1992b) (see Figure 2). It was shown that **11**, unlike RA, was able to sustain cell growth in lymphoblastoid line 5/2.

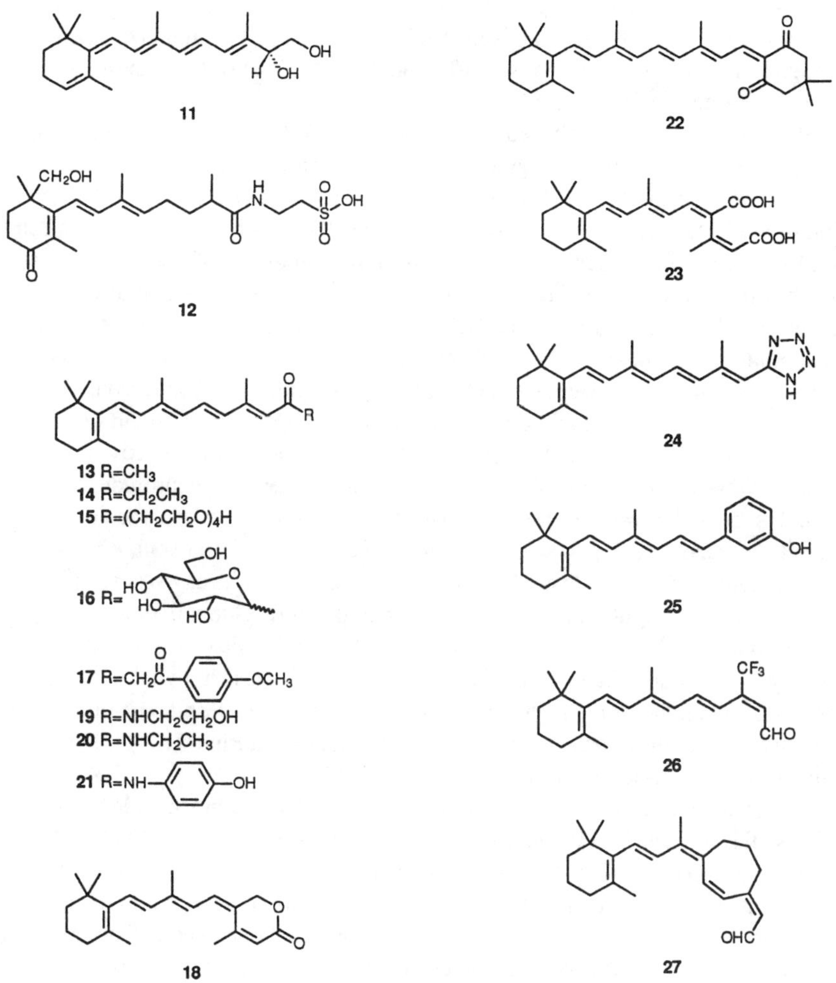

Figure 2. Structures of retinoids.

C. Conjugation

Conjugation is an enzymatic reaction whereby a functional group is masked by the addition of a new group. In the case of RA, the terminal carboxylic acid moiety is the group that is often masked. The most prevalent conjugate of RA is retinoyl-β-glucuronide (**10**). It was the first metabolite of RA to be identified (Dunagin et al., 1965, 1966; Zachman et. al., 1966). Retinoyl-β-glucuronide is secreted into the bile when RA is given orally to rats (Dunagin et al., 1966; Swanson et al., 1981; Skare and DeLuca, 1983; Frolik, 1984) and represents up to 48 percent of the metabolites present 2 hours post dosing. Retinoyl-β-glucuronide has also been detected in the urine (Nath and Olson, 1967) and in the intestinal mucosa (Zile et. al., 1982). Plasma concentrations of **10** range from 5-17 nmol/L in fasting human subjects (Barua and Olson, 1986).

Biosynthesis of **10** is believed to take place in the liver, intestine, kidney, and other tissues by a microsomal glucuronosyl transferase (Lippel and Olson,1968; Miller and DeLuca, 1986). The process involves enzymatic transfer of uridine diphosphoglucuronic acid to RA. It is unclear whether this metabolite plays a role in RA activity or if it is merely acting as a detoxifying product that becomes excreted. Retinoyl-β-glucuronide has been shown to be as active as RA in promoting growth of retinoid deficient rats (Nath and Olson, 1967) and as an inducer of cellular differentiation both *in vitro* (Zile et al., 1987; Gallup et al., 1987; Janick-Buckner et al.,1991) and *in vivo* (Sietsema and DeLuca, 1982b). Retinoyl-β-glucuronide does not bind to the cellular retinoid binding proteins or to the nuclear retinoid receptors (Mehta et al., 1992; Sani et al., 1992) which leaves doubt as to the active form of this somewhat unstable metabolite. It is not known whether the glucuronide is acting through some unknown mechanism or whether it is becoming hydrolyzed to RA to elicit its apparent activity. The low toxicity of **10** with respect to skin, embryonic development, and cells in tissue culture as compared to RA may reside to some extent in its water solubility (Janick-Buckner et al., 1991; Gunning et al., 1993).

Retinotaurine (**12**) is another conjugated metabolite that was isolated and identified in rat bile (Skare et al., 1982a). The biosynthesis is not well understood but it is known to be formed from conjugation with taurine. Despite the biological activity found for the glucuronide conjugate of RA, retinotaurine is not biologically active in vitamin A-deficient rats (Skare et al., 1982b).

Retinoylation or conjugation with proteins is another mode in which RA may act on cells (Takahashi and Breitman, 1989a; 1989b; 1991; Takahashi et al., 1991a; 1991b; Pipkin et al.,1991a; 1991b). The metabolic intermediate governing the retinoyllation has been suggested to be retinoyl-CoA which can transfer the retinoyl moiety to proteins (Renstrom and DeLuca, 1989). Hypothetically, other metabolic intermediates may be involved including retinoyl-β-glucuronide. Whether retinoyl-CoA is a true biosynthetic intermediate is not clear. Also, the amino acid residues that are altered and the functional effects of retinoylation on proteins warrant further attention to determine what role retinoylation plays in the biological actions of RA.

Finally, the metabolites discussed above are a small fraction of the total metabolites discovered from RA metabolism. The significant metabolites in terms of function and of quantity have been discussed. There are many other metabolites of RA that have been discovered but not identified. A more comprehensive discussion of metabolism and other aspects of retinoid biochemistry can be found in the reviews by Sporn and colleagues (Sporn et al., 1984, 1994) and Ganguly (1989).

III. RETINOIC ACID ANALOGUES

The discovery of the nuclear retinoic acid receptor (RARs) proteins and the relatively selective distribution of their subtypes in different tissues and cells (Giguere, 1994; Mangelsdorf et al., 1994) has contributed to a significant increase in efforts to synthesize new retinoid analogues. These efforts are now based on the hope that disease or tissue specific retinoids may be developed and/or that retinoid toxicities can be selectively ameliorated. Because of this substantial synthetic activity and for the sake of brevity, we will primarily restrict our considerations to the lines of investigation that appear to be leading to the most important advances. Consideration will be divided somewhat arbitrarily into discussion of analogues that are relatively simple modifications of the RA skeleton and those which represent a greater structural departure from the parent molecule. A more extensive compilation of the range of both active and inactive retinoid structures being generated has recently been published (Dawson and Hobbs, 1994).

A. Modifications of Retinoic Acid

The parent structure of RA can be divided into three major areas for the purpose of discussion of its structural modifications: (1) the polar, ionizable terminus; (2) the polyene chain; and, (3) the hydrophobic ring end.

Esters of the carboxyl terminus have been prepared using many different alcohols. While many of these esters can show biological activity, it is generally believed that these compounds behave as prodrugs that must be hydrolyzed back to the parent carboxylate to be effective. In addition to simple methyl (**13**) and ethyl (**14**) esters, more hydrophilic esters such as the polyoxyethyl ester **15** have been prepared (Gross, 1989). The hydrophilic glucosyl ester **16** has also been prepared (Barua and Olson, 1991) during chemical studies of the natural ester metabolite retinoyl-β-glucuronide (**10**). Earlier evidence of the biological activity of the simple esters **13** and **14** can be seen in the compilation of data by Newton and colleagues (Newton et al., 1980) on the ability of retinoids to reverse squamous metaplasia in the cultured vitamin A-deficient hamster trachea. While the glucoside **16** has also shown useful retinoic acid-like activity, most of the other RA esters have been described in the patent literature and it is difficult to find reports of pharmacological activity. One class of esters which is in this category and is apparently in pharmaceutical development is typified by the phenacyl ester **17** (Parish and Purcell, 1989; 1991). Members of this class have been stated to show utility in the treatment of acne, skin aging, and skin cancer. It will be interesting to see how these esters are developed.

While it requires some additional structural modification, the internal ester **18**, which is formally a cyclic ester of 11,13-*dicis*-12-hydroxymethylretinoic acid, has shown very favorable efficacy as an inhibitor of hyperkeratinization and epidermal ornithine decarboxylase activity. Although a few analogues of **18** have shown somewhat improved activity in these regards, only this unusual lactone **18** shows the desired improved therapeutic index for topical use relative to RA (Lewin et al., 1995).

Amide derivatives of the carboxyl terminus have also been extensively investigated, especially since the early observations that simple amides such as **19** show reduced toxicity relative to RA (Hixson and Denine, 1979). Of particular interest because of this early evidence of reduced toxicity and of useful retinoidal activity (Newton et al., 1980) have been the N-ethylamide (**20**), N-(2-hydroxyethyl)amide (**19**) and N-(4-hydroxyphenyl)amide (**21**). This last amide **21**, often referred to as 4-HPR, has been much studied because of its low toxicity and chemopreventive efficacy, particularly in animal models of mammary cancer (Moon et al., 1979). Success in these studies has resulted in an ongoing clinical evaluation of **21** as a breast cancer chemopreventive agent in Milan (Formelli et al., 1989; Costa et al., 1993). No doubt because of the success of retinamides such as **21**, an extensive number of amides derived from other amines (Shealy et al., 1984) or amino acids (Shealy et al., 1988) have been synthesized. While a number of these compounds have shown useful retinoidal activity, none have yet been developed as extensively as 4-HPR. Interestingly, while it is often assumed that these amides are also RA prodrugs, this issue has not been satisfactorily resolved in the case of 4-HPR. That is, there is limited evidence that has been obtained both for (Shih et al., 1988) and against (Hultin et al., 1986) the hydrolysis of 4-HPR to RA *in vivo*. The lack of certainty in this regard is perhaps more critical given our recent demonstration (Curley et al., 1996) that 4-HPR does not have significant affinity for the RA receptors. This could raise some questions about the molecular mechanism of action of **21**.

The vast majority of these ester and amide derivatives of RA are prepared by reacting the desired alcohol or amine with an appropriately activated, electrophilic acyl analogue of RA. Common strategies for the activation of RA include formation of its relatively unstable acid chloride (Moon et al., 1979) or synthesis of the more stable retinoyl fluoride (Barua and Olson, 1985) and retinoyl imidazolide (Dawson et al., 1980).

Relatively few other replacements of the carboxyl terminus of RA have resulted in significant new compounds, although at one time there was some interest in retinylidene dimedone **22** (Acton et al., 1980) and biscarboxylate analogues of RA such as **23** (Lewin et al., 1982).

Other isosteric/bioisoteric replacements of the carboxylate group have not generally been very effective. These substitutions include the tetrazole **24** (Dawson et al., 1980) and phenol **25** (Dawson et al., 1980). A number of other substitutions such as with sulfonic and phosphonic acid residues have usually been carried out on more heavily modified retinoid skeletons.

The majority of the relatively simple chemical modifications made to the RA polyene chain have not resulted in important active molecules, although many of these same changes have been of significance in more extensively modified analogues.

Some of the more common modifications have included substitution of vinyl protons or side chain methyl groups with halogen atoms (F, Cl, or Br), removal or homologation of side chain methyl groups, and reduction of specific side chain double bonds to produce dihydroRAs. A useful source to review the range of these structures prepared is the chapter by Frickel (1984).

Replacement of side chain methyl groups with trifluoromethyl groups, such as in **26**, have resulted in compounds useful for the study of retinal chemistry during rhodopsin and bacteriorhodopsin photocycles (Asato et al., 1982). Likewise, restricting rotation about the side chain bonds by incorporating them into aliphatic rings (**27**) has been especially exploited for $\Delta^{11,12}$ double bond restriction in studies of rhodopsin (Akita et al., 1980).

The use of acetylene moiety's in the polyene chain has long had a role in retinoid synthesis, most notably in the industrial synthesis of retinyl acetate developed at Hoffman-LaRoche (Isler et al., 1947). A modest number of acetylenic RA analogues have also been made, for example the (all-E)- and (13Z)-7,8-didehydroretinoic acids **28** and **29** (Attenburrow et al., 1952; Baumann et al., 1979) (see Figure 3). While these compounds have not proven to be particularly important advances, adaptation of the strategy of using acetylenic side-chain units in more heavily modified retinoids has led to important new compounds (see below).

Modifications or replacement of the trimethylcyclohexenyl ring of RA have been, perhaps, more useful in developing important new analogues as well as in understanding the role of metabolism in the activation/inactivation of RA. Many of these molecular modifications have inadvertantly or intentionally been carried out to inhibit one or more of the oxidative metabolic processes that occur on the cyclogeranyl ring resulting in inactivation of RA.

In hindsight, a fairly obvious modification of the RA cyclohexene ring would be to replace it with aromatic, substituted aromatic, and heteroaromatic rings. Knowledge concerning the reduction of activity brought about by ring oxidation makes it likely one could predict that replacement of this ring with more metabolically stable aromatic moiety's might result in active analogues. One of the first materials of this type prepared was the important 4-methoxy-2,3,6-trimethylphenyl analogue acitretin (**30**) which is clinically employed in dermatology (Mayer et al., 1978) as its ethyl ester etretinate (**31**). Interestingly, no details are generally provided concerning the original rationale for the preparation of **31** (Bollag et al.,

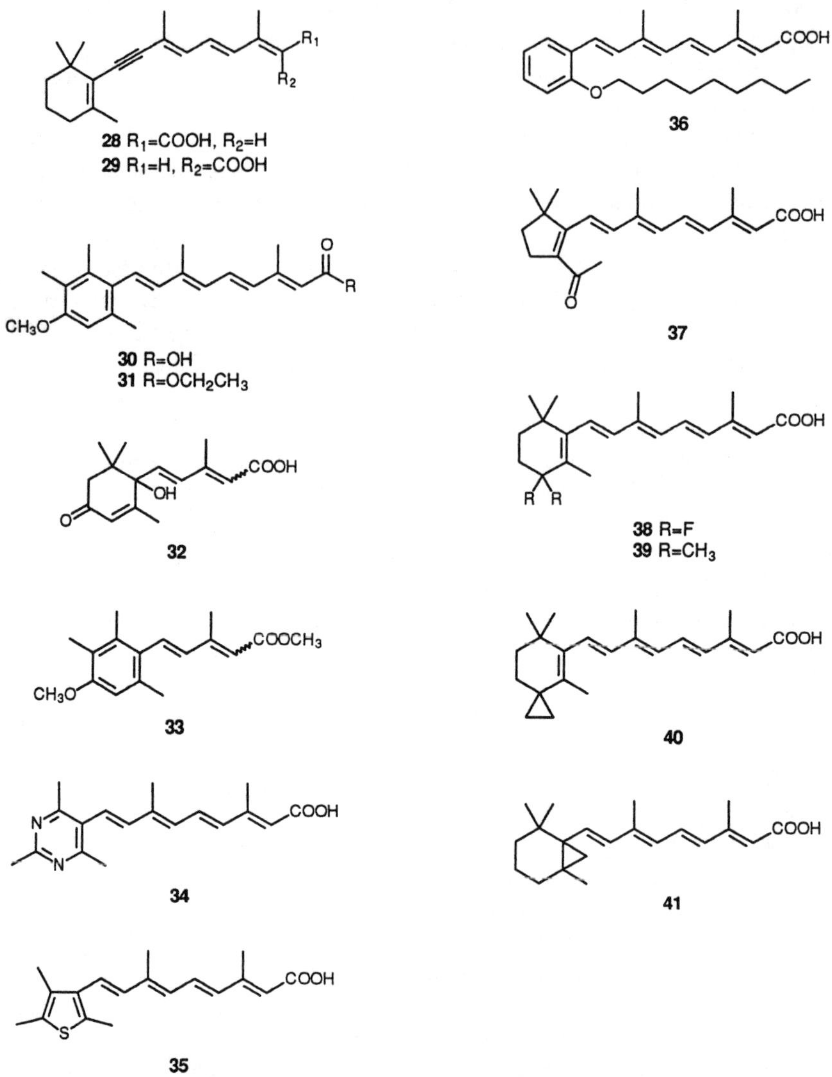

Figure 3. Structures of retinoids.

1975) and it may be that this effective retinoid was discovered somewhat more serendipitously. For example, in connection with some other research, we have found by chance (Sundaram and Curley, unpublished results) that the plant growth-promoter abscissic acid (**32**) rearranges to the etretinate-like compound **33** upon treatment with sulfuric acid in methanol. Nonetheless, etretinate (**31**) is a clinically useful retinoid which was originally identified to be more potent than RA in inhibiting carcinogen-induced rodent skin papilloma formation (Mayer et al., 1978) and has an exceptionally long half-life in humans leading to prolonged teratogenic risk (Lofberg et al., 1990a). While some of these effects may be due to the sequestration of the highly lipid soluble **31** in fat-storing tissues, even the active metabolite **30** also has a long biological half-life (Larsen et al., 1991), is teratogenic (Lofberg et al., 1990b), and is not extensively metabolized (Bugge et al., 1985).

The success of **31** has led to the preparation of a number of analogues, most prominently those in which halogen atoms, particularly chlorine, replace one or more of the methyl groups on the methoxytrimethylphenyl ring (Mayer et al., 1978). Furthermore, this success no doubt led to the preparation of heteroaromatic ring replacements including pyrimidinyl (**34**) (Trown et al., 1980) and thiophenyl (**35**) (Klaus and Pawson, 1981) rings as well as other ether substituted benzene rings such as **36** (Aig et al., 1986). While some of these other analogs are bioactive (Trown et al., 1980), none has proven to be as useful as etretinate or RA.

The size of the cyclohexenyl ring has also been successfully reduced to produce substituted cyclopentene **37** (Bollag et al., 1974), which has shown good bioactivity (Trown et al., 1980). Otherwise, many of the simpler chemical transformations that have been wrought on the trimethycyclohexenyl ring have been carried out to minimize the possibility of oxidative inactivation of RA, most notably at the allylic 4-position and 5,6-double bond.

Replacement of protons with halogen atoms, particularly fluorine, is an often used strategy to minimize metabolic hydroxylation of the carbons to which the protons are bound. To this end, Barua and Olson prepared the 4,4-difluororetinoic acid analogue **38** (Barua and Olson, 1984). Unfortunately, perhaps because of the allylic position of the fluorine atoms in **38**, these atoms showed some slight chemical lability implying the possibility of metaboic instability and perhaps contributing to the limited biological activity of **38** (Miller et al., 1985).

Other changes to the cyclohexene ring including preparing the 4,4-dimethyl (**39**), 4,4-cyclopropyl (**40**), 5,6-cyclopropyl (**41**), and norbornanyl (**42**) (see Figure 4) analogues might be expected to block one or. more of the known oxidative inactivation processes on the ring of RA. All of these changes have been reported to result in analogues with moderate to good biological activity (Dawson et al., 1981; Dawson et al., 1990).

An interesting confirmation that the inhibition of metabolic oxidative ring inactivation of RA may be useful has been obtained. That is, substituted imidazoles such as liarazole have been shown to have an *in vivo* retinoic acid-mimetic effect *specifically* by inhibiting the cytochrome P_{450} monooxygenase which hydroxylates the 4-position of RA (Wauwe et al., 1992). Nonetheless, while oxidation of retinoic acid to 4-oxoretinoic acid (**9**) is generally thought to result in inactivation of RA, **9** remains a potent teratogen in hamsters and mice (Howard et al., 1988; Kraft et al., 1989). Furthermore, Shealy and coworkers have recently shown that 3,3,-dialkyl

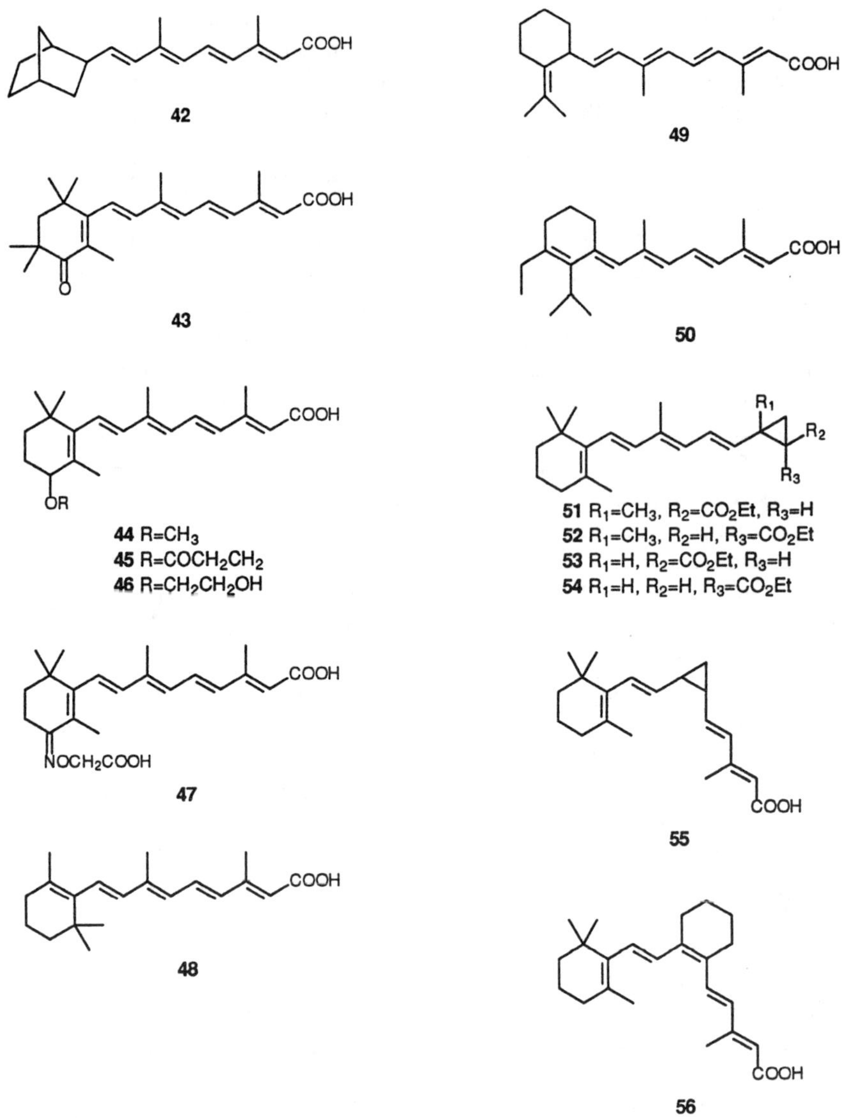

Figure 4. Structures of retinoids.

analogues of **9**, such as **43**, can show good retinoidal activity in cancer chemoprevention assays (Shealy et al., 1994).

We and others have also found that 4-oxygenated retinoids can have useful biological activity and/or be useful tools to study aspects of RA biochemistry. For example, we have found that 4-methoxyretinoic acid (**44**) and 4-acryloyloxyretinoic acid (**45**) show some skin cancer chemopreventive potential (Curley and Carson, 1992). In addition, we have also demonstrated that synthetic 4-(2-hydroxyethoxy)retinoic (**46**) acid can be coupled to epoxy-activated Sepharose 6B and used in an 8500-fold purification of rat testicular CRABP by affinity chromatography (Chapman and Curley, 1990). Zhou and coworkers have also prepared an oxime of **9** (**47**) and found it to be very useful for the generation of anti-RA antibodies when bound to chicken IgG (Zhou et al., 1991).

One extremely interesting recent research initiative has been to modify the trimethylcyclohexenyl ring of RA in efforts to determine whether the active conformer of retinoic acid has a 6-s-*cis* (see **1**) or 6-s-*trans* (**48**) ring-side chain relationship. This relationship is known to be important for the binding of retinal to opsin, but studies of this type have been neglected for retinoic acid. The team of Brouillette and Muccio and coworkers has prepared analogues such as **49** and **50**, which are fixed in a 6-s-*cis* and 6-s-*trans*-like conformation (Vaezi et al., 1994). The results of the biological activity of these compounds as well as their RA receptor binding affinity (Alam et al., 1995) strongly suggest it is the 6-s-*trans* conformation of RA (**48**) that is the active conformation. Similar types of studies are now underway with related 9-*cis* analogues to see if the same relationship holds for 9-*cis*-retinoic acid (Robinson et al., 1995).

A number of chemical studies have been engaged in to explore the importance of other metabolic processes that occur to RA, producing metabolites that may be active. 13-*cis*-RA (**4**) has long been employed in dermatology and is known to be a metabolite of RA. Despite the fact that no specific isomerase(s) has been found for this transformation, it remained possible that the bioactivity of **4** was due to its back isomerization to RA. Thus, cyclopropane analogues of the 13,14-double bond were prepared to prevent isomerization (Curley and DeLuca, 1984). While the methyl analogues (**51** and **52**) were not very effective, both desmethyl analogues **53** and **54** were equiactive in inducing epithelial differentiation, suggesting **4** may be active in its own right (Curley et al., 1985). This has become perhaps more interesting in light of the lack of demonstration of significant affinity of **4** for the RARs. We have just made possible a similar approach for the study of 9-*cis*-RA (**5**) isomerization via the synthesis of cyclopropane **55** (Wong and Curley, unpublished results). Others have found this to be a useful approach in studying more heavily modified analogues of **5** (see below) as well as the cyclohexenyl analogue **56** of 9-*cis*-RA (Brooks et al., 1996).

Much like retinoyl-β-glucuronide (**10**), the phenol O-glucuronide metabolite of 4-HPR is known (**57**) (Dawson and Hobbs, 1980) (see Figure 5). In a series of experiments, we have found that **57** is both more effective and less toxic than 4-HPR when employed as a mammary tumor chemopreventive agent (Abou-Issa et al., 1993). However, this metabolite is a substrate for β-glucuronidase and is susceptible to hydrolysis, therefore it was unclear whether this glucuronide serves as a prodrug or is active as the intact species. A series of stable carbon-linked analogues of the

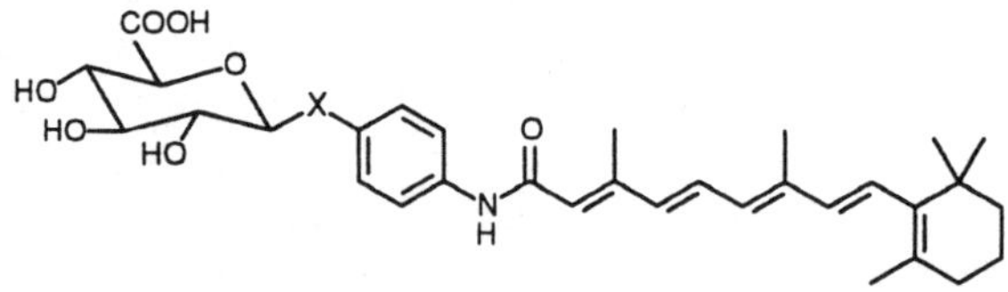

57 X=O
58 X=CH2

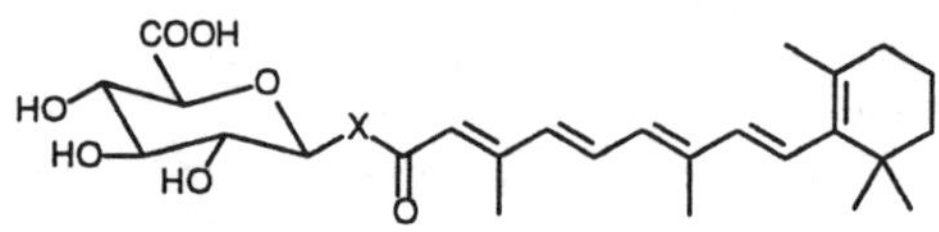

59 X=NH
60 X=CH2

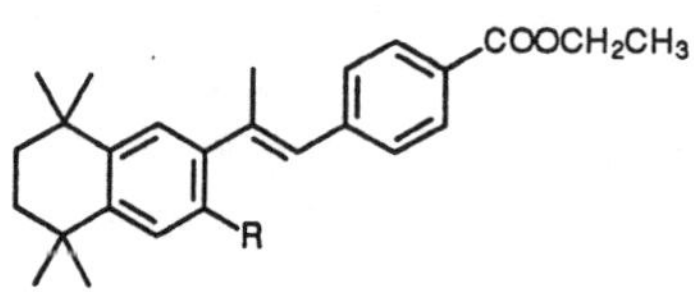

61 R=H
66 R=CH3
67 R=CH2CH3
68 R=CH2CH2CH3
69 R=CH(CH3)2
70 R=OCH3

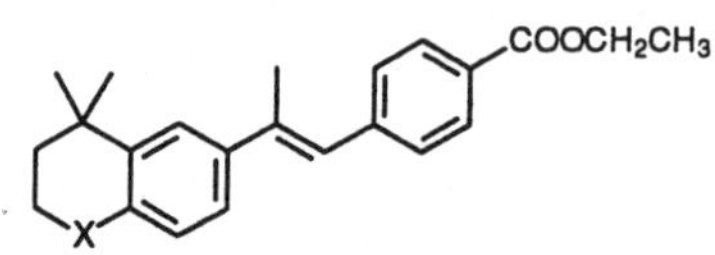

64 X=S
65 X=O

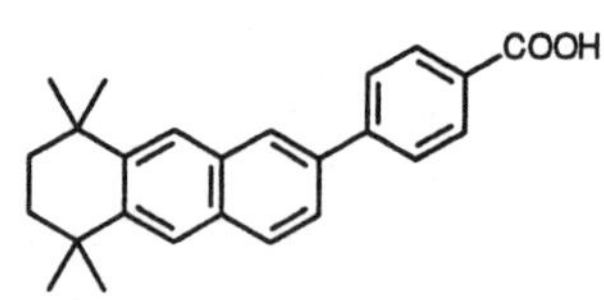

71

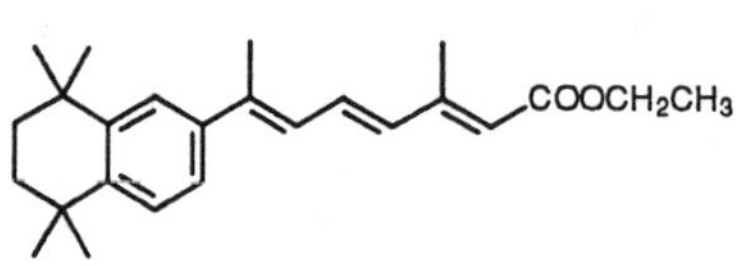

62

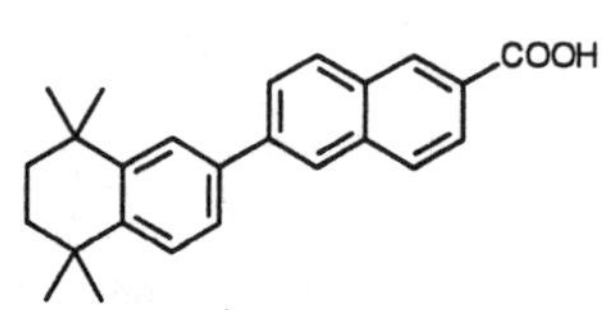

72

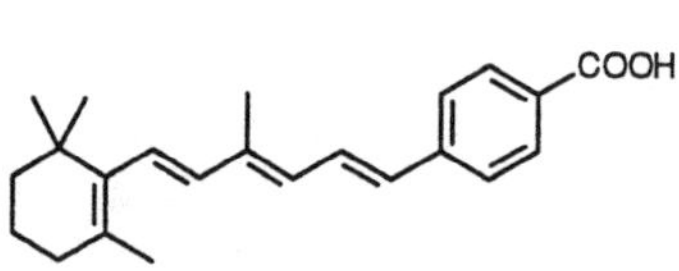

63

Figure 5. Structures of retinoids.

oxygen-linked glucuronide were thus prepared (Panigot et al., 1994) and analogue **58** was found to be an even more effective chemopreventive agent than **57** despite the fact that it too has low affinity for the RARs (Curley et al., 1996). Even greater instability is to be expected for the O-acylglucuronide metabolite **10**. Therefore, the somewhat more stable amide analogue **59** has been prepared and found to be more effective as a mammary tumor chemopreventive agent than retinoic acid. This retinamide **59** is also not a high affinity ligand for the RARs (Robarge et al., 1994). We have just completed the synthesis of the even more stable carbon-linked analogue of **10** and **59** (**60**) (Robarge and Curley, unpublished results) and hope to shortly complete its biological evaluation.

B. Synthetic Retinoids

One of the first highly active structural departures from the retinoid polyene skeleton was the naphthalenyl benzoic acid **61** known as ethyl TTNPB (ethyl *p*-[(E)-2-(5,6,7,8-tetrahydro-5,5,8,8-tetramethyl-2-naphthylenyl)-propenyl]-ben zoate). This compound, as well as analogues such as **62**, were prepared to fix the retinoid side chain stereochemistry with respect to rotation about the single bonds (Loeliger et al., 1980) and may also have shown high biological potency by structural inhibition of metabolic inactivation pathways. The potent activity of **61** in inhibiting the formation of carcinogen-induced papillomas on mouse skin led to a flurry of activity that involved structural modifications to restrict various sites of conformational mobility in RA. Somewhat simpler benzoic acids such as **63** have been synthesized and generally show moderate-to-good activity inhibiting phorbol ester-influenced ornithine decarboxylase induction in mouse skin (Dawson et al., 1981).

A number of analogues related to **61** and **62** have also been prepared in which a heteroatom (usually oxygen or sulfur) has been substituted for one of the dimethylalkyl residues in the saturated ring. For example, the analogues **64** and **65** have been prepared and referred to as "heteroarotinoids" (Waugh et al., 1985). In the hamster tracheal organ culture model, sulfur analogue **64** is particularly effective although oxidation to its sulfoxide appears to reduce potency.

Again, because of the potent activity of **61**, a series of 3-alkyl and alkoxy analogues were prepared (**66-70**) in which the R group on the naphthalene aromatic ring was varied (Strickland et al., 1983). While none of these 3-substituted analogues showed activity in the range of **61** in differentiating murine F9 teratocarcinoma cells, the 3-methyl analogue (**66**) was the most effective compound in differentiating human HL-60 promyelocytic leukemia cells. This earlier observation of the selective activity of **66** has turned out to have greater significance in the more recent past (see below).

Further evolution in this structural class has resulted in even more extensive restriction of side chain flexibility by preparation of molecules such as **71** and **72**. In particular, the tetrahydrotetramethyanthracenylbenzoic acid **71** (TTAB) has recently been shown to have comparable or improved binding affinity for the RARs and somewhat improved efficacy in the F9 cell assay relative to **1** (Charpentier et al., 1995). Changes in the aromatic ring structure of TTAB to include heteroaromatic rings, such as the furan **73** (see Figure 6), has also resulted in compounds

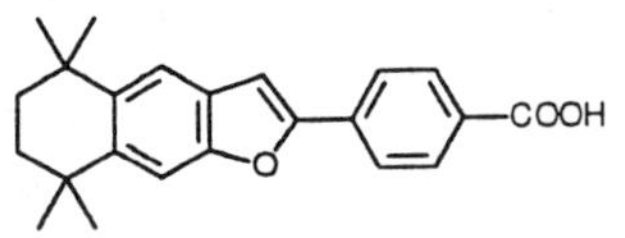

73

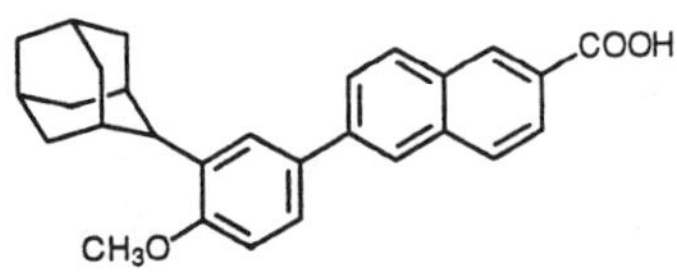

74

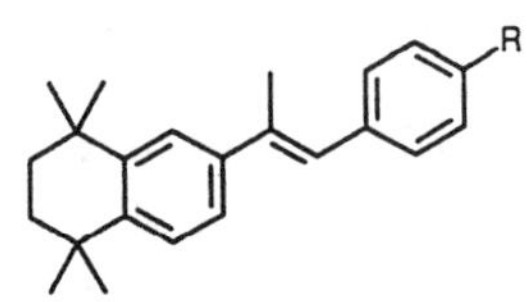

75 R=P(O)(OH)$_2$
76 R=SO$_3$H
77 R=SO$_2$H
78 R=SO$_2$CH$_2$CH$_3$

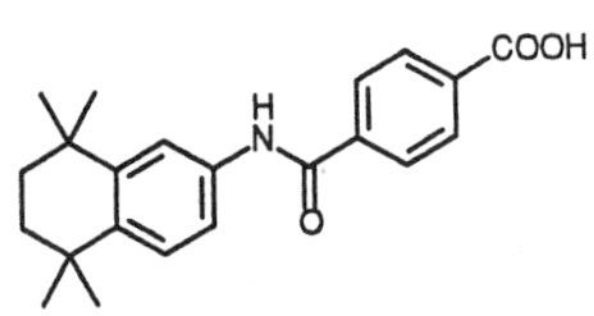

79

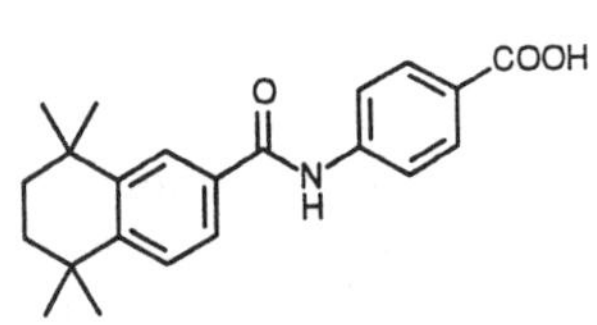

80

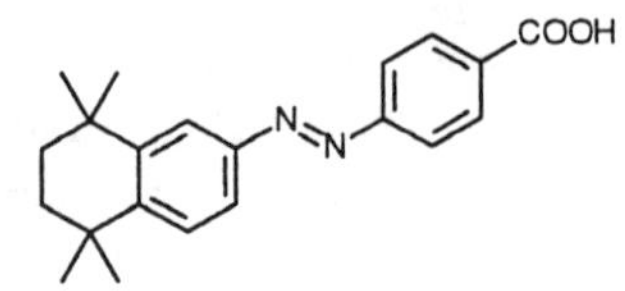

81

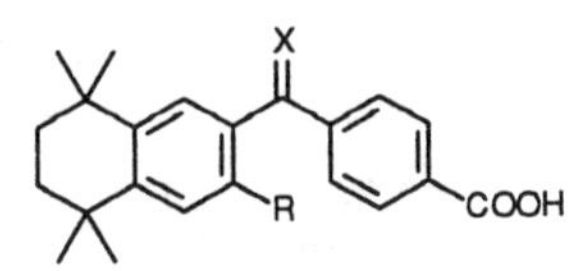

82 X=O, R=H
83 X=CH$_2$, R=H
84 X=O, R=CH$_3$
85 X=CH$_2$, R=CH$_3$

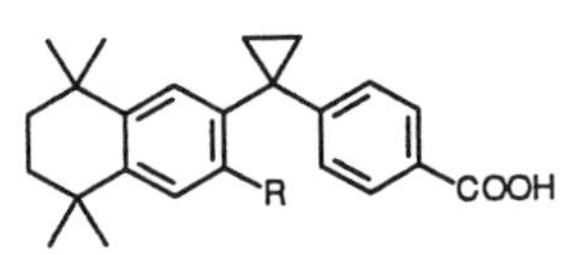

86 R=H
87 R=CH$_3$

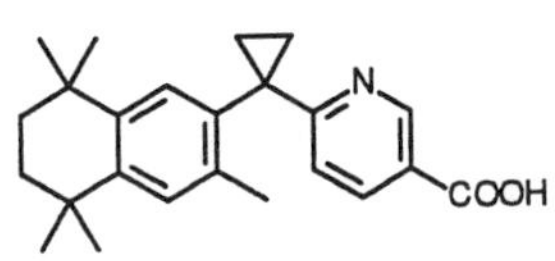

88

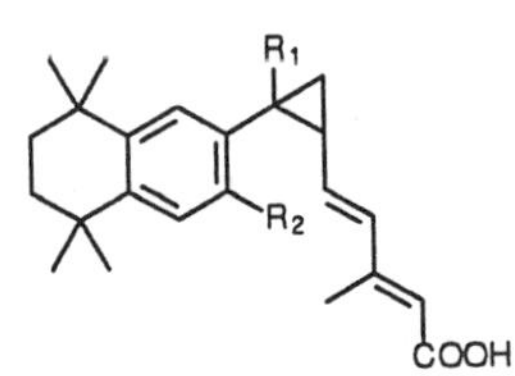

89 R$_1$=H, R$_2$=H
90 R$_1$=CH$_3$, R$_2$=H
91 R$_1$=CH$_3$, R$_2$=CH$_3$

Figure 6. Structures of retinoids.

with differentiation-inducing activity comparable or greater to TTNPB or RA (Kagechika et al., 1989). Subsequent development of this structural type has led to the observation that the tetrahydrotetramethyl ring in structures like TTNPB can be replaced by ortho substituents and retain excellent bioactivity. For example,

Charpentier and coworkers also recently disclosed (Charpentier et al., 1995) that the adamantyl- and methoxy-substituted analogue **74** shows high receptor affinity and differentiation-inducing activity. This ortho substitution pattern on TTNPB-like molecules (one moderately polar and one highly hydrophobic group) can also be replaced by two moderately hydrophobic residues, for example isopropyl groups.

As mentioned earlier, isosteric replacement of the retinoid carboxylate group with sulfur or phosphorous acids has only been popular with heavily modified retinoids such as TTNPB. For example, phosphonates such as **75** have been claimed (Wuest and Janssen, 1989) and sulfonic and sulfinic acid analogues **76** and **77** show good antipapilloma activity, although ethyl sulfone derivative **78** proved to be the most effective compound in this series (Klaus et al., 1983). It was suggested that this activity of **78** was dependent on oxidation to molecules like **76** and **77** *in vivo*. Considerable early interest in sulfone **78** was also based on its favorably low bone toxicity (Kistler et al., 1984), although other teratogenic malformations it causes may have slowed its development (Kistler et al., 1990).

An important area for further investigation, which has evolved from TTNPB, has been exploration of the required linker between the two aromatic rings. That is, is the propenyl group necessary for the activity of TTNPB-like molecules? Shudo and coworkers have convincingly shown this is not the case by synthesizing amides **79** and **80** (known as Am80 and Am580) (Kagechika et al., 1988). These amides show 3.5 and 7 times the efficacy of retinoic acid in inducing HL-60 cell differentiation. Even further evolution in the acceptable structural substitutes for the propenyl linker has been more recently demonstrated by the group of Shudo (Kagechika et al., 1989). That is, the propenyl group can be replaced by an azo linkage producing an azobenzene analog such as **81**, which has 1.3 times the activity of retinoic acid in inducing HL-60 cell differentiation.

Shortly after the discovery of the three major subtypes of the RAR (α,β,γ), a second class of retinoid receptor, termed RXR, was also discovered in three major subtypes (α,β,γ) (Giguere, 1994; Mangelsdorf et al., 1994). As mentioned earlier, it was subsequently proposed that the isomeric metabolite 9-*cis*-RA (**5**) is the natural ligand for the RXRs while RA preferentially binds to the RARs. While it is not yet completely clear what the significance of this variety of retinoid signalling pathways is, the differential tissue distribution of certain of these receptors has led to the hope that receptor or tissue selective retinoids might be discovered that would have selective actions or selectively reduced toxicities. With the discovery of the nuclear receptors, it was subsequently established that the potent retinoids TTNPB (**62**), TTAB (**72**), and the bisnaphthalenyl carboxylic acid TTNN (**73**) were relatively selective ligands for the RARs (Lehmann et al., 1991) and activators of RAR-selective pathways. Subsequently, it was recognized that molecules like 3-methyl TTNPB (**67**) have selective affinity for the RXRs (Beard et al., 1994; Boehm et al., 1994). This observation may account for the earlier noted selective action of **67** in differentiating F9 but not HL-60 cells (see above). Furthermore, it is likely this RXR-selectivity occurs because the presence of the 3-methyl group of **67** causes the more planar stilbene moiety of TTNPB (**62**) to now be twisted into a more 9-*cis*-RA-like conformation. Using this type of rationale, a number of research groups are now developing similar molecules with high selectivity for RXR receptors/pathways that resemble **67** but are pre-organized presumably into a more 9-*cis*-RA-like conformation. For example, Dawson and colleagues (Dawson et al., 1995) and Boehm and coworkers (Boehm et al., 1994; 1995) have been developing compounds such as **82-88** that show very high RXR-selectivity. In fact, **85**, designated as LGD1069, is the first RXR-selective retinoid to enter clinical trials for the treatment of certain cancers.

As mentioned earlier, because we have previously observed that 9-*cis*-RA is not a highly preferred RA isomer at the photostationary state (Curley and Fowble, 1988) and no isomerase for the production of **5** is known, we have been concerned that **5** may be isomerically unstable during assay conditions and have thus prepared and are evaluating **55**. Others have recently used a similar cyclopropane analog approach on more heavily modified TTNPB-like molecules. Apfel and coworkers have found (1995) that cyclopropane **89** is a very effective RXR ligand and RXR-pathway activator while Chandraratna's group have found similar results with the more substituted **90** and **91** (Vuligonda et al., 1996), perhaps lifting doubts about the importance of **5** acting as its 9-*cis* isomer.

Replacement of the benzoic acid ring of TTNPB and 3-methyl TTNPB with heteroaromatic rings has also recently been a successful approach in the development of new retinoids with unique properties. Certainly, this approach has resulted in the highly RXR-selective ligand **88**. In addition, a number of other 3-methyl TTNPB analogs with heteroaromatic rings have been described that show either RXR-specificity (**92-94**) (see Figure 7) or, interestingly, RXR and RARβ,γ selectivity (**95** and **96**) (Beard et al., 1995a; 1995b). The preparation of retinoids with receptor subtype specificity as opposed to RAR or RXR selectivity has also been identified as a goal because the selective tissue distribution of subtypes may permit development of tissue or disease-specific retinoids, perhaps with reduced toxicities. In addition to **95** and **96**, recent successes in this area have been primarily with the development of RARβ,γ-selective agents, although RARα-selective agents such as **80** are known. For example, the marriage of aromatic or heteroaromatic carboxylic acids with acetylenic side chain units have permitted Chandraratna and coworkers to develop molecules such as **97** and **98** that show excellent selective RARβ,γ binding/transcriptional activating activity and potency as topical antiproliferatives, which indicates some potential for treatment of psoriasis (Chandraratna et al., 1995).

Some conflicting results have recently been published, however, with regard to the possibility that receptor selective retinoids will be likely to have altered toxicity patterns. Chandraratna, Kochhar, and coworkers have recently shown that in mouse embryo limb bud cell cultures, RAR selective retinoids such as **1**, **62** (TTNPB) and **80** show potent teratogenic potential while RXR-selective compounds such as **83**

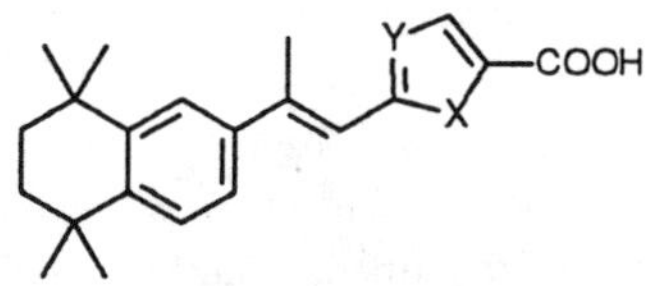

92 X=O, Y=CH_2
93 X=NH, Y=N
95 X=S, Y=CH_2

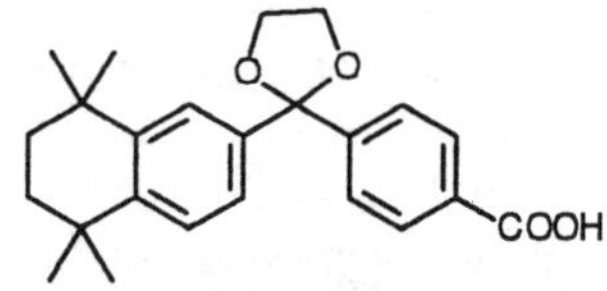

99

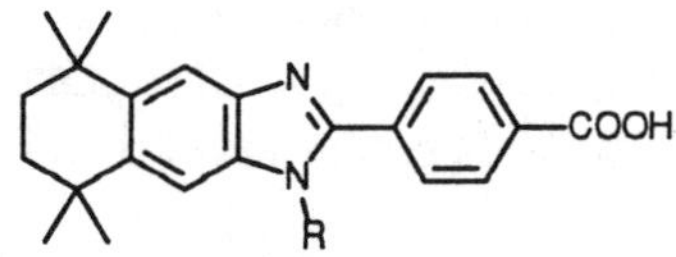

100 R=Ph
101 R=CH_2Ph

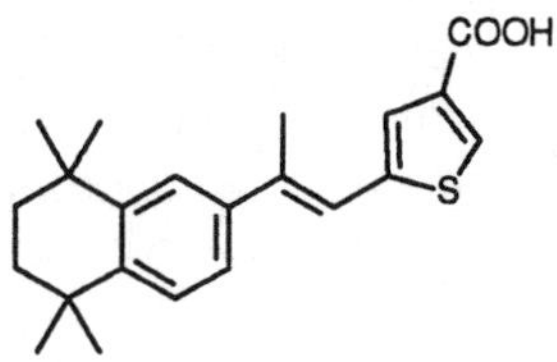

94

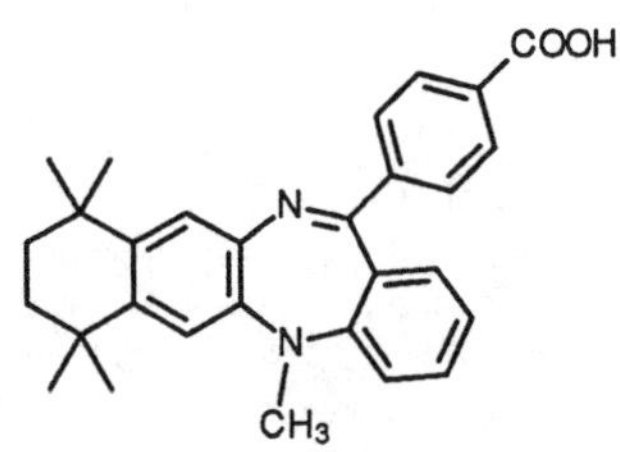

102

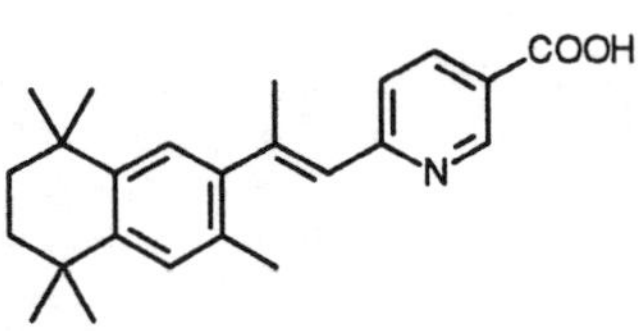

96

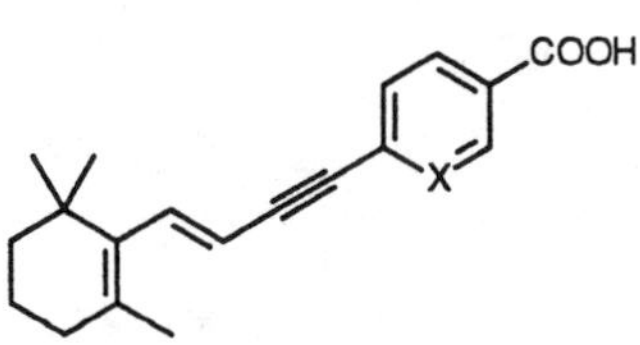

97 X=CH
98 X=N

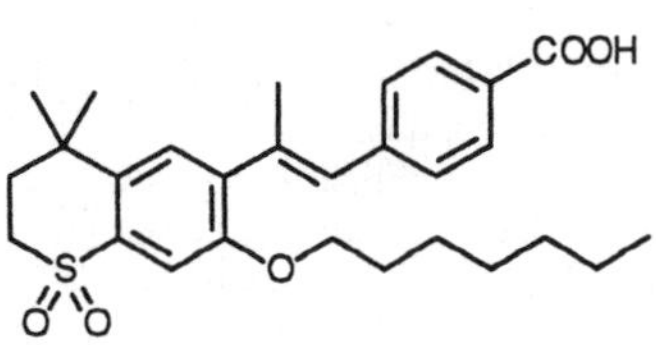

103

Figure 7. Structures of retinoids.

and **99** do not (Jiang et al., 1995). On the other hand, it has also recently been disclosed that RXR-selective ligand **99** causes embryonic malformations in *Xenopus* but not in zebrafish (Minucci et al., 1996). While these results may be due to the different nature of the models employed, it does indicate some caution need be excercised in interpreting some claims made for receptor-selective retinoids.

When specific receptors are discovered for natural substances that are potent regulators of biological responses, it is often found useful to identify receptor antagonists for use as pharmacological research tools and potential therapeutic agents. This has begun to occur in the retinoid research area. Shudo's group has evolved some heteroaromatic analogs from molecules such as **79** that function as antagonists of the binding of RAR ligands as well as inhibiting the ability of these ligands to induce HL-60 differentiation (Eyrolles et al., 1994). Among these molecules are the structurally interesting benzimidazoles **100** and **101** and the dibenzodiazepine **102**. That retinoid receptor antagonists might have some future important practical interest is suggested by the demonstration that RO 41-5253 (**103**) is a specific RARα antagonist that appears to counteract the side effects, especially teratogenic, of the RARα agonist **80** (Apfel et al., 1992; Eckhardt and Schmitt, 1994).

Given the explosion of recent knowledge about the retinoid receptors and their biology, and the beginnings of some real potential for developing receptor-selective retinoids with unique therapeutic/toxicity profiles, it seems likely that the effort to develop new synthetic retinoid analogues will continue at its current vigorous pace for the next number of years.

REFERENCES

Abou-Issa, H., Curley, R.W., Jr., Panigot, M.J., Wilcox, K.A., & Webb, T.E. (1993). *In vivo* use of N-(4-hydroxyphenyl)retinamide-O-glucuronide as a breast cancer chemopreventive agent. Anticancer Res. 13, 1431-1436.

Acton, N., Brossi, A., Newton, D.L., & Sporn, M.B. (1980). Potential prophylactic antitumor activity of retinylidene 1,3-diketones. J. Med. Chem. 23, 805-809.

Aig, E.R., Coffey, J.W., Lovey, A.J., & Rosenberger, M. (1986). Phenylnonatetraenoic acid derivatives and their pharmaceutical use. Chem. Abstr. 105, P42486e.

Akita, H., Tanis, S.P., Adams, M., Balogh-Nair, V., & Nakanishi, K. (1980). Nonbleachable rhodopsins retaining the full natural chromophore. J. Am. Chem. Soc. 102, 6370-6372.

Alam, M., Zhestkov, V., Sani, B.P., Venepally, P., Levin, A.A., Kazmer, S., Li, E., Norris, A.W., Zhang, X.-K., Lee, M.-O., Hill, D.L., Lin, T.-H., Brouillette, W.J., & Muccio, D.D. (1995). Conformationally defined 6-s-*trans*-retinoic acid analogs. 2. Selective agonists for nuclear receptor binding and transcriptional activity. J. Med. Chem. 38, 2302-2310.

Apfel, C., Bauer, F., Crettaz, M., Forni, L., Kamber, M., Kaufmann, F., LeMotte, P., Pirson, W., & Klaus, M. (1992). A retinoic acid receptor α antagonist selectively counteracts retinoic acid effects. Proc. Natl. Acad. Sci. USA 89, 7129-7133.

Apfel, C.M., Kamber, M., Klaus, M., Mohr, P., Keidel, S., & LeMotte, P.K. (1995). Enhancement of HL-60 differentiation by a new class of retinoids with selective activity on retinoid X receptor. J. Biol. Chem. 270, 30765-30772.

Asato, A.E., Mead, D., Denny, M., Bopp, T.T., & Liu, R.S.H. (1982). 19,19,19-Trifluororetinal and 20,20,20-trifluororetinal. J. Am. Chem. Soc. 104, 4979-4981.

Attenburrow, J., Cameron, A.F.B., Chapman, J.H., Evans, R.M., Hems, B.A., Jansen, A.B.A., & Walker, T. (1952). Synthesis of vitamin A from cyclohexanone. J. Chem. Soc., 1094-1111.

Barua, A. B., & Olson, J. A. (1986). Retinoyl β-glucuronide: An endogenous compound of human blood. Am. J. Clin. Nutr. 50, 370-374.

Barua, A. B., & Ghosh, M. C. (1972). Preparation and properties of 4-oxoretinoic acid and its methyl ester. Tetrahedron Lett. 1823-1825.

Barua, A.B., & Olson, J.A. (1985). Preparation of retinamides by use of retinoyl fluoride. J. Lipid Res. 26, 258-262.

Barua, A.B., & Olson, J.A. (1991). All-*trans*-retinoyl β-glucose: Chemical synthesis, growth-promoting activity, and metabolism in the rat. Int. J. Vitam. Nutr. Res. 61, 258-263.

Barua, A.B., & Olson, J.A. (1984). Synthesis of 4,4-difluoro analogs of retinol and retinoic acid. J. Lipid Res. 25, 304-309.

Baumann, M., Hoffman, W., & Nuerrenbach, A. (1979). Syntheses and some reactions of 2,6,6-trimethyl-2-cyclohexenone. Liebigs Ann. Chem., 1945-1951.

Bavik, C.O., Ericksson, U., Allen, R.A., & Peterson, P.A. (1991). Identification and partial characterization of retinal pigment epithelial membrane receptor for plasma retinol-binding protein. J. Biol. Chem. 266, 14978-14985.

Beard, R.L., Gil, D.W., Marler, D.K., Henry, E., Colon, D.F., Gillett, S.J., Arefieg, T., Breen, T.S., Krauss, H., Davies, P.J.A., & Chandraratna, R.A.S. (1994). Synthesis and structure-activity relationships of novel retinoid X receptor-selective retinoids. J. Med. Chem. 37, 2930-2941.

Beard, R.L., Colon, D.F., Klein, E.S., Vorse, K.A., & Chandraratna, R.A.S. (1995a). Differential RXR and RAR activity of stilbene retinoid analogs bearing thiazole and imidazole carboxylic acids. Bioorg. Med. Chem. Lett. 5, 2729-2734.

Beard, R.L., Chandraratna, R.A.S., Colon, D.F., Gillett, S.J., Henry, E., Marler, D.K., Song, T., Denys, L., Garst, M.E., Arefieg, T., Klein, E., Gil, D.W., Wheeler, L., Kochhar, D.M., & Davies, P.J.A. (1995b). Synthesis and structure-activity relationships of stilbene retinoid analogs substituted with heteroaromatic carboxylic acids. J. Med. Chem. 38, 2820-2829.

Blaner, W.S., & Olson, J.A. (1994). Retinol and retonic acid metabolism. In: The Retinoids: Biology, Chemistry, and Medicine, 2nd Ed. (Sporn, M.B., Roberts, A.B., & Goodman, D.S., eds.), pp. 229-255. Wiley, New York.

Blomhoff, R., Green, M.H., Green, J.B., Berg, T., & Norum, K.R. (1991). Vitamin A metabolism: New perspectives on absorption, transport, and storage. Physiol. Rev. 71, 951-990.

Boehm, M.F., McClurg, M.R., Pathirana, C., Mangelsdorf, D., White, S.K., Hebert, J., Winn, D., Goldman, M.E., & Heyman, R. (1994). Synthesis of high specific activity [^{3}H]-9-*cis*-retinoic acid and its application for identifying retinoids with unusual binding properties. J. Med. Chem. 37, 408-414.

Boehm, M.F., Zhang, L., Badea, B.A., White, S.K., Mais, D.E., Berger, E., Suto, C.M., Goldman, M.E., & Heyman, R.A. (1994). Synthesis and structure-actvity relationships of novel retinoid X receptor-selective retinoids. J. Med. Chem. 37, 2930-2941.

Boehm, M.F., Zhang, L., Zhi, L., McClurg, M.R., Berger, E., Wagoner, M., Mais, D.E., Suto, C.M., Davies, P.J.A., Heyman, R.A., & Nadzan, A.M. (1995). Design and synthesis of retinoid X receptor selective ligands that induce apoptosis in leukemia cells. J. Med. Chem. 38, 3146-3155.

Bok, D., & Heller, J. (1976). Transport of retinol from blood to retina: Autoradiographic study of pigment epithelial cell surface receptors for plasma retinol-binding protein. Exp. Eye Res. 22, 395-402.

Bollag, W., Rigassi, W., & Schwieter, U. (1974). Polyene compounds. Chem. Abstr. 80, P121150x.

Bollag, W., Ruegg, R., & Ryser, G. (1975). Polyene compounds. Chem. Abstr. 82, P111788x.

Brooks, S.C., III, Kazmer, S., Levin, A.A., & Yen, A. (1996). Myeloid differentiation and retinoblastoma phosphorylation changes in HL-60 cells induced by retinoic acid receptor- and retinoid X receptor-selective retinoic acid analogs. Blood 87, 227-237.

Buck, J., Myc, A., Garbe, A., & Cathomas, G. (1991a). Differences in the action and metabolism between retinol and retinoic acid in B lymphocytes. J. Cell. Biol. 115, 851-859.

Buck, J., Derguini, F., Levi, E., Nakanishi, K., & Hämmerling, U. (1991b). Intracellular signaling by 14-hydroxy-4,14-*retro*-retinol. Science 254, 1654-1655.

Bugge, C.J.L., Rodriguez, L.C., & Vane, F.M. (1985). Determination of isotretinoin, or etretinate and their major metabolites in human blood by reversed-phase high-performance liquid chromatography. J. Pharm. Biomed. Anal. 3, 269-277.

Canada, F. J., Law, W. C. Rando, R. R.,Yamamoto, T., Derguini, F., & Nakanishi, K. (1990). Substrate specificities and mechanism in the enzymatic processing of vitamin A to 11-cis-retinol. Biochemistry 29, 9690-9697.

Chandraratna, R.A.S., Gillett, S.J., Song, T.K., Attard, J., Vuligonda, S., Garst, M. E., Arefieg, T., Gil, D.W., & Wheeler, L. (1995). Synthesis and pharmacological activity of conformationally restricted, acetylenic retinoid analogs. Bioorg. Med. Chem. Lett. 5, 523-527.

Chapman, J.M., & Curley, R.W., Jr. (1990). Affinity purification of retinoic acid-binding proteins using immobilized 4-(2-hydroxyethoxy)retinoic acid. Protein Expression Purif. 1, 63-69.

Charpentier, B., Bernardon, J.-M., Eustache, J., Millois, C., Martin, B., Michel, S., & Shroot, B. (1995). Synthesis, structure-affinity relationships, and biological activities of ligands binding to retinoic acid receptor subtypes. J. Med. Chem. 38, 4993-5006.

Connor, M.J., & Smit, M.H. (1987). Terminal-group oxidation of retinol by mouse epidermis. Biochem. J. 244, 489-492.

Costa, A., Sacchini, V., & Decensi, A. (1993) Retinoids and tamoxifen in breast cancer chemoprevention. Int. J. Clin. Lab. Res. 23, 53-55.

Curley, R.W., Jr., & DeLuca, H.F. (1984). Ethyl 13,14-dihydro-13,14-methyleneretinoates: Analogues of all-*trans*- and 13-*cis*-retinoic acid. J. Org. Chem. 49, 1944-1946.

Curley, R.W., Jr., Silva, D.P., Jr., & DeLuca, H.F. (1985). The biological activity of cyclopropyl analogues of all-*trans*- and 13-*cis*-retinoic acid in the rat vaginal smear assay. Arch. Biochem. Biophys. 238, 484-489.

Curley, R.W. Jr., & Fowble, J.W. (1988). Photoisomerization of retinoic acid and its photoprotection in physiologic-like solutions. Photochem. Photobiol. 47, 831-836.

Curley, R.W., Jr., Abou-Issa, H., Panigot, M.J., Repa, J.J., Clagett-Dame, M., & Alshafie, G. (1996). Chemopreventive activities of C-glucuronide/glycoside analogs of retinoid-O-glucuronides against breast cancer development and growth. Anticancer Res. 16, in press.

Curley, R.W., Jr., & Carson, D.L. (1992). 4-Oxygenated retinoids: Unexpected chemopreventive potential for analogues originally synthesized as affinity labels. Bioorg. Med. Chem. Lett. 2, 1133-1136.

Dawson, M.I., & Hobbs, P.D. (1994). The synthetic chemistry of retinoids. In: *The Retinoids: Biology, Chemistry, and Medicine,* 2nd Ed. (Sporn, M.B., Roberts, A.B., & Goodman, D.S., eds.), pp. 5-178. Wiley, New York.

Dawson, M.I., Jong, L., Hobbs, P.D., Cameron, J.F., Chao, W.-r., Pfahl, M., Lee, M.-O., Shroot, B., & Pfahl, M. (1995). Conformational effects on retinoid receptor selectivity. 2. Effects of retinoid bridging group on retinoid X receptor activity and selectivity. J. Med. Chem. 38, 3368-3383.

Dawson, M.I., Hobbs, P.D., Chan, R.L., Chao, W.-R., & Fung, V.A. (1981). Aromatic retinoic acid analogues. Synthesis and pharmacological activity. J. Med. Chem. 24, 583-592.

Dawson, M.I., Hobbs, P.D., Kuhlmann, K., Fung, V.A., Helmes, C.T., & Chao, W.-R. (1980). Retinoic acid analogues. Synthesis and potential as cancer chemopreventive agents. J. Med. Chem. 23, 1013-1022.

Dawson, M.I., Chao, W.-R., Hobbs, P.D., & Delair T. (1990). The inhibitory effects of retinoids on the induction of ornithine decarboxylase and the promotion of tumors in mouse epidermis. In: *Chemistry and Biology of Synthetic Retinoids* (Dawson, M.I., & Okamura, W.H., eds.) pp. 385-466. CRC Press, Boca Raton.

Dawson, M.I., & Hobbs, P.D. (1980). Synthesis of the D-glucuronic acid conjugates of N-(4-hydroxyphenyl)- and N-(2-hydroxethyl)-retinamides. Carbohydr. Res. 85, 121-129.

Dawson, M.I., Hobbs, P.D., Chan, R.L.-S, & Chao, W.-R. (1981). Retinoic acid analogues with ring modifications. Synthesis and pharmacological activity. J. Med. Chem. 24, 1214-1223.

De Leenheer, A.P., Lambert, W.E., & Claeys, I. (1982). All-*trans*-retinoic acid: Measurement of reference values in human serum by high performance liquid chromatography. J. Lipid Res. 23, 1362-1367.

Dunagin, P. E., Jr., Meadows, E.H. & Olson, J.A. (1965). Retinoyl β-glucuronic acid; a major metabolite of vitamin A in rat bile. Science 148, 86-87.

Dunagin, P. E., Jr., Zachman, R. D., & Olson, J. A. (1966). The identification of metabolites of retinol and retinoic acid in rat bile. Biochim. Biophys. Acta 124, 71-85.

Eckhardt, K., & Schmitt, G. (1994). A retinoic acid receptor α antagonist counteracts retinoid teratogenicity *in vitro* and reduced incidence and/or severity of malformations *in vivo*. Toxicol. Lett. 70, 299-308.

Eckhoff, C., Collins, M.D., & Nau, H. (1991). Human plasma all-*trans*, 13-*cis*, and 13-*cis*-4-oxoretinoic acid profiles during subchronic vitamin A supplementation: Comparision to retinol and retinyl ester plasma levels. J. Nutr. 121, 1016-1025.

Eyrolles, L., Kagechika, H., Kawachi, E., Fukasawa, H., Iijima, T., Matsushima, Y., Hashimoto, Y., & Shudo, K. (1994). Retinobenzoic acids. 6. Retinoid antagonists with a heterocyclic ring. J. Med. Chem. 37, 1508-1517.

Fiorella, P.D., & Napoli, J.L. (1991). Expression of cellular retinoic acid binding protein (CRABP) in *Escherichia coli*. J. Biol. Chem. 266, 16572-16579.

Formelli, F., Carsana, R., Costa, A., Buranelli, F., Campa, T., Dossena, G., Magni, A., & Pizzichetta, M. (1989). Plasma retinol level reduction by the synthetic retinoid fenretinide: A one year follow-up study of breast cancer patients. Cancer Res. 48, 6149-6152.

Frickel, F. (1984).Chemistry and physical properties of retinoids. In: *The Retinoids*, Vol. 1. (Sporn, M.B., Roberts, A.B., & Goodman, D.S., eds.), pp. 8-145. Academic Press, Orlando.

Frolik, C.A. (1984). Metabolism of retinoids. In: *The Retinoids,* Vol. 2. (Sporn, M.B., Roberts, A.B., & Goodman, D.S., eds.), pp. 177-208. Wiley, New York.

Frolik, C. A., Tavela, T. E., Newton, D. L., & Sporn, M. B. (1978). *In vivo* metabolism and biological activity of all-*trans*-retinoic acid and its metabolites in hamster trachea. J. Biol. Chem. 253, 7319.

Frolik, C. A., Roberts, A. B., Tavela, T. E., Roller, P. P., Newton, D. L., & Sporn, M. B. (1979). Isolation and identification of 4-hyroxy- and 4-oxoretinoic acid. *In vivo* metabolites of all-*trans*-retinoic acid in hamster trachea and liver. Biochemistry 18, 2092-2097.

Gallup, J. M., Barua, A. B. Furr, H. C., & Olson, J. A. (1987). Effects of retinoid β-glucuronide and N-retinoyl amines on the differentiation of HL-60 cells *in vitro*. Proc. Soc. Exp. Biol. Med. 186, 269-274.

Ganguly, J. (1989). Biochemistry of Vitamin A. CRC Press, Boca Raton, FL.

Giguere, V. (1994). Retinoic acid receptors and cellular retinoid binding proteins: Complex interplay in retinoid signaling. Endocrine Rev. 15, 61-79.

Gross, G. (1989). Preparation of retinoid acid esters for treatment of skin disorders. Chem. Abstr. 110, P95572q.

Gunning, D. B., Barua, A. B., & Olson, J.A. (1993). Comparative teratogenicity and metabolism of all-*trans* retinoic acid, all-*trans*-retinoyl-β-glucose, and all-*trans*-retinoyl-β-glucuronide in pregnant Sprague-Dawley rats. Teratology 47, 29-36.

Hänni, R., Bigler, F., Meister, W., & Englert, G. (1976). Isolation and identification of three urinary metabolites of retinoic acid in the rat. Helv. Chim. Acta 59, 2221-2227.

Hänni, R., & Bigler, F. (1977). Isolation and identification of three major metabolites of retinoic acid from rat feces. Helv. Chim. Acta 60, 881-887.

Heyman, R. A., Mangelsdorf, D. J., Kyck, J. A., Stein, R. B., Eichele, G., Evans, R. E., & Thaller, C. (1992). 9-*cis* retinoic acid is a high affinity ligand for the retinoid X receptor. Cell 66, 397-406.

Hixson, E.J., & Denine, E.P. (1979). Comparative subacute toxicity of retinyl acetate and three synthetic retinamides in swiss mice. J. Natl. Cancer Inst. 63, 1359-1364.

Howard, W.B., Willhite, C.C., Dawson, M.I., & Sharma, R.P. (1988). Structure-activity relationships of retinoids in developmental toxicology III. Contribution of the vitamin A β-cyclogeranylidene ring. Toxicol. Appl. Pharmacol. 95, 122-138.

Hubbard, R., & Wald, G. (1952). *Cis-trans* isomers of vitamin A and retinene in vision. Science 115, 60-63.

Hultin, T.A., May, C.M., & Moon, R.C. (1986). N-(4-Hydroxyphenyl)-all-trans-retinamide pharmacokinetics in female rats and mice. Drug Metab. Disp. 14, 714-717.

Hussain, M.M., Mahley, R.W., Boyles, J.K., Fainaru, M., Brecht, W.J., & Lindquist, P.A. (1989). Chylomicron-chylomicron remnant clearance by liver and bone marrow in rabbits. J. Biol. Chem. 244, 9571-9582.

Isler, O., Huber, W., Ronco, A., & Kofler, M. (1947). Synthesis of vitamin A. Helv. Chim. Acta 30, 1911-1927.

Janick-Buckner, D., Barua, A. B., & Olson, J. A. (1991). Induction of HL-60 cell differentiation by water soluble and nitrogen-containting conjugates of retinoic acid and retinol. FASEB J. 5, 320-325.

Jiang, H., Penner, J.D., Beard, R.L., Chandraratna, R.A.S., & Kochhar, D.M. (1995). Diminished teratogenicity of retinoid X receptor-selective synthetic retinoids. Biochem. Pharmacol. 50, 669-676.

Kagechika, H., Himi, T., Namikawa, K., Kawachi, E., Hashimoto, Y., & Shudo, K. (1989). Retinobenzoic acids. 3. Structure-activity relationships of retinoidal azobenzene-4-carboxylic acids and stilbene-4-carboxylic acids. J. Med. Chem. 32, 1098-1108.

Kagechika, H., Kawachi, E., Hashimoto, Y., Himi, T., & Shudo, K. (1988). Retinobenzoic acids. 1. Structure-activity relationships of aromatic amides with retinoidal activity. J. Med. Chem. 31, 2182-2192.

Kakkad, B.P., & Ong, D.E. (1988). Reduction of retinaldehyde bound to cellular retinol-binding protein (type II) by microsomes from rat small intestine. J. Biol. Chem. 263, 12916-12919.

Kistler, A., Galli, B., & Howard, W.B. (1990). Comparative teratogenicity of three retinoids: the arotinoids Ro 13-7410, Ro 13-6298 and Ro 15-1570. Arch. Toxicol. 64, 43-48.

Kistler, A., Sterz, H., & Teelmann, K. (1984). Ro 15-1570, a new sulfur-containing retinoid devoid of bone toxicity in rats. Arch. Toxicol. 56, 117-122.

Klaus, M., Bollag, W., Huber, P., & Kueng, W. (1983). Sulfur-containing arotinoids, a new class of retinoids. Eur. J. Med. Chem. 18, 425-429.

Klaus, M.J., & Pawson, B.A. (1981). Polyene compounds. Chem. Abstr. 95, P169542a.

Kraft, J. C, Löfberg, B., Chahoud, I., Bochert, G., & Nau, H. (1989). Teratogenicity and placental transfer of all-*tran*-, 13-*cis*, 4-oxo-all-*trans*-, and 4-oxo-13-*cis*-retinoic acid after administration of a low oral dose during organogenesis in mice. Toxicol. Appl. Pharmacol. 100, 162-176.

Kutner, A., Renstrom, B., Schnoes, H.K., & DeLuca, H.F. (1986). Synthesis of coenzyme A ester of retinoic acid: Intermediate in vitamin A metabolism. Proc. Natl. Acad. Sci. USA 83, 6781-6784.

Larsen, F.G., Jakobsen, P., Eriksen, H., Gronhoj, J., Kragballe, K., & Nielsen-Kudsk, F. (1991). The pharmacokinetics of acitretin and its 13-cis-metabolite in psoriatic patients. J. Clin. Pharmacol. 31, 477-483.

Lehmann, J.M., Dawson, M.J., Hobbs, P.D., Husbann, M., & Pfahl, M. (1991). Identification of retinoids with nuclear receptor subtype-selective activities. Cancer Res. 61, 4804-4809.

Leo, M.A., Iida, S., & Lieber, C.S. (1984). Retinoic acid metabolism by a system reconstituted with cytochrome P-450. Arch. Biochem. Biophys. 234, 305-312.

Leo, M. A., Lasker, J.M., Raucy, J. L., Kim, C-I., Black, M., & Lieber, C. S. (1989). Metabolism of retinol and retinoic acid by human liver cytochrome P450IIC8. Arch. Biochem. Biophys. 269, 305-312.

Levin, A. A., Sturzenbecker, L. J., Kazmer, S., Bosakowski, T., Huselton, C., Allenby, G., Speck, J., Kratzeisen, C. L., Rosenberger, M., Lovey, A., & Grippo, J. F. (1992). 9-*cis* retinoic acid stereoisomer binds and activates the nuclear receptor RXRα. Nature 355, 359-361.

Lewin, A.H., Whaley, M.G., Parker, S.R., & Carroll, F.I. (1982). 12-Carboxyretinoic acids. Synthesis and structure. J. Org. Chem. 47, 1799-1807.

Lewin, A.H., Black, S.L., Bos, M.E., Goehring, R.R., Nair, X., Whiting, G., Bouquin, P., Tetrault, G., & Carroll, F.I. (1995). Evaluation of retinoid lactones as topical therapeutic agents in dermatology. Pharm. Res. 12, 983-992.

Lippel, K., & Olson, J.A. (1968). Biosynthesis of β-glucuronides of retinol and of retinoic acid *in vivo* and *in vitro*. J. Lipid Res. 9, 168-175.

Loeliger, P., Bollag, W., & Mayer, H. (1980). Arotinoids, a new class of highly active retinoids. Eur. J. Med. Chem. 15, 9-15.

Lofberg, B., Chahoud, I., Bochert, G., & Nau, H. (1990b). Teratogenicity of the 13-cis and all-trans-isomers of the aromatic retinoid etretin: Correlation to transplacental pharmacokinetics in mice during organogenesis after a single oral dose. Teratology 41, 707-716.

Lofberg, B., Reiners, J., Spielmann, H., & Nau, H. (1990a). Teratogenicity of steady-state concentrations of etretinate and metabolite acitretin maintained in maternal plasma and embryo by intragastric infusion during organogenesis in the mouse: A possible model for the extended elimination phase in human therapy. Dev. Pharmacol. Ther. 15, 45-51.

Mangelsdorf, D.J., Umesono, K., & Evans, R.M. (1994). The retinoid receptors. In: *The Retinoids: Biology, Chemistry, and Medicine*, 2nd Ed. (Sporn, M.B., Roberts, A.B., & Goodman, D.S., Eds.), pp.319-349. Wiley, New York.

Mangelsdorf, D. J., Borgmeyer, U., Heyman, R. A., Zhou, J. Y., Ong, E. S., Oro, A. E., Kakizuka, A., & Evans, R. M. (1992). Characterization of three RXR genes that mediate the action of 9-*cis*-retinoic acid. Genes Dev. 6, 329-344.

Mayer, H., Bollag, W., Hanni, R., & Ruegg, R. (1978). Retinoids, a new class of compounds with prophylactic and therapeutic activities in oncology and dermatology. Experientia 34, 1105-1119.

McCormick, A. M., Napoli, J. L., Schnoes, H.K., & DeLuca, H. F. (1978). Isolation and identification of 5,6-epoxyretinoic acid: A biologically active metabolite of retinoic acid. Biochemistry 17, 4085-4090.

McCormick, A. M., & Napoli, J. L. (1982). Identification of 5,6-epoxyretinoic acid as an endogenous retinol metabolite. J. Biol. Chem. 257, 1730-1735.

Mehta, R., Barua, A. B., Moon, R. C., & Olson, J. A. (1992). Interactions between retinoid β-glucuronides and cellular retinol- and retinoic acid-binding proteins. Int. J. Vit. Nutr. Res. 62, 143-147.

Miller, D. A. & DeLuca, H. F. (1986). Biosynthesis of retinoyl β-glucuronide, a biologically active metabolite of all-*trans*-retinoic acid. Arch. Biochem. Biophys. 244, 179-186.

Miller, D.A., Stephens-Jarnagin, A., & Deluca, H.F. (1985). The epithelial differentiating activity *in vivo* of (E)-4-[2-(5,6,7,8-tetrahydro-5,5,8,8-tetramethyl-2-naphthalenyl)-1-propenyl]benzoic acid and 4,4,-difluororetinoic acid. Biochem. J. 227, 311-316.

Minucci, S., Saint-Jeannet, J.-P., Toyama, R., Scita, G., DeLuca, L.M., Taira, M., Levin, A.A., Ozato, K., & Dawid, I.B. (1996). Retinoid X-receptor-selective ligands produce malformations in *Xenopus* embryos. Proc. Natl. Acad. Sci. USA 93, 1803-1807.

Moon, R.C., Thompson, H.J., Becci, P.J., Grubbs, C.J., Gander, R.J., Newton, D.L., Smith, J.M., Phillips, S.L., Henderson, W.R., Mullen, L.T., Brown, C.C., & Sporn, M.B. (1979). N-(4-Hydroxyphenyl)retinamide, a new retinoid for prevention of breast cancer in the rat. Cancer Res. 39, 1339-1346.

Napoli, J.L. (1986). Retinoid metabolism in LLC-PK_1 cells. Characterization of retinoic acid synthesis by an established mammalian cell line. J. Biol. Chem. 261, 13594-13597.

Napoli, J.L., & Race, K.R. (1988). Biogenesis of retinoic acid from β-carotene. J. Biol Chem. 263, 17372-17377.

Napoli, J. L., Khalil, H., & McCormick, A. M. (1982). Metabolism of 5,6-epoxyretinoic acid *in vivo:* Isolation of a major intestinal metabolite. Biochemistry 21, 1942-1949.

Nath, K., & Olson, J. A. (1967). Natural occurrence and biological activity of vitamin A derivatives in rat bile. J. Nutr. 93, 461-469.

Newton, D.L., Henderson, W.R., & Sporn, M.B. (1980). Structure-activity relationships of retinoids in hamster tracheal organ culture. Cancer Res. 40, 3413-3425.

Norum, K.R., & Blomhoff, R. (1992). Vitamin A absorption, transport, cellular uptake, and storage. Am. J. Clin. Nutr. 56, 735-744.

Noy, N., & Blaner, W.S. (1991). Interactions of retinol with binding proteins: Studies with rat cellular retinol-binding protein and with rat retinol-binding protein. Biochemistry 30, 6380-6386.

Olson, J.A. (1989). Provitamin A function of carotenoids: The conversion of β-carotene into vitamin A. J. Nutr. 119, 105-108.

Ong, D.E., McDonald, P.N., & Gubitosi, A.M. (1988). Esterification of retinol in rat liver. Possible participation by cellular retinol-binding protein and cellular retinol-binding protein II. J. Biol. Chem. 263, 5789-5796.

Panigot, M.J., Humphries, K.A., & Curley, R.W., Jr. (1994). Preparation of 4-retinamidophenyl- and 4-retinamidobenzyl-C-glycosyl and C-glucuronosyl analogues of the glucuronide of 4-hydroxyphenylretinamide as potential stable cancer chemopreventive agents. J. Carbohydr. Chem. 13, 303-321.

Parish, H.A., & Purcell, W.P. (1989). Preparation of esters of 13-trans-retinoic acid for treatment of acne and pharmaceutical compositions containing them. Chem. Abstr. 111, P115646t.

Parish, H.A., & Purcell, W.P. (1991). Skin preparations containing nonirritating retinoids. Chem. Abstr. 114, P108705v.

Pipkin, J. L., Hinson, W. G., Anson, J. F., Schol, H. M., Lyn-Cook, L. E., Burns, E. R., & Casciano, D. A. (1991a). Synthesis and biochemical characteristics of nucleoproteins following a toxic dose of retinoic acid in cycling and differentiating HL-60 cells. Appl. Theor. Electrophor. 2, 31-41.

Pipkin, J. L., Hinson, W. G., Anson, J. F., Lyn-Cook, L. E., Burns, E. R., & Casciano, D. A. (1991b). Synthesis and phosphorylation of nuclear matrix proteins following a toxic dose of retinoic acid in cycling and differentiating HL-60 cells. Appl. Theor. Electrophor. 2, 17-29.

Rando, R. R., Bernstein, P. S., & Barry, R. J. (1991). New insights in the visual cycle. In: *Progress in Retinal Research* (Osborne, N. & Chader, G., Eds.), pp. 161-178. Pergamon Press, Oxford.

Rando, R. R. (1990). The chemistry of vitamin A and vision. Angew. Chem. Int. Ed. Engl. 29, 461-480.

Renstrom, B., & DeLuca, H.F. (1989). Incorporation of retinoic acid into proteins via retinoyl-CoA. Biochim. Biophys. Acta 998, 69-74.

Rietz, P., Wiss, O., & Weber, F. (1974). Metabolism of vitamin A and the determination of vitamin A status. Vitam. Horm. (N.Y.) 32, 237-249.

Robarge, M.J., Repa, J.J., Hanson, K.K., Seth, S., Clagett-Dame, M., Abou-Issa, H., & Curley, R.W., Jr. (1994). N-Linked analogs of retinoid O-glucuronides: Potential cancer chemopreventive/chemotherapeutic agents. Bioorg. Med. Chem. Lett. 4, 2117-2122.

Roberts, A. B. & Frolik, C. A. (1979). Recent advances in the *in vivo* and *in vitro* metabolism of retinoic acid. Fed. Proc., Fed. Am. Soc. Exp. Biol. 38, 2524-2527.

Roberts, A. B., Lamb, L. C., & Sporn, M. B. (1980). Metabolism of all-*trans* retinoic acid in hamster liver microsomes: Oxidation of 4-hydroxy to 4-keto-retinoic acid. Arch. Biochem. Biophys. 199, 374-383.

Roberts, A.B., Nichols, M.D., Newton, D.L., & Sporn, M.B. (1979). *In vitro* metabolism of retinoic acid in hamster intestine and liver. J. Biol. Chem. 254, 6296-6302.

Roberts, E. S., Vaz, A.D. N., & Coon, M. J. (1992). Role of isozymes of rabbit microsomal cytochrome P-450 in the metabolism of retinoic acid, retinol, and retinal. Mol. Pharmacol. 41, 427-433.

Robinson, C.Y., Waterhous, D.V., Muccio, D.D., & Brouillette, W.J. (1995). Conformationally constrained precursors to retinoic acid analogs which stabilize the 9Z-configuration. Bioorg. Med. Chem. Lett. 9, 953-958.

Sani, B. P., Barua, A. B., Hill, D. L., Shih, T.-W., & Olson, J. A. (1992). Retinoyl β- glucuronide: Lack of binding to receptor proteins of retinoic acid as related to biological activity. Biochem. Pharmacol. 43, 919-922.

Shealy, Y.F., Frye, J.L., & Schiff, L.J. (1988). N-(Retinoyl)amino acids. Synthesis and chemopreventive activity *in vitro*. J. Med. Chem. 31, 190-196.

Shealy, Y.F., Frye, J.L., O'Dell, C.A., Thorpe, M.C., Kirk, M.C., Coburn, W.C., Jr., & Sporn, M.B. (1984). Synthesis and properties of some 13-*cis*- and all-*trans*-retinamides. J. Pharm. Sci. 73, 745-751.

Shealy, Y.F., Hosmer, C.A., Riordan, J.M., Wille, J.W., Rogers, T.S., & Hill, D.L. (1994). Cancer chemopreventive 3-substituted-4-oxoretinoic acids. J. Med. Chem. 37, 3051-3056.

Shih, T.-W., Shealy, Y.F., & Hill, D.L. (1988). Enzymatic hydrolysis of retinamides. Drug Metab. Disp. 16, 337-340.

Sietsema, W. K., & DeLuca, H. F. (1982). A new vaginal smear assay for vitamin A in rats. J. Nutr. 112, 1481-1489.

Skare, K.L., Schnoes, H.K., & DeLuca, H.F. (1982a). Biliary metabolites of all-trans-retinoic acid in the rat: Isolation and identification of a novel polar metabolite. Biochemistry 21, 3308-3317.

Skare, K. L., & DeLuca, H. F. (1983). Biliary metabolites of all-*trans*-retinoic acid in the rat. Arch Biochem. Biophys. 224, 13-18.

Skare K. L., Schnoes H. K., & DeLuca H. F. (1982b). The biological activity of retinotaurine. J. Nutr. 112, 1626-1630.

Soprano, D.R., & Blaner, W.S. (1994). Plasma retinol-binding protein. In: *The Retinoids: Biology, Chemistry, and Medicine*, 2nd Ed. (Sporn, M.B., Roberts, A.B., & Goodman, D.S., Eds.), pp. 257-281. Wiley, New York.

Sporn, M. B., Roberts, A. B., & Goodman, D. S., Eds. (1984). *The Retinoids*, Vol. 1 & 2. Academic Press, Orlando.

Sporn, M. B., Roberts, A. B., & Goodman, D. S., Eds. (1994). *The Retinoids: Biology, Chemistry, and Medicine*. Raven Press, New York.

Strickland, S., Breitman, T.R., Frickel, F., Nuerrenbach, A., Hadicke, E., & Sporn, M.B. (1983). Structure-activity relationships of a new series of retinoidal benzoic acid derivatives as measured by induction of differentiation of murine F9 teratocarcinoma cells and human HL-60 promyelocytic leukemia cells. Cancer Res. 43, 5268-5272.

Surekha Rao, M. S., John, J., & Cama, H. R. (1972). Vitamin A_2. Preparation, properties, metabolism, and biological activity of 4-oxoretinoic acid. Int. J. Vit. Nutr. Res. 42, 368-378.

Swanson, B. M., Frolik, C. A., Zaharevitz, D. W., Roller, P. P., & Sporn, M. B. (1981). Dose-dependent kinetics of all-*trans* RA in rats. Biochem. Pharmacol. 30, 107-113.

Takahashi, N., & Breitman, T.R. (1989a). Retinoic acid acylation (retinoylation) of a nuclear protein in the human acute mycloid leukemia cell line HL 60. J. Biol. Chem. 264, 5159-5163.

Takahashi, N., & Breitman, T. R. (1989b). Covalent binding of 17 beta-estradiol and retinoic acid to proteins in the human breast cancer cell line MCF-7. *In Vitro* Cell. Dev. Biol. 25, 1199-1200.

Takahashi, N., & Breitman, T. R. (1991). Retinoylation of proteins in leukemia, embryonal carcinoma, and normal kidney cell lines: Differences associated with differential responses to retinoic acid. Arch. Biochem. Biophys. 285, 105-110.

Takahashi, N., Jetten, A. M., & Breitman, T. R. (1991a). Retinoylation of cytokeratins in normal human epidermal keratinocytes. Biochem. Biophys. Res. Commun. 180, 393-400.

Takahashi, N., Liapi, C., Anderson, W. B., & Breitman, T. R. (1991b). Retinoylation of the cAMP-binding regulatory subunits of type I and type II cAMP-dependent protein kinases in HL60 cells. Arch. Biochem. Biophys. 290, 293-302.

Tang, G., & Russell, R. M. (1990). 13-*cis* Retinoic acid is an endogenous compound in human serum. J. Lipid Res. 30, 175-182.

Tang, G., & Russell, R. M. (1991). Formation of all-*trans*-retinoic acid and 13-*cis*-retinoic acid from all-*trans*-retinyl palmitate in humans. J. Nutr. Biochem. 2, 210-213.

Thaller, C., & Eichele, G. (1990). Isolation of 3,4-didehydroretinoic acid, a novel morphogenetic signal in the chick wing bud. Nature 345, 815-819.

Trown, P.W., Palleroni, A.V., Bohoslawec, O., Richelo, B.N., Halpern, J.M., Gizzi, N., Geiger, R., Lewinski, C., Machlin, L.J., Jetten, A., & Jetten, M.E.R. (1980). Relationship between binding affinities to cellular retinoic acid-binding protein and *in vivo* and *in vitro* properties for 18 retinoids. Cancer Res. 40, 212-220.

Vaezi, M.F., Alam, M., Sani, B.P., Rogers, T.S., Simpson-Herren, L., Wille, J.J., Hill, D.F., Doran, T.I., Brouillette, W.J., & Muccio, D.D. (1994). A conformationally defined 6-s-*trans*-retinoic acid isomer: Synthesis, chemopreventive activity, and toxicity. J. Med. Chem. 37, 4499-4507.

Vuligonda, V., Lin, Y., & Chandraratna, R.A.S. (1996). Synthesis of highly potent RXR-specific retinoids: The use of a cyclopropyl group as a double bond isostere. Bioorg. Med. Chem. Lett. 6, 213-218.

Wald, G. (1934-1935). Vitamin A in eye tissues. J. Gen. Physiol. 18, 905-915.

Wald, G. (1935-1936). Carotenoids and the visual cycle. J. Gen. Physiol. 19, 351-371.

Wald, G. (1968). Molecular basis of visual excitation. Science 162, 230-239.

Waugh, K.M., Berlin, K.D., Ford, W.T., Holt, E.M., Carrol, J.P., Schomber, P.R., Thompson, M.D. & Schiff, L.J. (1985). Synthesis and characterization of selected heteroarotinoids. Pharmacological activity as assessed in vitamin A deficient hamster tracheal organ cultures. Single crystal x-ray diffraction analysis of 4,4-dimethylthiochroman-6-yl methyl ketone 1,1-dioxide and ethyl (*E*)-*p*-[2-(4,4-dimethylthiochroman-6-yl)propenyl]benzoate. J. Med. Chem. 28, 116-124.

Wauwe, J.V., Nyen, G.V., Coene, M.-C., Cools, W., Goossens, J., Borghgraef, P., & Janssen, P.A.J. (1992). Liarozole, an inhibitor of retinoic acid metabolism, exerts retinoid- mimetic effects *in vivo*. J. Pharmacol. Exp. Ther. 261, 773-779.

Wouters, W., van Dun, J., Dillen, A., Coene, M.-C., Cools, W., & De Coster, R. (1992). Effects of liarozole, a new antitumoral compound, on retinoic acid-induced inhibition of cell growth and on retinoic acid metabolism in MCF-7 human breast cancer cells. Cancer Res. 52, 2841-2846.

Wuest, H.-H., & Janssen, B. (1989). Preparation of arylphosphonates, -phosphinates, and - phosphine oxides for treatment of dermatologic conditions and tumors. Chem. Abstr. 111, P39583c.

Yoshizawa, T., & Wald, G. (1963). Prelumirhodopsin and the bleaching of visual pigments. Nature 197, 1279-1286.

Zachman, R. D., Dunagin, P. E., Jr., & Olson, J. A. (1966). Formation and enterohepatic circulation of metabolites of retinol and retinoic acid in bile duct-cannulated rats. J. Lip. Res. 7, 3-9.

Zhou, H.-R., Cullum, M.E., Gerlach, T., Gage, D.A., & Zile, M.H. (1991). Immunogens for the preparation of retinoic acid antibodies. J. Nutr. Biochem. 2, 122-131.

Zile, M.H., Cullum , M.E., Simpson, R.U., Barua, A.B., & Swartz, D.A. (1987). Induction of differentiation of human promyelocytic leukemia cell line HL-60 by retinoyl glucuronide, a biologically active metabolite of vitamin A. Proc. Natl. Acad. Sci. USA 84, 2208-2219.

Zile, M. H., Emerick, R. J., & DeLuca, H. F. (1967). Identification of 13-*cis* retinoic acid in tissue extracts and its biological activity in rats. Biochim. Biophys. Acta, 141, 639-641.

Zile, M. H., Inhorn, R.C., & DeLuca, H. F. (1982). Metabolism *in vivo* of all-*trans* retinoic acid. J. Biol. Chem., 257, 3544-3550.

Zile, M., & Deluca, H. F., (1968). Retinoic acid: Some aspects of growth-promoting activity in the rat. J. Nutr. 94, 302-308.

Zile, M. H., Inhorn, R. C., & DeLuca, H. F. (1980). The biological activity of 5,6- epoxyretinoic acid. J. Nutr. 110, 2225-2230.

MOLECULAR MECHANISMS OF RETINOID FUNCTION

Christopher P.F. Redfern

Abstract . . . 36
I. Introduction . . . 36
II. Retinoids in the Extracellular Compartment . . . 38
III. Retinoids in the Cytoplasmic Compartment . . . 40
A. Cellular Retinoid Binding Proteins . . . 41
B. The Metabolism of Retinol and Retinoic Acid . . . 44
IV. Retinoids in the Nuclear Compartment . . . 47
A. Retinoic Acid Receptors (RARs) . . . 47
B. Retinoid X Receptors (RXRs) . . . 49
C. Developmental Function of RARs and RXRs . . . 50
D. Retinoid Response Elements and Receptor Dimerization . . . 52
E. Transcriptional Regulation by Retinoic Acid Receptors . . . 54
F. Receptor-Specific Ligands . . . 58
V. Other Mechanisms of Action of Retinoic Acid . . . 59
VI. An Overview of the Developmental Function of Retinoic Acid . . . 60
References . . . 63

Advances in Organ Biology
Volume 3, pages 35-77.

ISBN: 0-7623-0285-2

ABSTRACT

Retinoids, derivatives of vitamin A, influence many aspects of cell and developmental biology, and may be important intracellular regulators or signaling molecules. This chapter attempts to outline current knowledge of how retinoids áffect cell and tissue development with respect to their sites of absorption in the gut and journey to their final (but not necessarily sole) target-receptors within the cell nucleus that function mainly, although not exclusively, as ligand-dependent transcriptional regulators. While the emphasis is on all-*trans* retinoic acid, the potential biological role of the 9-*cis* isomer is stressed and the importance of other retinoids and mechanisms of retinoid action are outlined. Since retinoic acid affects many aspects of cell function, nuclear retinoic acid receptors are important targets for the development of new drugs and therapeutic strategies in medicine.

I. INTRODUCTION

Retinoids are compounds derived from or structurally related to retinol (Fig. 1), otherwise known as vitamin A. The developmental importance of vitamin A was indicated by early studies on rats (Wolbach and Howe, 1925) that demonstrated a marked effect of vitamin A deficiency on the characteristics of epithelial tissues. It was not until the pioneering studies of Fell and colleagues in the 1950s (Fell, 1957; Fell and Mellanby, 1953) that the ability of retinol to promote the formation of mucous metaplasia in embryonic chick skin was described, an observation that has had an important role in stimulating subsequent research on the molecular and cellular mechanisms of action of retinoids in epithelial tissues.

The active "principle" of retinol is usually its derivative, all-*trans* retinoic acid. However, the possibility that retinol itself or metabolites other than retinoic acid may have a direct effect on cell behaviour and metabolism must always be borne in mind (see below). Apart from the clear and opposing effects of retinol excess and deficiency on epithelial tissues, retinol or retinoic acid excess during embryonic development is markedly teratogenic, particularly with respect to the development of the brain, craniofacial structure, and upper limb girdle (see Chapter 4). The teratogenicity of retinoic acid and its ability to induce limb duplications in experimentally-manipulated embryos, mimicking the properties of the Zone of Polarizing activity (ZPA) (Tickle et al., 1982), has led to it being viewed as a morphogen during embryological development. A morphogen is defined as a diffusable molecule imparting positional information and, while there is evidence for a gradient in retinoic acid concentration across the chick wing bud (Thaller and Eichele, 1987) and some embryonic tissues *in vitro* appear to release retinoic acid (Hogan et al., 1992; Wagner et al., 1992), it is not easy to imagine that such a lipophilic molecule would diffuse between cells. The fact that a concentration gradient in limb buds exists is not in itself evidence that this concentration gradient was produced by retinoic acid diffusing out from a source. It may be more appropriate to view retinoic acid as an intracellular regulatory

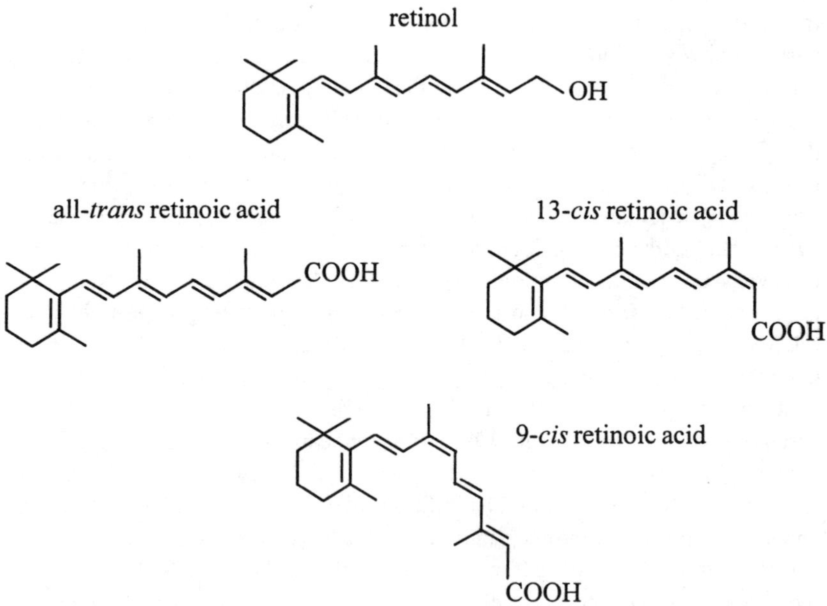

Figure 1. Basic retinoid structures: all-*trans* retinol (Vitamin A), all-*trans* retinoic acid, 13-*cis* retinoic acid and 9-*cis* retinoic acid.

molecule. This view is underlined by the fact that retinoic acid has effects on a wide range of different cell types apart from epithelial tissues. Vitamin A and its derivatives may be important for promoting and maintaining cellular control and thus may be key components of fundamental cellular control pathways.

In outline, the main molecular mechanism of action of retinoic acid is via specific interactions with nuclear retinoic acid receptors (RARs). These are ligand-dependent transcriptional regulators that interact as dimers with specific DNA sequences or "retinoic acid response elements" (RAREs) within gene promoters and enhancers, and provide a mechanism whereby retinoic acid can control the expression of specific genes. Up until recently all-*trans* retinol and all-*trans* retinoic acid were the forms used experimentally and considered to be the natural biologically-relevant isomers of retinol and retinoic acid, with the exception of 11-*cis* retinal, the cofactor of the visual pigment rhodopsin. Within the last few years nuclear receptors called retinoid X receptors (RXRs) that bind a minor retinoic acid isomer, 9-*cis* retinoic acid, have been identified and appear to function both as RXR homodimers binding to retinoid X receptor response elements (RXREs) and as heterodimers with RARs.

The importance of retinoids in medicine stems from their ability to induce differentiation or modify cell phenotype in a wide variety of cell types from different tissue systems, including the nervous system, the immune system, epithelial tissues and tissues associated with reproduction. Various retinoids, but in particular 13-*cis* retinoic acid, have been used in the treatment of skin diseases, and for the treatment of various types of cancer such as melanoma, neuroblastoma, bladder and stomach cancer, and squamous cell carcinoma. There is no doubt of the efficacy of retinoic acid in the treatment of skin diseases, but in cancer treatment retinoic acid has not had a great impact, with the notable exception of acute promyelocytic leukaemia resulting from t15:17 translocations. The potential relevance of retinoic acid-regulated pathways to cancer development or progression has received very little attention, despite their evident importance in developmental processes during embryogenesis. However, inappropriate expression of cytosolic retinoid binding proteins and nuclear retinoid receptors may be associated with oncogenesis. As an example, transgenic animals overexpressing cellular retinoic acid binding protein I (CRABP I) have high rates of spontaneous pancreatic carcinoma (Perez-Castro et al., 1993). Reduced expression of nuclear retinoic acid receptor-β is associated with head and neck cancer (Xu et al., 1994), and increased expression of cytosolic retinoid binding proteins is associated with high grade endometrial adenocarcinoma (Siddiqui et al., 1995). Given that exogenous retinoic acid has antiproliferative effects and may also reduce the ability of cells to metastasize through the basement membrane, if retinoic acid has these effects *in vivo* then perturbations in intracellular retinoic acid control pathways may have important consequences for oncogenesis. To understand the mechanism and importance of retinoids in biology and medicine, it is necessary to consider the derivation of retinoids from the diet, transport within the body, entry into cells and tissues and the molecular mechanisms within cells leading to the effects of retinoids on development, differentiation, and teratogenesis.

II. RETINOIDS IN THE EXTRACELLULAR COMPARTMENT

Vitamin A is derived from the diet, either directly as retinol or retinyl esters from animal tissues, or in the form of plant carotenoids. Dietary retinyl esters are hydrolysed to retinol by luminal retinyl ester hydrolases from the intestinal brush border (Rigtrup et al., 1994), and free retinol is transferred into enterocytes via a specific carrier-mediated mechanism (Dew and Ong, 1994). Carotenoids are directly internalized and subsequently broken down to retinol and retinoic acid either in the gut or at target tissues. Since retinol is important as a precursor for retinoic acid and other biologically active forms of vitamin A, there have to be mechanisms to transfer dietary retinoids from the sites of intestinal absorption to target tissues. Retinol is esterified in the intestine via acyl-CoA-independent enzymes acting on retinol bound to cellular retinol binding protein II (CRBP II) as substrate (Ong et al., 1987), and the retinyl esters are packaged into chylomicrons and transferred to the liver via the lymphatic system. Retinol is mobilized from the liver bound to serum retinol binding protein (RBP), a 182 amino-acid-residue (Rask et al., 1979) protein (21kDa) synthesized by hepatocytes. RBP consists of eight antiparallel β-strands folded to form a flattened barrel with amino- and carboxy-terminal extensions (Cowan et al., 1990; Newcomer et al., 1984). The barrel, closed at one end, completely encapsulates a retinol molecule with the hydroxyl group towards the open end and accessible to solvent (Cowan et al., 1990; Newcomer et al., 1984). Secretion of RBP from the liver is dependent on the presence of bound retinol, and conformational differences between apo- and holo-RBP may play an important part in facilitating the secretion of holo-RBP from hepatocytes (Kaji and Lodish, 1993). Once in the serum, RBP circulates as a complex with transthyretin. Retinol is unloaded as a result of interactions between holo-RBP and cell surface receptors; the resulting apo-RBP has reduced affinity for the receptor and transthyretin and is removed by glomerular filtration (Rask et al., 1979; Sivaprasadarao and Findlay, 1988).

The mechanism of entry of retinol into cells is uncertain. It has been suggested that, on the basis of the physical properties of retinol and its affinity for RBP, a transport mechanism to allow retinol to enter cells is not necessary (Noy and Xu, 1990). On the other hand, there is evidence for the involvement of specific cell-surface receptors in the uptake of retinol into cells (Macdonald et al., 1990; Rask and Peterson, 1976; Shingleton et al., 1989; Sivaprasadarao and Findlay, 1988). A specific cell surface receptor for RBP has now been isolated and cloned from retinal pigment epithelium (RPE) and comprises a terminally glycosylated 63 kDa membrane protein noncovalently associated with a high molecular weight complex (Bavik 1991; 1992; 1993). Although the expression of this protein is apparently confined to RPE, related proteins may be expressed in other tissues. Bavik and colleagues have shown that retinol bound to RBP is more effective than free retinol at inhibiting keratinocyte differentiation, an effect blocked by an antibody that recognizes the RPE cell surface receptor for RBP (Bavik et al., 1995). Although conventional opinion would be that RBP is the major carrier of retinol in the blood, it is also possible that serum albumins play a role in retinol transport. Retinol has a lower avidity for serum albumin, but since albumin is present at much higher concentrations than RBP it could therefore be a significant carrier of retinol (Noy and Xu, 1990) while lacking the specificity of cellular transfer afforded by specific cell-surface receptors for retinol-bound RBP.

Although liver is one of the major sites of RBP synthesis, this protein is secreted by other tissues such as the retinal pigment epithelium, sertoli cells and the peritubular cells of the seminiferous epithelium (Davis and Ong, 1992; 1995; Ong et al., 1994). In these tissues, RBP secretion may serve to transport retinol across blood-tissue barriers. Retinol is important for the later stages of spermatogenesis and Davis and Ong (1995) have suggested that a receptor-mediated process is responsible for the uptake of retinol from the serum by the peritubular cells of the seminiferous tubules; retinol is then exported to the seminiferous epithelium via new RBP synthesized by the peritubular cells. Extracellular retinoid transport in the male reproductive system is not confined to RBP secretion. The epididymus secretes a retinoic acid binding protein, E-RABP, which, like RBP, has an 8-stranded β-sheet structure folded into an open-ended barrel (Newcomer, 1993). Unlike other retinoid binding proteins in which the retinoid ligand is bound with the polyene tail fully extended, E-RABP binds retinoic acid in a bent conformation and will bind both all-*trans* and 9-*cis* retinoic acid (Newcomer et al., 1993). Although both all-*trans* and 9-*cis* retinoic acid have been identified in epididymal extracts, the function of this protein and the relevance of both retinoic acid isomers the function or maintenance of the male reproductive tract is obscure.

Carotenoids from plant materials are one of the major sources of vitamin A: retinol is generated in the intestine by central cleavage of carotenoids to retinal which is subsequently reduced to retinol by retinal reductase in a process mediated by CRBP II (Wardlaw and Ong, 1996). However, carotenoids are also present in the serum, bound to low density lipoproteins (Clevidence and Bieri, 1993), and can be directly metabolized, via retinal, to retinoic acid by a variety of tissues (Napoli and Race, 1988). The entry of serum carotenoids into cells is presumably mediated by LDL receptors but the biological significance of serum carotenoids as a source of retinoic acid at "target" tissues is unknown. However, unlike serum retinol levels, serum carotenoids increase with increased dietary intake of carotenoids (Carughi and Hooper, 1994; Hughes et al., 1994), and a potentially increased availability of retinoic acid resulting from high serum carotenoid levels may be *part* of the explanation for the reduced cancer risk associated with high carotenoid intakes (Block et al., 1992).

The identification of 9-*cis* retinoic acid as a ligand for RXRs and RARs raises the question of the source of this retinoic acid isomer *in vivo*. Serum or tissue 9-*cis* β-carotenoids derived from the diet are a *potential* source of 9-*cis* retinoic acid (Hebuterne et al., 1995; Nagao and Olson, 1994; Wang et al., 1994). In the formation of retinoic acid from retinol, the initial dehydrogenation from retinol to retinal is carried out by retinol dehydrogenase, which discriminates against 9-*cis* retinol (Boerman and Napoli, 1995). Conversely, a cytosolic retinal dehydrogenase converting the aldehyde into retinoic acid is active with both the all-*trans* and 9-*cis* isomers (El Akawi and Napoli, 1994). Since retinal and not retinol is the intermediate in the metabolism of carotenoids to retinoic acid, 9-*cis* carotenoids in the serum or in tissues such as the liver (Stahl et al., 1993) may be important in relation to the intracellular regulation of 9-*cis* retinoic acid concentrations.

III. RETINOIDS IN THE CYTOPLASMIC COMPARTMENT

The entry of retinol into cells serves at least two recognized purposes: storage as retinyl esters or the production of metabolites that regulate nuclear retinoid receptors. Although retinol reportedly activates RARs (Repa et al., 1993), one of its main functions may be to act as a substrate for the synthesis of retinoic acid, thought to be the main ligand for RARs. Nevertheless, recent studies on embryonal carcinoma cells have suggested that the 4-oxo and 4-hydroxy metabolites of retinol may also be effective ligands for activating RARs (Achkar et al., 1996). Furthermore, studies on B- and T-cells have indicated that retinol, and not retinoic acid, is important for permitting B-cell proliferation (Buck et al., 1990) and T-cell activation (Garbe et al., 1992), apparently via metabolism to relatively unknown metabolites 14-hydroxy-*retro*-retinol (Buck et al., 1991) or 13,14-dihydroxy-*retro*-retinol (Derguini et al., 1995). 14-hydroxy-*retro*-retinol can also be formed directly from retinyl esters without retinol (Eppinger et al., 1993).

The discovery that all-*trans* retinoic acid was the ligand for nuclear receptors mediating the effects of vitamin A has emphasized the importance of understanding the way in which intracellular concentrations of all-*trans* retinoic acid are regulated. Consequently, recent work on vitamin A metabolism has been focused on the synthesis of all-*trans* retinoic acid from retinol, and the subsequent metabolic degradation of retinoic acid. The importance of other retinol metabolites such as 4-oxo-retinol (Achkar et al., 1996) and the *retro*-retinol derivatives (Buck et al., 1991; Derguini et al., 1995) in pathways of cellular control is unknown, as are the biochemical pathways and their control for generating these newly identified retinol metabolites.

A. Cellular Retinoid Binding Proteins

The synthesis and degradation of retinoic acid involve a number of different enzymes. Cellular binding proteins for both retinol and retinoic acid play a key role in these enzymatic pathways. Biochemical studies by David Ong and Frank Chytil in the mid-1970s (for review see Chytil and Ong, 1984) demonstrated the existence of cytosolic binding proteins for retinol (CRBP) and retinoic acid (CRABP). These proteins, initially isolated from bovine testis, were found to be low molecular weight, acidic proteins, and bound retinol or retinoic acid, respectively, with high affinity. Subsequent studies established the existence of a separate, intestinal form of CRBP, referred to as CRBP II (Ong, 1984), and two distinct forms of CRABP, referred to as CRABP I and CRABP II (Bailey and Siu, 1988; 1990). These cytosolic retinoid binding proteins are members of a family of related lipid-binding proteins, including the fatty-acid binding proteins and myelin basic protein 2, in which the ligand is enclosed within a deep hydrophobic pocket formed by the 10-stranded β-barrel polypeptide structure (Cowan et al., 1993; Winter et al., 1993). One end of the barrel is closed by the N-terminal end of the protein and the other end capped by a helix-turn-helix motif. Unlike serum RBP, the ligand for CRBPs and CRABPs lies with the polar hydroxyl or acid group in the interior of the barrel (Cowan et al., 1993; Winter et al., 1993). Structure determinations indicate that there is no aperture large enough to admit the ligand into its binding site in the interior of the barrel, and this suggests that a conformational change involving the cap is necessary for the ligand to obtain access to the hydrophobic internal pocket (Jamison et al., 1994).

The genes for the cytosolic retinoid binding proteins have all been cloned, and analyzed to varying extents. Although the genes vary in size from 6 kb for human CRABP II to 19.5 kb for rat CRBP II, they all have a similar 4-exon structure that is common to other members of this protein family (Astrom et al., 1992; Demmer et al., 1987; Macgregor et al., 1992; Nilsson et al., 1988; Wei et al., 1990). In the human genome, CRBP I and II are closely linked on 3q21-q22 (CRBP I: [Rocchi et al., 1989]) and 3q21-qter (CRBP II: (Demmer et al., 1987)), whereas CRABP I and CRABP II are on 15q22-qter (van Kessel et al., 1991) and 1q21 (Elder et al., 1992), respectively.

Cellular Retinol Binding Proteins

Although widely expressed in different tissues, CRBP I has temporally and spatially restricted patterns of expression during development (Dollé et al., 1990; Maden et al., 1990; Perez-Castro et al., 1989; Ruberte et al., 1991; 1993). Conversely, the expression of CRBP II is, in the adult animal at least, mainly confined to the small intestine where it is an abundant protein, representing 1 percent of soluble protein in rat jejunal mucosa (Ong, 1984; Schaefer et al., 1989). In neonatal rats, CRBP II is also found in liver (Ong, 1984). The factors responsible for regulating the expression of CRBPs during development are unknown. However, analysis of the promoter regions of these two genes has shown that the CRBP I promoter, in common with many "house-keeping" genes, lacks a TATA box and has GC-rich sequences characteristic of Sp1 response elements (Smith et al., 1991). In addition, putative binding sites for Krox-24 and a DR-2 RARE (see part IV, section D for details of RAREs) have been identified and may be involved in developmental control of CRBP I expression (Smith et al., 1991). The CRBP II promoter, on the other hand, contains TATA and CAAT boxes, and a series of DR-1 RXREs (see part IV, section D) and it has been suggested that tissue specificity in CRBP II expression is achieved, in part, by the co-expression of RXR-α at high levels in cells expressing CRBP II (Mangelsdorf et al., 1991).

CRBP I and II have > 50 percent amino-acid sequence identity and both bind retinol in a planar configuration (Cowan et al., 1993; Macdonald and Ong, 1987). However, CRBP I has an affinity for retinol approximately 100-fold higher than does CRBP II with a K_d in the subnanomolar range (Li et al., 1991). Unlike CRBP I, CRBP II binds both retinol and retinal with comparable affinity (Li et al., 1991). Since CRBP II is 1000-fold more abundant than CRBP I in intestinal mucosa (Ong, 1984), this protein is likely to play a central role in the processing and intracellular transport of both retinol and the retinal produced by cleavage of dietary carotenoids.

Cellular Retinoic Acid Binding Proteins

CRABPs are expressed in most cells and tissues although there can be considerable variability in their relative levels of expression: some cell lines and tissues express both forms, whereas in others only one of the two CRABPs is detectably expressed (Dollé et al., 1990; Ruberte et al., 1991; 1993; Siddiqui et al., 1994; 1995). In the nervous system, both CRABPs are expressed in a rhombomere-specific pattern, but in different nervous-system components (Lyn and Giguere, 1994). CRABP I and II differ substantially in promoter structure; the CRABP I promoter, like CRBP I, lacks a TATA box and has GC-rich sequences in common with housekeeping genes (Wei et al., 1990). Methylation of these GC-rich sequences may play an important part in regulating the tissue and developmental expression of CRABP I (Wei and Lee, 1994). The CRABP II promoter has a conventional TATA box, and putative binding sites for a number of transcription factors including Sp1, AP1, AP2, and Krox-24 (Astrom et al., 1992; Macgregor et al., 1992). Although both CRABP I and II are reported to be inducible in response to retinoic acid, the mechanism of induction is different. Retinoic acid induction of CRABP I is dependent on the synthesis of other protein factors (Wei et al., 1989; Wei and Lee, 1994), whereas the CRABP II promoter, unlike CRABP I, has several distinct RAREs that are likely to mediate a direct transcriptional induction via RARs (Astrom et al., 1992; 1994).

While the differences in promoter structure between CRABP I and II provide a mechanism for regulating developmentally- and tissue-regulated patterns of expression, the reason for this tissue-specific variability is unknown and reflects our lack of understanding of the precise roles of these two proteins in cellular retinoid pathways. It has been suggested that the function of CRABP is to facilitate the transport of retinoic acid to the nucleus and deliver it to specific chromatin sites in order to regulate gene expression (Takase et al., 1986). However, although the hypothesis that CRABP transports retinoic acid to the nucleus cannot be discounted, there is rather more evidence to support an alternative (or additional) function of CRABPs in mediating the metabolic degradation of retinoic acid. Over-expression of CRABP I in embryonal carcinoma cells increases both the resistance of these cells to the differentiation-inducing effects of retinoic acid and the production of retinoic acid metabolites (Boylan and Gudas, 1991; 1992). *In vitro* studies using recombinant CRABP I in *E.coli* also indicate that CRABP mediates the metabolic degradation of retinoic acid (Fiorella and Napoli, 1991), perhaps by mediating an interaction with cytochrome P450 enzymes that are important in retinoic acid metabolism. It should be stressed that CRABPs could conceivably function both as a transport protein (to the nucleus) and in facilitating the metabolic degradation of retinoic acid. Furthermore, it is not clear why there should be two forms of CRABP. Since CRABP II has a similar spectrum of specificity for different retinoids as CRABP I but at a 3-fold lower level, it has been suggested that both forms operate to regulate intracellular retinoic acid concentrations but to different set points (Fiorella et al., 1993). It is important to note that CRABPs discriminate between retinoic acid isomers: 13-*cis* and 9-*cis* retinoic acid have a lower affinity for CRABPs (Daly and Redfern, 1988; Fogh et al., 1993; Redfern and Wilson, 1993). Conversely, some modifications to the retinoid cyclohexenyl ring do not greatly affect binding affinity and the 4-oxo and 4-hydroxy metabolites have an affinity for CRABPs comparable to all-*trans* retinoic acid (Daly and Redfern, 1988; Fiorella et al., 1993).

Thus, a high level of expression of CRABP may be predicted in tissues in which a low level of "free" retinoic acid is required and it has been suggested that an opposing gradient of expression of CRABP in embryonic tissues may be a mechanism to steepen the concentration gradient of "free" retinoic acid (Maden et al., 1988). A low level of effective retinoic acid would be achieved either by the sequestration of retinoic acid by CRABP, or sequestration and subsequent metabolism to inactive metabolites. The concept of "free" retinoic acid remains poorly defined. Since retinoic acid is a relatively lipophilic molecule, it may never exist "free" in an aqueous environment but be transported through the cell and from compartment to compartment via intracellular membranes, or by binding proteins. Since proteins such as microsomal cytochrome P450 enzymes may interact with CRABP, it is possible that the prevalence of such interactions determines whether retinoic acid is sequestered, transported to the nucleus, or degraded to inactive metabolites. This emphasizes a potentially important function of CRABPs in cellular homeostasis: regulating the concentration of ligand (retinoic acid) available to the nuclear receptors. The fact that CRABP II is rapidly inducible in response to retinoic acid could thus be an important component of cellular homeostasis and a mechanism to ensure that retinoic acid is maintained at a low level at critical points in development or to maintain cellular phenotype. Whether this model is valid requires experimental verification. In studies of CRABP II and CRABP I expression in the murine nervous system, Lyn and Giguere have found that in response to teratogenic doses of retinoic acid, although the expression of CRABP II was increased (CRABP I was unaffected), there was no alteration in the normal spatial or temporal boundaries of CRABP II expression (Lyn and Giguere, 1994). Thus, if CRABP II is involved in cellular homeostasis, it is only in certain tissues. Transgenic mice homozygous null for either CRABP I, CRABP II or both CRABP I and II are viable and fertile with no apparent major defects (Garry et al., 1994; Fawcett et al., 1995; Lampron et al., 1995), and this clearly raises the issue of how important CRABPs are in development and whether or not CRABP II does indeed have an important role in neural development, as suggested by expression studies (Lyn and Giguere, 1994). The only defect detectable in the CRABP II-nulls was an additional post-axial digit (Fawcett et al., 1995). Inappropriate expression of CRABPs can also have developmental consequences, as illustrated by defective lens differentiation and pancreatic tumourigenesis in transgenic mice overexpressing CRABP I (Perez-Castro et al., 1993).

B. The Metabolism of Retinol and Retinoic Acid

In view of the fact that there are nuclear receptors for retinoic acid that appear to play a key role in the developmental regulation of gene expression, the main focus of studies in retinoid metabolism is directed towards understanding the way in which intracellular concentrations of retinoic acid are controlled. Most information on the pathways of retinoic acid synthesis and metabolism has come from studies on a limited range of tissues, predominantly liver. It is thus important to recognize that, in other tissues, these pathways and their mechanisms of control may differ in detail or importance with respect to controlling retinoic acid levels. Furthermore, other metabolites of retinol and retinoic acid may have important biological functions, perhaps mediated by RARs/RXRs or as-yet-unidentified nuclear receptors.

There are two main routes by which retinoic acid may be generated intracellularly: synthesis from retinol and by central or excentric cleavage from carotenoids. More is known about the former, and as yet the importance of serum carotenoids with respect to the regulation of intracellular levels of retinoic acid is unknown. As summarized in part II, current evidence favors the idea that retinol delivery to cells is achieved by a specific mechanism involving cell surface receptors for RBP. The 63 kDA component of these receptors has no obvious transmembrane regions, and the mechanism of retinol transfer across the plasma membrane is open to speculation. Whether CRBP plays an active role in retinol uptake by specific interaction with a translocation mechanism, or picks up retinol nonspecifically from the plasma membrane lipid bilayer, is also unknown.

CRBP I plays a central role in pathways of retinol metabolism and the predominant form of retinol *in vivo* is as a complex with CRBP I (Napoli et al., 1991). Synthesis of retinoic acid from retinol is a two stage process involving retinol dehydrogenases to convert retinol to retinal, and retinal dehydrogenases to convert retinal to retinoic acid. The first stage is the rate-limiting step carried out by microsomal NADP-dependent retinol dehydrogenases (RoDH) that recognize holo-CRBP I as substrate (Napoli et al., 1992; Posch et al., 1991). At least three retinol dehydrogenases have been identified so far—all are members of the short-chain alcohol dehydrogenase family and expressed with some degree of tissue specificity: RoDH III is expressed only in liver whereas RoDH I and II, while predominant in liver, are expressed more widely (Chai et al., 1995a; 1996; 1995b). Holo-CRBP I is also a substrate for Lecithin-retinol acetyl transferase (LRAT) which catalyzes the formation of retinyl esters (Ong et al., 1988), but this activity is inhibited by apo-CRBP I (Herr and Ong, 1992). In addition, apo-CRBP I stimulates retinyl ester hydrolysis (Boerman and Napoli, 1991) and thus, under conditions where holo-CRBP I is low, plays a key role in maintaining intracellular levels of retinoic acid synthesis. Although microsomal RoDHs are the main source of retinal for retinoic acid biosynthesis, cytosolic NAD- and NADP-dependent retinol dehydrogenases have been reported (Boerman and Napoli, 1996). The function of these enzymes is uncertain. However, their activity is inhibited by apo-CRBP I and it has been speculated that their primary function may be as retinal reductases, reducing retinal to retinol; under conditions of vitamin A depletion this activity would be inhibited by apo-CRBP I, thus allowing the conversion of retinal (derived, for example, from carotenoids) to retinoic acid (Boerman and Napoli,

1996). Oxidation of retinal to retinoic acid can be achieved via cytosolic "retinal dehydrogenases" (Napoli and Race, 1987), or by microsomal cytochrome p450 isoenzymes (Raner et al., 1996; Roberts et al., 1992; Tomita et al., 1993), or retinal (aldehyde) oxidase (Tsujita et al., 1994).

The emphasis of much of the work on retinoic acid metabolism has been on the synthesis and degradation of all-*trans* retinoic acid. Although 9-*cis* retinoic acid is now thought to be a biologically important retinoid, pathways for its synthesis *in vivo* are unknown and largely a matter of speculation. As discussed earlier, the cleavage of 9-*cis*-β-carotenoids to 9-*cis* retinal and subsequent oxidation to 9-*cis* retinoic acid by a "retinal dehydrogenase" that does not discriminate between all-*trans* and 9-*cis* isomers, could be an important route for the production of 9-*cis* retinoic acid. Isomerization *in vivo*, either of all-*trans* retinoic acid or β-carotene precursors, is an additional route that could be important for 9-*cis* retinoic acid formation, and may occur via microsomal enzymes (Samokyszyn and Marnett, 1993) or nonspecifically by protein thiol groups (Urbach and Rando, 1994), but no specific isomerase has yet been identified.

While CRBP I plays a key role in the synthesis of retinoic acid and its regulation, holo-CRABP I and II have a correspondingly important, but less well-defined, role in the degradation or metabolic inactivation of retinoic acid. Although it is clear that increasing CRABP expression reduces sensitivity to retinoic acid and increases the formation of metabolites (Boylan and Gudas, 1991; 1992), all-*trans* retinoic acid can also be oxidized by microsomal enzymes independently of CRABP (Fiorella and Napoli, 1994). Moreover, there are multiple and potentially tissue-specific pathways of retinoic acid metabolism, involving multiple P450 isozymes (e.g., Muindi and Young, 1993) and hydroperoxide-dependent mechanisms. Cytochrome P450 enzymes mediate an initial C4-hydroxylation step, resulting in the formation of 4-hydroxy- and 4-oxo-retinoic acid, although hydroxylation at C18 can also occur (Fiorella and Napoli, 1994). C4-oxidized metabolites, particularly 4-oxo-retinoic acid, are potentially able to activate RARs (Pijnappel et al., 1993) and bind to CRABPs with an affinity only slightly lower than that of all-*trans* retinoic acid (Fiorella et al., 1993). However, unlike all-*trans* retinoic acid, C4-oxidized metabolites are reported to be metabolized slowly, if at all, when bound to CRABP (Fiorella and Napoli, 1994). Recent work on the metabolism of 9-*cis* retinoic acid suggests that, in addition to 4-oxidation, reduction to 13,14-dihydro-9-*cis* retinoic acid may be an initial step leading to β-oxidation (Shirley et al., 1996).

The peroxidase activity of microsomal prostaglandin H synthase acting on lipid hydroperoxides is an indirect route of retinoic acid catabolism, leading to the formation of C4-oxidized and 5,6-epoxy metabolites of retinoic acid (Samokyszyn et al., 1995; Samokyszyn and Marnett, 1987). The relative importance of prostaglandin H synthase in retinoic acid metabolism and the role of CRABP in this pathway is unknown. P450-dependent metabolism may be an important route of degradation of all-*trans* retinoic acid, since P450 inhibitors such as the imidazole antimycotics ketoconazole and fluconazole can increase the sensitivity of cells to retinoic acid (Williams and Napoli, 1985) and *in vivo* may maintain plasma retinoic acid levels in patients undergoing retinoic acid therapy (Schwartz et al., 1995), although these drugs also inhibit lipoxygenase activity.

Despite the existence of CRABP-independent routes of retinoic acid metabolism, it is possible that CRABP is the main route of catabolism of endogenous all-*trans* retinoic acid. This idea is based on a number of observations: first, CRABP II is inducible by retinoic acid, suggesting that this is an important homeostatic mechanism regulating retinoic acid availability; second, clinical resistance to all-*trans* retinoic acid may be associated with increased expression of CRABP (Cornic et al., 1994), although inducible P450 isozymes may also be an important component of resistance; third, isomers such as 13-*cis* and 9-*cis* retinoic acid that do not bind to CRABP (Fogh et al., 1993; Redfern and Wilson, 1993) are metabolized more slowly (Adamson, 1994; Adamson et al., 1995; Duell et al., 1996; Muindi et al., 1994). Since prostaglandin H synthase-mediated oxidation does not discriminate between 13-*cis* and all-*trans* retinoic acid (Samokyszyn et al., 1995; Samokyszyn and Marnett, 1987), this implies that the degradation of all-*trans* retinoic acid is mediated by specific mechanisms in which CRABP mediates the direct enzymatic catabolism of bound all-*trans* retinoic acid (Fiorella and Napoli, 1991). However, there is still a great deal to be learned about the microsomal enzymes involved, the relative importance of P450 and prostaglandin H synthase routes of catabolism, the precise roles of CRABPs in sequestering and promoting the metabolism of all-*trans* retinoic acid, and the way in which retinoic acid degradation is integrated and regulated in different tissues.

IV. RETINOIDS IN THE NUCLEAR COMPARTMENT

A. Retinoic Acid Receptors (RARs)

On the basis of biochemical studies with purified or partially purified CRBP and CRABP it was originally suggested that retinol and retinoic acid interacted with specific sites on the chromatin, via CRBP and CRABP (Liau et al., 1981; 1985; Takase et al., 1986). Although it does not seem likely that the cellular retinoid binding proteins directly deliver retinoids to chromatin, the basic idea that retinoids control gene expression by binding to "chromatin" (in a broad sense) is essentially correct. Specific nuclear retinoic acid binding proteins or receptors were identified on the basis of biochemical experiments (Daly and Redfern, 1987) and cloned independently at about the same time (Giguere et al., 1987; Petkovich et al., 1987). Predicted amino acid sequences of RAR-α, the first RAR type to be identified (Giguere et al., 1987; Petkovich et al., 1987), suggested that the protein would fold into the discrete A/B, C (DNA-binding), D, E (ligand-binding), and F domains characteristic of steroid and thyroid hormone receptors, and RARs are therefore included with the steroid receptor superfamily. The isolation of RAR-α (Giguere et al., 1987; Petkovich et al., 1987) was followed by the discovery of two additional

RARs, RAR-β (Benbrook et al., 1988; Brand et al., 1988) and RAR-γ (Krust et al., 1989; Zelent et al., 1989). Each RAR type is encoded by a distinct gene; in humans, these have been mapped to 17q21.1 (RAR-α), 3p24 (RAR-β), and 12q13 (RAR-γ) (Mattei et al., 1991).

The identification of three distinct RAR types goes some way to explaining the diversity of cellular responses to retinoic acid. The fact that each type is highly conserved between species suggests that the different RARs have important and distinct functions, and this is borne out by temporally- and tissue-specific patterns of RAR expression during animal development (Dollé et al., 1989; 1990; Ruberte et al., 1991; 1990). Further complexity is indicated by the existence of RAR isoforms within each RAR type (Kastner et al., 1990; Leroy et al., 1991a; Zelent et al., 1991). These isoforms are generated by transcription from different promoters and alternative splicing, resulting in mature transcripts with 5'UTR and A-domains derived from different exons within each *RAR* gene (Lehmann et al., 1991b; Leroy et al., 1991a; Zelent et al., 1991). Two major isoforms for RAR-α (Leroy et al., 1991a) and RAR-γ (Kastner et al., 1990; Lehmann et al., 1991b), and four for RAR-β (Nagpal et al., 1992b; Zelent et al., 1991) have been identified, although a range of minor isoforms of uncertain significance has also been described for RAR-α and RAR-γ (Kastner et al., 1990; Leroy et al., 1991a). For a given RAR type, mRNA for each isoform has a different 5'UTR and A-domain while sharing common B, C, D, E, and F domains. RAR isoforms show tissue-specific patterns of expression (Leroy et al., 1991a; Zelent et al., 1991) and since the A-domains have transcriptional activation functions (Nagpal et al., 1993; 1992a), this suggests some specificity in interactions with components of the preinitiation transcription complex. Unlike all the other major RAR isoforms, RAR-β4 lacks all except four amino acid of the A-domain (Nagpal et al., 1992b), suggesting that this isoform could acts as a dominant-negative receptor able to bind to a retinoic acid response element in a ligand-dependent manner but unable to activate transcription.

Transcription of each *RAR* gene can be driven from two promoters, one upstream of the 5' UTR and coding exons, and an internal promoter that drives transcription of the RAR-α2 (Leroy et al., 1991a; 1991b), RAR-γ2 (Lehmann et al., 1991b; 1992b) and RAR-β2/β4 isoforms (Nagpal et al., 1992b; Zelent et al., 1991). RAR-β expression is induced by all-*trans* retinoic acid (de The et al., 1989; Redfern et al., 1990) and characterization of the RAR-β2 promoter revealed a retinoic acid response element (RARE) consisting of two direct repeats (DR) of the sequence (G/A)GTTCA separated by 5 bases (de The et al., 1990). Although the expression of RAR-α and RAR-γ is less sensitive to retinoic acid and inducible only by high levels of the ligand, both the RAR-α2 and RAR-γ2 promoters contain similar RAREs (Lehmann et al., 1992b; Leroy et al., 1991b). The RAR-α2 RARE has a lower affinity for RARs than does the RAR-β2 RARE (Leroy et al., 1991b), thus providing an explanation for the differential sensitivity of these RAREs to retinoic acid. Neither the RAR-α2 or RAR-γ2 promoters contain a consensus TATA box (Lehmann et al., 1992b; Leroy et al., 1991b), but both contain putative binding sites for the Sp1 transcription factor, a feature common to many housekeeping genes. The upstream promoters of the *RAR* genes have been less well studied, with the exception of the RAR-α1 promoter, which lacks TATA or CAAT box elements but has several putative Sp1 binding sites and a Krox-20 enhancer element (Brand et al., 1990). The Krox-20 enhancer may regulate RAR-α expression in embryonic rhombomeres, since both Krox-20 and RAR-α are coexpressed at similar times during brain development (Ruberte et al., 1991).

B. Retinoid X Receptors (RXRs)

RARs control gene transcription as ligand-dependent receptors and it was initially assumed that RARs worked as homodimers, as is the case for glucocorticoid and other steroid hormone receptors. However, proteins present in nuclear extracts were found to increase the activity of RARs on their response elements (Glass et al., 1990). A nuclear receptor similar to RARs, but sufficiently distinct on sequence criteria to warrant classification as a separate nuclear receptor subfamily, was cloned subsequently (Mangelsdorf et al., 1990) and shown to act as an auxiliary factor to RARs and other nuclear receptors such as vitamin D receptors (VDR), thyroid hormone receptors (TR), and peroxisome proliferator-activated receptors (PPAR) (Issemann et al., 1993; Kliewer et al., 1992; Yu et al., 1991; Zhang et al., 1992a). This new receptor did not have appreciable affinity for all-*trans* retinoic acid but was called a retinoid X receptor or RXR because the structure of the putative ligand binding domain suggested that a retinoid or related molecule could act as a ligand. Further work revealed a family of related RXRs, designated as RXR-α, RXR-β, and RXR-γ (Mangelsdorf et al., 1992), differentially expressed in adult and embryonic tissues (Dolle et al., 1994; Mangelsdorf et al., 1992). These RXRs are encoded by separate genes that, in humans, have been mapped to 9q34.3 (RXR-α), 6p21.3 (RXR-β), and 1q22-q23 (RXR-γ) (Almasan et al., 1994). As with the RARs, different isoforms of the RXRs have been identified, at least for RXR-β and RXR-γ. In the mouse, two major isoforms of RXR-β have been identified, corresponding to 2.6 and 2.8 kb mRNAs expressed tissue-specifically and transcribed from separate promoters (Nagata et al., 1994). Like RAR isoforms, the two RXR-β isoforms have common B through to F domains and differ at the A domains, which are encoded by different exons: mRXR-β1 has a N-terminal extension of 72 amino-acid residues relative to mRXR-β2 (Nagata et al., 1994). Two major isoforms of RXR-γ have also been characterized in mouse tissues: mRXR-γ2 differs from mRXR-γ1 with respect to the 5'UTR and A-domain; translation of the mRXR-γ2 mRNA probably starts within the B-domain sequence, yielding a receptor protein lacking an A-domain and with a shorter B-domain (Liu and Linney, 1993). Unlike RARs and the other RXRs, RXR-α is expressed as a single large transcript of c. 5.6 kb (Mangelsdorf et al., 1992).

As predicted from the structure of its ligand-binding domain, the ligand for RXRs is a retinoid and turns out to be the 9-*cis* isomer of retinoic acid (Heyman et al., 1992; Levin et al., 1992). 9-*cis* retinoic acid also binds with high affinity to RARs (Allegretto et al., 1993; Allenby et al., 1993; 1994), but the binding site within the RAR ligand binding domain is apparently distinct from that of all-*trans* retinoic acid (Tate et al., 1994; Tate and Grippo, 1995). Unlike 9-*cis* , all-*trans* retinoic acid binds only to RARs and not to RXRs. Equilibrium dissociation constants (K_d) are within the range 0.1 to 1 nM for the binding of all-*trans* retinoic acid to RARs and 9-*cis* retinoic acid to RARs and RXRs (Allegretto et al., 1993). However, more recent studies suggest that 9-*cis* retinoic acid has a significantly higher affinity for RARs than for RXRs (Kersten et al., 1996). Two hydrophobic pockets have been identified from structural studies within the ligand binding domain of RXRs and one of these, which has some similarity to the binding pocket of E-RABP, is thought to be the binding site for 9-*cis* retinoic acid (Bourguet et al., 1995).

In Zebrafish, RXRs corresponding to the human and murine RXR-α and RXR-γ have been identified; however, two additional RXR types, characterized by 14 amino-acid insertions in the ligand binding domain have also been found. These novel Zebra fish RXRs, designated RXR-δ and RXR-ε, do not bind 9-*cis* retinoic acid but seem to form heterodimers with TRs although RXR-δ acts as a dominant negative repressor of transcription (Jones et al., 1995).

C. Developmental Function of RARs and RXRs

RARs, particularly RAR-β and RAR-γ, have discrete patterns of expression during embryonic development and the mature animal also has tissue-specific patterns of RAR expression (Rees et al., 1989; Zelent et al., 1989). The high degree of RAR sequence conservation between different species, the existence of three distinct RAR types, each with N-terminally distinct isoforms, and the temporally and spatially restricted patterns of RAR expression during development all point to important roles for RARs and RXRs in mediating retinoic acid signaling during development. To characterize the function of RARs, homologous recombination between mutant and endogenous wild-type *RAR* genes in embryonic stem cells has been used to generate transgenic mice homozygous for null mutations in one or more of the *RAR* genes. These experiments, in general, confirm the developmental importance of RARs, RXRs, and retinoic acid, but an unexpected finding is that there is partial functional redundancy between RAR types and isoforms. For example, transgenic mice homozygous null for RAR-γ2 (Lohnes et al., 1993), RAR-α1 (Lufkin et al., 1993), or RAR-β2 (Mendelsohn et al., 1994) all appear normal, implying functional redundancy between RAR isoforms. RAR-β null homozygotes are apparently viable and fertile, with no externally apparent abnormalities (Luo et al., 1995) and this implies functional redundancy between RAR types. Despite the wide distribution of expression of RAR-α, mice homozygous null for this RAR display a limited range of defects consisting of early post-natal lethality and testis generation (Lufkin et al., 1993). RAR-γ nulls display growth deficiency, early lethality and male sterility due to squamous metaplasia of seminal vesicles and prostate (Lohnes et al., 1993). None of the single RAR-null mutants display the multiple congenital abnormalities characteristic of Vitamin A-deficiency Syndrome (VAD) (Wilson et al., 1953) whereas RAR-null double mutants do mimic VAD.

The fact that RAR-β2 is induced in many tissues (de The et al., 1989; Lovat et al., 1993; Redfern et al., 1990; Redfern and Todd, 1992; Viallet et al., 1991) by exogenous all-*trans* retinoic acid suggest that this RAR may mediate the teratogenic effects of retinoic acid, or be part of a homeostatic mechanism to abrogate its effects. However, mice embryos homozygous null for RAR-β2 (Mendelsohn et al., 1994) or RAR-β (Luo et al., 1995) are comparable to wild-type in their susceptibility to the teratogenic effects of retinoic acid. The lack of effect of RAR-β null mutations, an indication of functional redundancy between RAR types, raises the question of the function of RAR-β. A partial answer has been obtained from the generation of double-null mutants in which the compensatory RAR (presumably RAR-α) is removed: homozygous null double mutants for RAR-α1/RAR-β show a number of defects not apparent with RAR-α1/RAR-β2 homozygous nulls, implying that the RAR-β1/3 isoforms have distinct functions to RAR-β2 (Luo et al., 1996). In addition, expression of a RAR-β4-like transgene in mice predisposes towards hyperplasia and neoplasia, in contrast to RAR-β2, which shows tumor suppresser activity (Berard et al., 1994).

Gene-knockout experiments with embryonal carcinoma cells *in vitro* complement the transgenic mouse studies and show that each RAR may regulate different subsets of retinoic acid-responsive genes (Boylan et al., 1993; 1995). Over-expression experiments on cells *in vitro* also demonstrate partial functional redundancy (Taneja et al., 1995), which may vary with the nature of the target gene (Roy et al., 1995) and in relation to cell type (Taneja et al., 1996). However, this type of experiment should be interpreted with caution: other proteins such as CBP and E1a-like proteins (see part IV, section E) link hormone-receptor dimers with the transcription complex (Berkenstam et al., 1992; Folkers and Vandersaag, 1995; Kamei et al., 1996; Kruyt et al., 1993) and swamping the cell with large amounts of RARs may have indirect "squelching" effects on other receptor-mediated processes by reducing the availability of these proteins.

Experiments to establish the developmental function of RXRs are not so well advanced. Unlike mice lacking a single RAR-α or -β gene, mice homozygous null for RXR-α suffer embryonic lethality with hypoplastic development of the heart and delayed liver development (Sucov et al., 1994). The cardiac defects in RXR-α null mice represent a subset of the defects seen in vitamin A deficiency and presumably result from a lack of RXR-α to form heterodimers with RARs. As with RAR-null mice, the fact that defects are so limited, despite the apparent requirement for RXRs to heterodimerize with RARs and the widespread distribution of RXR-α expression, is surprising. One explanation could be that RXRs may be required in these other tissues for mature tissue function, rather than for morphogenesis (Sucov et al., 1994). Further work on these mice have shown that the RXR-α-null embryos are resistant to the teratogenic effects of excess retinoic acid at specific stages in pregnancy (Sucov et al., 1995). This suggests that retinoic acid teratogenicity is mediated by RXR-RAR heterodimers and since heterozygotes are less resistant, RXR-α dosage is clearly important in limiting effective concentrations of ligand-activated RXR-RAR heterodimers (Sucov et al., 1995). As yet, the developmental function of RXR-β and RXR-γ and the degree of functional redundancy between RXR types is unknown.

D. Retinoid Response Elements and Receptor Dimerization

Steroid and thyroid hormone receptors bind (usually as dimers) to hormone response elements consisting of specific DNA sequences associated with gene promoters or enhancers. DNA binding occurs via two α-helices of the C-domain, arranged perpendicular to each other and with the N-terminal end of each helix held in a loop ("Zinc finger") by the coordination of a zinc ion with conserved cysteine residues at the loop base (Schwabe et al., 1990). The same is true for RARs and RXRs and, following the characterization of the RAR-β2 RARE (de The et al., 1990), it is now accepted that direct repeats (DR) of the hexanucleotide PuGGTCA (the consensus half-site) separated by 1-5 neutral nucleotides (DR1-DR5) form the basis of response elements for RARs, RXRs, and heterodimers between RXRs and other nuclear receptors such as VDRs, TRs, and PPARs (Stunnenberg, 1993; Umesono et al., 1991). Different response elements are bound by different receptor hetero- and homodimers. For example, RXR-RAR heterodimers bind DR1, DR2, and DR5 elements, although heterodimer binding to DR5 elements mediate transcriptional induction, whereas heterodimer binding to DR1 elements is reported to repress transcription (Kurokawa et al., 1994; Vudac et al., 1996). RXR homodimers bind to DR1 response elements, such as that associated with the CRBP II promoter (Mangelsdorf et al., 1991), which has been used extensively in experimental studies. In addition to these DR1-DR5 response elements, there is some flexibility in the orientations and spacing of the repeat that can be recognized. For example, the γF-crystalin gene promoter contains an everted paldromic repeat of two consensus RARE half-sites separated by 8 nucleotides (Tini et al., 1993) and the human immunodeficiency virus-1 (HIV-1) LTR contains an inverted palidromic RARE, recognized by RXR-RAR heterodimers and RXR homodimers, consisting of two RARE half-sites separated by 9 nucleotides (Lee et al., 1994). Response elements may be located close to the site of assembly of the transcription complex on the TATA box, as in the case of the RAR-β2 RARE, or some distance away, such as in the human *CRABP II* gene where a DR5 RARE is approximately 5.6 kb upstream of the transcription start site (Astrom et al., 1994).

The characterization of RAREs as direct repeats consisting of two consensus half-sites is important confirmation that RARs and RXRs work as dimers, each partner binding to a half-site of the response element. However, some nuclear receptors such as TR and NGFI-β can also bind efficiently as monomers because the α-helix at the C-terminal end of the DNA-binding domain (DBD) forms DNA contacts 5' to the half-site (Rastinejad et al., 1995). The binding of RXR-heterodimers to a response element occurs with a specific polarity, the RXR partner binding to the upstream 5' half-site and the other partner to the 3' half-site (Kurokawa et al., 1993; Perlmann et al., 1993; Predki et al., 1994; Zechel et al., 1994). Dimerization involves distinct domains within the DNA-binding domain (DBD) and ligand binding domains of RARs, RXRs, and other nuclear receptors (Nagpal et al., 1993). The dimerization domains within the DBDs of TR and RXR have been characterized in some detail (Rastinejad et al., 1995): heterodimers of RXR-α and TR-β bind to DR4 response elements with the RXR-α partner on the 5' half-site (Perlmann et al., 1993). In solution, isolated DBDs of these receptors dimerize only weakly but in the presence of DNA there is a cooperative assembly of RXR-α -TR-β DBDs to properly spaced half-sites (Rastinejad et al., 1995). The specificity of binding depends on the dimerization interface between monomers and the long A-box helix at the C-terminal end of the TR DBD, which plays a key role in the discrimination of half-site spacing by producing a sterically unfavorable interface with half-site spacings of less than 4 bp (Rastinejad et al., 1995). The ligand binding domain (LBD) of RARs and RXRs also contains a dimerization domain (Nagpal et al., 1993) important for high-affinity DNA binding and heterodimerization. It has been suggested that heterodimerization via the DBD and LDB dimerization domains occurs sequentially: heterodimerization via the LBD dimerization interface in solution results in heterodimers which can only bind with high affinity to specific direct repeat elements via the formation of the second dimerization interface within the DBD (Perlmann et al., 1996).

Early studies on RXRs demonstrated marked 9-*cis* retinoic acid-dependent positive cooperativity in the binding of RXR homodimers to retinoid response elements (Zhang et al., 1992b). More recent data show that RXRs form tetramers in solution (Kersten et al., 1995a) and that these tetramers dissociate to form RXR dimers in response to 9-*cis* retinoic acid (Kersten et al., 1995b). Surprisingly, at low levels of ligand saturation, the affinity of 9-*cis* retinoic acid for RXRs has been found to be much weaker than at high levels (Kersten et al., 1995c). These reports provide an explanation for the different dose-dependent effects on gene activation of all-*trans* and 9-*cis* retinoic acid in neuroblastoma cells: unlike all-*trans* retinoic acid, 9-*cis* has less effect on the induction of RAR-β and CRABP II than at higher concentrations where the situation is reversed and 9-*cis* produces a greater transcriptional activation (Redfern et al., 1994; 1995). Thus, at low 9-*cis* retinoic acid concentrations, RXRs would remain largely as apo-RXR tetramers; at higher 9-*cis* concentrations, RXRs would dissociate into holo-RXR dimers and monomers with the ability to regulate gene transcription either as holo-RXR homodimers or heterodimers between holo-RXR and other nuclear receptors such as LXRs, NGFI-β, and NURR-1. This model implies that RXR homodimers may be mediators of the effects of 9-*cis* retinoic acid on cells and tissues *in vivo* and *in vitro*.

However, it is not clear if 9-*cis* retinoic acid has a role in activating RXR-RAR heterodimers by binding to the RAR partner. Structural studies on the RXR-α DBD have shown that a third helix of this domain, referred to as the T-box, forms a homodimerization interface with the second zinc finger of another RXR monomer and also has DNA binding activity (Lee et al., 1993). There is an additional homodimerization domain, at least in RXR-β, localized within the last 30 residues at the C-terminal end of the protein and this is distinct from the C-terminal heterodimerization domain (Leng et al., 1995). Clearly, RXR homodimerization and DNA binding may involve different mechanisms to heterodimerization. Although it has been suggested that RXR tetramers may bind to the CRBP II promoter, which consists of a tandem array of DR1 RXREs, and that this interaction is moderately enhanced by 9-*cis* retinoic acid (Chen and Privalsky, 1995), it is not clear if this is relevant to transcriptional regulation *in vivo*.

E. Transcriptional Regulation by Retinoic Acid Receptors

The regulation of transcription is a complex process that is only beginning to be understood, and occurs at the level of chromatin packing and the regulation of transcription complex assembly and function. Assembly of the preinitiation transcription complex on DNA occurs via binding of the TATA-box binding protein (TBP), a component of transcription factor TFIID, to a core promoter element (such as the TATA box). This DNA-TFIID complex is probably very stable (Zawel and Reinberg, 1995). RNA polymerase II is then recruited via additional bridging factors such as TFIIB. For transcription to occur, RNA polymerase II must then be released from the preinitiation complex and this is dependent on other transcription factors such as TFIIH. At the same time, TFIIB or other bridging factors that recruited RNA polymerase II may themselves be released from the preinitiation complex. One of the functions of transcriptional activators, such as ligand-dependent protein dimers bound to hormone response elements, is apparently to retain TFIIB or other bridging factors in the preinitiation complex, via interactions between the TBP-associated factors (TAFs) constituting TFIID and/or the RNA polymerase II holoenzyme (Chao and Young, 1996; Walker et al., 1996), thus allowing multiple reinitiation events (review: Ho et al., 1996; Zawel and Reinberg, 1995). Protein-protein interactions between hormone-dependent receptors bound to response elements and the pre-initiation complex may be facilitated by a physical bending of DNA at the response element induced by the binding of receptors (Lu et al., 1993).

A number of different modes of transcriptional regulation by RARs and RXRs have been described. Most attention has been focused on transcriptional induction where the binding of RXR-RAR heterodimers to DR5 response elements results in increased transcription in response to ligand. The ligand in this case is all-*trans* retinoic acid, which is sufficient to activate RXR-RAR heterodimers by binding to the RAR partner (Xiao et al., 1995). Activation results from ligand-dependent conformational changes in an "activation function" (AF) within the LBD. In addition to the LBD AF, referred to as AF-2, there is a ligand-independent AF (AF-1) associated with the N-terminal or A-domain that synergizes with AF-2 (Nagpal et al., 1993; 1992a). The AFs of different RARs are distinct, showing specificity for different response elements and the context of the promoter (Nagpal et al., 1993). RXRs also have AF1 and AF2: AF2 in RXR-α and RXR-β is an α-helical domain that is inactive in the absence of ligand (Bourguet et al., 1995); proteolysis (Leng et al., 1995) and fluorescence anisotropy experiments (Kersten et al., 1995b) indicate that the binding of ligand results in a conformational change such that holo-RXR is more compactly folded (Kersten et al., 1995b).

Although studies have confirmed that activation of the RAR partner of RXR-RAR heterodimers is sufficient to confer transcriptional activation, the role of 9-*cis* retinoic acid in RXR-RAR activation is unclear. 9-*cis* retinoic acid binds to RARs with an affinity comparable to all-*trans* retinoic acid (Allegretto et al., 1993), but apparently to a different site within the LBD (Tate et al., 1994; Tate and Grippo, 1995). Whether this results in transcriptional activation is difficult to answer with certainty because of the likelihood of significant isomerization to all-*trans* retinoic acid that may occur during the incubations (12 hours or more) used for transient transfection experiments. Despite the isomerization of 9-*cis* to all-*trans* retinoic acid that occurs in cells *in vitro* (Urbach and Rando, 1994), the fact that 9-*cis* does not bind appreciably to CRABPs (Fogh et al., 1993; Redfern and Wilson, 1993) and the slower metabolism of 9-*cis* retinoic acid compared to all-*trans*, it is still frequently concluded that 9-*cis* activates RARs (e.g., Jong et al., 1993). Recent data that 9-*cis* retinoic acid has a higher affinity for RARs compared with RXRs (Kersten et al., 1996) implies that treating cells with a low concentration of 9-*cis* will result in occupation of RARs first. At higher concentrations, 9-*cis* will bind to the RXR partner of RXR-RAR heterodimers and the levels of RXR homodimers and RXR monomers will increase as a result of the dissociation of RXR tetramers (Kersten et al., 1995c). Two important questions are: (1) does the binding of 9-*cis* retinoic acid to RARs result in transcriptional activation? (2) does the binding of 9-*cis* retinoic acid to the RXR partner make any contribution, either synergistic or additive, to the activation of heterodimers? An answer to the first question has not been obtained unequivocally, and comparative dose-response studies using 9-*cis* and all-*trans* retinoic acid do not support the idea that 9-*cis* activates the RAR partner, at least in human neuroblastoma cells (Redfern et al., 1994).

The use of RAR- and RXR-specific agonists and antagonists has given some answers to the question of whether each partner in RXR-RAR heterodimers contributes to transcriptional regulation. As with many facets of retinoid biology, a different answer may be obtained in different experimental systems. For example, the human cytomegalovirus promoter contains a DR2 and two DR5 RAREs; RXR-RAR heterodimers bind to all elements with equilibrium dissociation constants in the range 5-20 nM and RAR-specific ligands such as TTNPB (all-E-4[2-(5,6,7,8-tetrahydro-5,5,8,8-tetramethyl-2-naphthalenyl)-1-propenyl]benzoic acid) are sufficient to achieve full activation. RXR-specific ligands on their own do not activate and do not synergize with RAR-specific ligands (Angulo et al., 1996). Conversely, in NB4 acute promyelocytic leukaemia cells, RAR- and RXR-specific ligands are reported to synergize in promoting differentiation, and the effects of the RXR-specific ligand are dependent on the presence of ligand, whether agonist or antagonist, bound to the RAR partner (Chen et al., 1996b). These conflicting results presumably depend on the differential expression of transcriptional coactivators in each cell type studied.

Many advances in our understanding of nuclear receptors have been achieved using transient transfection techniques to introduce receptor and reporter constructs into cells. Despite their elegance as an experimental approach to studying the regulation of transcription, extrapolating the results of transient transfection experiments to normal, intact cells requires some caution since reporter and receptor constructs may be expressed in excess over that normally achieved by cellular regulatory mechanisms. *In vivo*, ligand-dependent activation of different genes via RXRs, RARs, or other nuclear receptors depends on the presence of other factors or linking proteins that interact either directly or indirectly with ligand-activated RXR, or ligand-activated RAR, or both. Initial studies demonstrated that the ligand-dependent activation of RXR-RAR heterodimers from a DR5 response element required either E1a or an E1a-like activity expressed in target cells (Berkenstam et al., 1992; Folkers and Vandersaag, 1995; Kruyt et al., 1993). The E1a-like activity interacts with the AF2 domain of the RAR partner to facilitate transcriptional induction, but does not appear to interact with the RXR partner (Folkers and Vandersaag, 1995). This provides an explanation for the observation that activation of the RAR heterodimer partner by all-*trans* retinoic acid is sufficient for transcriptional induction from a DR5 response element. In this context the RXR would be acting as a ligand-independent auxiliary factor, but still leaves open the question of whether 9-*cis* retinoic acid can synergize with all-*trans* retinoic acid via RXR.

In their experiments on the dependence of E1a-like activity for the ligand-dependent activation of RXR-RAR heterodimers, Folkers and Vandersaag (1995) suggest that E1a stabilizes the preinitiation complex by interacting with both RAR and TBP, and propose that a third protein may be necessary for the up-regulation of E1a activity. The suggestion that a third protein may be necessary for receptor function has been confirmed by the recent identification of a number of proteins which act as coactivators or corepressors of nuclear receptors. One of these is CREBP binding protein (CBP), a protein coactivator linking cyclic AMP response-element binding protein (CREBP) homodimers with the transcription complex. CBP and a related protein, p300, interact with E1a (Arany et al., 1995; Lundblad et al., 1995) and mediate transcriptional activation via RXR-RAR heterodimers and RXR homodimers (Chakravarti et al., 1996; Kamei et al., 1996). CBP and P300 also mediate transcriptional induction by other transcription factors, including TR, GR, and AP1, and may function as general integrators for extracellular and intracellular signaling pathways (Chakravarti et al., 1996; Kamei et al., 1996). Other classes of proteins interacting with nuclear receptors have been found: receptor-interacting proteins (RIP) of M_r 140 and 160 interact with the AF2 domain of the oestrogen receptor (Cavailles et al., 1994; 1995) and also appear to be coactivators of RXR-RAR heterodimers (Kurokawa et al., 1995). TIF1 is a ligand-dependent coregulator via the LDBs of RARs, RXRs, and other members of the nuclear receptor family but does not associate tightly with TFIID, TFIIB, or TBP (Le Douarin et al., 1995). SUG1 interacts with RAR in a ligand- and AF2-dependent manner (Baur et al., 1996). RARs and RXRs may show specificity in their interactions with these various coactivators: for example, both RAR-α and RXR-α interact efficiently with TIF1 but only RAR-α interacts efficiently with both TIF1 and SUG1 (Baur et al., 1996). Thus, the function of RXR-RAR in cells expressing TIF1 could be synergistically mediated via interactions with either or both the RAR or RXR partner, whereas in cells only expressing SUG1, ligand-dependent activation would be via the RAR heterodimer partner. Clearly, whether the RXR partner of receptor heterodimers functions as a ligand independent auxiliary factor or is capable of activation by 9-*cis* retinoic acid will depend on cell type and the pattern of expression of nuclear receptor coactivators.

As well as requiring cofactors such as CBP and other interacting proteins (TIF1, RIPs, SUG1, E1a-like proteins), transcriptional activation by RXR-RAR heterodimers also requires the ligand-dependent dissociation of co-repressors. For example, a 270 kDa protein called N-CoR and isolated from CV1 cells (Horlein et al., 1995) functions as a polarity-specific repressor, dissociating from the RAR partner in the presence of ligand when the heterodimer is bound to a DR5 response element with the RXR bound to the 5' half-site, but remaining bound when the heterodimer is bound in the reverse orientation to DR1 response elements (Kurokawa et al., 1995). Similarly, two T3 Receptor-associating cofactors (TRACs) have been isolated from human fetal liver and interact differentially with RARs and RXRs (Sande and Privalsky, 1996). TRAC-1 and TRAC-2 are apparently derived from the same gene by alternative splicing, and although they interact with both RARs and RXRs, the interaction with RARs is inhibited by ligand (Sande and Privalsky, 1996). Thus, in the absence of ligand, TRACs bind to receptor heterodimers and repress transcription. These proteins thus mediate the ligand-independent silencing function of RARs and TRs. TRAC-2 has also been isolated independently as SMRT (silencing mediator for retinoic acid and thyroid hormone receptors) from human lymphocytes and HeLa cells (Chen and Evans, 1995) and has since been shown to be a ubiquitous nuclear protein related to N-CoR (Chen et al., 1996a). Since the receptor-interacting domain of SMRT/TRAC-2 acts as anti-repressor, Chen and colleagues also point out that splicing variants or isoforms of SMRT may be important modulators of retinoid signalling (Chen et al., 1996a).

Transcriptional repression via RARs and RXRs may be important in cellular and developmental regulation, and may be either direct (DNA-dependent) or indirect (DNA-independent). Direct repression may be ligand-independent, mediated via TRACs and functionally related proteins, or, in the case of Zebrafish

RXR-δ, a dominant-negative receptor unable to bind ligand. The competition of dominant-negative receptors lacking AF1 (possibly RAR-β4, for example) for a RARE, or the competition of RXR-heterodimers with other transcription factors for composite response elements are ligand-dependent mechanisms leading to direct transcriptional repression. Indirect repression may occur as a result of DNA-independent interactions between RARs and RXRs and other transcription factors. Perhaps the best-characterized mechanism for indirect transcriptional repression is through the ligand-dependent interaction of RARs and RXRs with the *fos* and *jun* protein components of the AP-1 transcription factor, resulting in a ligand-enhanced inhibition of AP-1 binding to its response element (Salbert et al., 1993). In this context, RARs and RXRs may be active as monomers (Salbert et al., 1993) although it has also been suggested that heterodimers inhibit AP-1 activity (Soprano et al., 1996). Other steroid hormone receptors, notably the glucocorticoid receptor (GR), also inhibit AP-1 activity although this is not accompanied by a reduction in AP-1 binding-activity in gel shift assays; in this case it has been suggested recently that the inhibition of AP-1 activity may result from a reduction in available CBP via ligand-dependent interactions between GR and CBP (Kamei et al., 1996). This mechanism may also be responsible for some forms of AP-1 repression by RARs and RXRs (Kamei et al., 1996). Since AP-1 is an important transcription factor promoting cell proliferation (Angel and Karin, 1991), its down-regulation may be part of the explanation for the antiproliferative effects of retinoic acid. Interestingly, the PML-RAR-α fusion protein expressed in acute promyelocytic leukemia cells carrying the chromosome 15:17 translocation activates AP-1 by cooperating with *fos* and *jun* in the presence of ligand, yet is an AP-1 inhibitor in the absence of ligand (Doucas et al., 1993).

F. Receptor-Specific Ligands

The identification of RARs and RXRs has opened the way for the design of new compounds that would bind to the nuclear receptors yet have distinct biochemical or biological properties. It is now possible to use recombinant receptors, purified in large quantities, to study their affinities for different ligands, and to use transient transfection methodology with specific receptor/reporter combinations to assay the ability of different compounds to activate transcription. Initial studies were directed at identifying compounds that would discriminate between RAR-α, RAR-β and RAR-γ and resulted in the discovery of relatively specific agonists for RAR-α (Delescluse et al., 1991; Hashimoto et al., 1990), mixed RAR-β/RAR-γ agonists (Charpentier et al., 1995; Delescluse et al., 1991; Lehmann et al., 1991a), and agonists selective for RAR-γ (Bernard et al., 1992). Useful RAR antagonists have been hard to find, but an RAR-α-specific antagonist has been described (Apfel et al., 1992), and in a recent report, two further relatively selective RAR antagonists have been identified and these may produce their biological effects by competing with agonists for binding to RARs (Lee et al., 1996).

Although, like RARs, three distinct classes of RXR are expressed in mammalian tissues, compounds that distinguish between these different RXR classes have not been identified. This reflects the greater concentration of effort on characterizing the functional significance of RAR types and our limited understanding of the biological role and importance of the different RXR types. However, there are now a number of compounds available that are selective for either RARs or RXRs. These have been extremely useful as research tools to understand the role of heterodimers in gene regulation and facilitate the design of novel therapeutic strategies based on activating specific receptors. All-*trans* retinoic is selective for RARs but may be converted to 9-*cis* by isomerization. Conversely, TTNPB has been known as a retinoid analogue for some time and is RAR-specific, metabolically more stable, and not subject to isomerization (Jong et al., 1993). RXR-selective retinoids have been developed more recently (Boehm et al., 1994; Dawson et al., 1995; Jong et al., 1993; Lehmann et al., 1992a) and a general approach to the design of these analogs is to constrain the double bond positions of natural retinoids to a particular configuration by the inclusion of aromatic rings (Dawson et al., 1995). Perhaps the demonstration that these conformationally restricted analogues of 9-*cis* retinoic acid have greater RXR-specificity than 9-*cis* itself is an indication that the apparent activation of RARs by 9-*cis* may be due either to isomerization to all-*trans* or greater conformational flexibility. While these compounds are reported to selectively activate RXR-homodimers (Lehmann et al., 1992a), their ability to activate the RXR-partner of RXR-RAR heterodimers may depend on the presence of an agonist or antagonist bound to the RAR partner (Chen et al., 1996b), and on the presence of appropriate coactivators able to interact with ligand-occupied RXRs. Furthermore, there is evidence that RXR-specific ligands may enable RXR-RAR heterodimers to act as repressors on DR1 RAREs (Kurokawa et al., 1994).

Transcriptional repression by retinoid receptors may be mediated via DNA-independent ligand-dependent protein-protein interactions between RARs and/or RXRs and AP1 components (part IV, section E). This appears to be a distinct receptor function with different ligand requirements and it has been possible to identify retinoid analogues that specifically promote ligand-dependent interaction with AP1 components but are unable to facilitate transcriptional induction (Fanjul et al., 1994). These analogues inhibit proliferation (Fanjul et al., 1994; 1996) and AP1 interactions may be a general mechanism underlying the antiproliferative effects of retinoic acid rather than through differential gene induction or DNA-dependent transcriptional repression.

V. OTHER MECHANISMS OF ACTION OF RETINOIC ACID

So far, I have stressed a "conventional" mechanism of action of retinoic acid as a key component in a pathway of nuclear receptor activation that is regulated by altering patterns of nuclear receptor expression and the levels of activating ligand.

However, retinoic acid may have other mechanisms of action—perhaps by altering membrane fluidity (Meeks et al., 1981) and permeability (Stillwell et al., 1982), or by acting on other cellular enzyme systems. The possibility that retinoic acid alters membrane fluidity may have consequences for the properties of the plasma membrane and internal membrane systems. For example, it has recently been shown that retinoic acid reduces membrane fluidity in human fibroblasts (Varani et al., 1996), affecting Ca^{2+} transport across the plasma membrane, and disrupts the structure and function of the golgi apparatus, thus potentiating the action of immunotoxins (Wu et al., 1994). These effects are likely to be relatively nonspecific with respect to retinoic acid isomers, as is the case for the effects on the golgi apparatus in which 13-*cis* and all-*trans* retinoic acid are equally effective (Wu et al., 1994), and independent of changes in gene expression.

Retinoic acid may also interact directly with cellular proteins other than nuclear receptors and cellular binding proteins. Many cellular proteins can be "retinoylated": the formation of thioester bonds between retinoic acid and protein (Takahashi and Breitman, 1990a; 1990b; 1991; 1994; Takahashi et al., 1991a). Since the RI and RII regulatory subunits of cyclic AMP-dependent protein kinase can also be retinoylated (Takahashi et al., 1991b), this may account for the observation that retinoic acid increases cyclic AMP-dependent protein kinase activity (Plet et al., 1985; 1986). Retinoylation of the RII subunit could modify the association of RII with $p34^{cdc2}$ or other regulatory proteins (Tournier et al., 1996) and account for some of the effects of retinoic acid on cell proliferation. However, although these effects are apparent after treating cells with retinoic acid, a physiological role for retinoylation has not been established. A further intriguing observation concerns the ability of 13-*cis* retinoic acid to irreversibly and stereospecifically inhibit thioredoxin reductase, a key enzyme in the nucleotide salvage pathway (Schallreuter and Woods, 1989). Some cell types such as melanoma cells express high levels of this enzyme and thus 13-*cis* retinoic acid may have unique effects on particular cell types not shared by the other retinoic acid isomers.

VI. AN OVERVIEW OF THE DEVELOPMENTAL FUNCTION OF RETINOIC ACID

The role of retinoic acid in development and differentiation should be considered at two levels: which genes are regulated in response to retinoic acid, and how the expression of these genes is regulated by the degree of RAR activation. Now that we have a better idea of how receptor dimers might interact with the transcription machinery, it is clear that the capacity for a given gene to be regulated in response to retinoic acid depends not only on the presence of RARs, RXRs, and a suitable RARE but also on the expression of a number of coactivators. An important point of control of RAR and RXR activity will be the availability of ligand and it seems likely that retinoic acid may be an important intracellular regulator and an intracellular messenger. Both these functions can only be carried out effectively if intracellular concentrations of retinoic acid can be regulated by external factors. Given the role of CRBP and CRABP in retinoid metabolism, regulating the balance of expression of these retinoid binding proteins could be an important control point. This has to be considered with respect to tissue-specific but invariant patterns of CRBP and CRABP expression required to maintain retinoic acid concentrations at specific levels necessary for correct tissue function, and changes in the expression of either or both proteins in response to external stimuli as part of programs of differentiation or development.

There have been few studies on differentiation-related changes in CRABP or CRBP expression during processes of renewal in mature tissues. Studies on human endometrium, a tissue that undergoes regular cycles of coordinated proliferation and differentiation, have shown that CRBP expression in epithelial and stromal cells remains constant during the cycle, in contrast to the expression of CRABP II, which shows a marked decrease in expression during the secretory differentiation phase of the endometrium (Loughney et al., 1995). Endometrial stromal cells from the secretory phase are also more sensitive to the antiproliferative effects of all-*trans* retinoic acid than at other times (Loughney and Redfern, 1995). These results suggest that cellular levels of all-*trans* retinoic acid may be elevated in both the epithelial cells and stromal cells during the mucosecretory differentiation of the endometrial epithelial cells. Since RARs and RXRs are expressed in endometrial epithelial and stromal cells at approximately constant levels throughout the menstrual cycle (Kumarendran et al., 1996; 1994; Prentice et al., 1992), it is likely, that the activity of these nuclear receptors may be regulated by ligand availability as a result of decreases in CRABP expression. This work on human endometrium complements experimental studies on the role of retinoic acid in promoting glandular differentiation in other epithelia (Covant and Hardy, 1990; Hardy et al., 1990) and suggests that elevation of retinoic acid via changes in the expression of CRABP or CRBP may be an important signal specifying the mucosecretory differentiation of epithelia.

There are many aspects of the biological effects of retinoic acid in general that remain obscure. For example, the mechanism of entry of retinoic acid into the nucleus is not known. If retinoic acid levels are regulated by changes in CRABP/CRBP expression, how is the expression of these proteins regulated? Furthermore, is this the only mechanism by which cells may regulate retinoic acid synthesis and catabolism, or is regulation at the level of mRNA translation, P450 activity, retinol dehydrogenase activity, or CRABP/CRBP availability also an important mechanism?

Many of the recent studies on the developmental effects of retinoic acid have emphasized the potential role of RARs in controlling gene expression in combination with RXRs. RXRs in this context presumably function as ligand-independent auxiliary factors. Given that 9-*cis* retinoic acid is a high affinity ligand for RXRs and also binds to RARs, does this isomer of retinoic acid have a developmental or regulatory role *in vivo*? Despite the fact that 9-*cis* retinoic acid has been identified in Xenopus embryos (Kraft et al., 1994), there is as yet no unequivocal evidence that it has a regulatory role *in vivo*. Apart from the potential metabolism of 9-*cis* carotenoids to 9-*cis* retinoic acid (Hebuterne et al., 1995; Nagao and Olson, 1994), no specific mechanism for its formation *in vivo* has been identified, although thiol-group-mediated isomerization (Urbach and Rando, 1994), or, for surface tissues, light-mediated isomerization of all-*trans* retinoic acid are nonspecific mechanisms that could generate 9-*cis in vivo*. Since 9-*cis* retinoic acid does not bind to CRABPs (Fogh et al., 1993; Redfern and Wilson, 1993), there is as yet no certain mechanism for regulating its degradation or stabilisation *in vivo*. However, like all-*trans* retinoic acid, oxidative metabolism of 9-*cis* is probably achieved by cytochrome P450-mediated reactions, and additional pathways of reduction followed by β-oxidation or taurine conjugation have been suggested (Shirley et al., 1996). Overall, these considerations suggest that 9-*cis* may not have a biological role *in vivo*. However, it should be borne in mind that some tissues such as liver and gut have high levels of RXR expression, and in these tissues there may be as yet unidentified mechanisms for generating and regulating 9-*cis* retinoic acid concentrations. If 9-*cis* retinoic acid is present in cells and serves to regulate RXRs, does it activate heterodimers between RXRs and other nuclear receptors such as LXRs, NGFI-β, and NURR-1 in which these receptors take on the role of ligand-independent auxiliary factors, or does it function to promote the formation of RXR homodimers that may promote transcription from DR1-type RAREs? Both functions are, of course, possible, and may depend on the abundance of RXRs relative to other nuclear factors. There is clearly a great deal of research to be done before these questions can be answered.

Despite this uncertainty over the regulatory function of 9-*cis* retinoic acid in relation to RXRs, it is clear from a number of studies that this retinoic acid isomer may have an important place in clinical therapy for some types of cancer. For example, in a rat model, 9-*cis* retinoic acid was more effective than all-*trans* in preventing chemically induced breast cancers (Anzano et al., 1994). Acute promyelocytic leukemia (M3 subtype), characterized by chromosome 15:17 translocations involving the RAR-α and PML loci, responds to all-*trans* retinoic acid with disease remission in a very high proportion of cases. Since relapse and resistance to retinoic acid develop, possibly mediated by increased CRABP expression (Cornic et al., 1994), 9-*cis* retinoic acid may be a more appropriate retinoid to use for therapy. In addition, *in vitro* studies have shown that 9-*cis* retinoic acid is more potent at inducing gene expression and cell differentiation in neuroblastoma cell lines (Han et al., 1995; Redfern et al., 1994; 1995), and thus may be a more appropriate retinoid to use than the all-*trans* or 13-*cis* isomers used in past clinical trials. Since the mechanism of action of 9-*cis* may be different to all-*trans*, and involve RXR-homodimers or heterodimers between RXRs and nuclear receptors other than RARs, 9-*cis* may have qualitatively as well as quantitatively different effects.

Although there have been rapid advances in our understanding of retinoid biology over the last decade, many unanswered questions remain. There can be little doubt that 9-*cis* retinoic acid, whatever its natural biological role, holds considerable promise as a pharmacological agent for the treatment of many diseases. Even if this is not realized, the development of synthetic RAR- and RXR-specific agonists and antagonists will open up new avenues of research and new ways in which to treat disease based on a rational understanding of gene regulation by nuclear receptors.

REFERENCES

Achkar, C. C., Derguini, F., Blumberg, B., Langston, A., Levin, A. A., Speck, J., Evans, R. M., Bolado, J., Nakanishi, K., Buck, J., & Gudas, L. J. (1996). 4-Oxoretinol, a new natural ligand and transactivator of the retinoic acid receptors. Proc. Natl. Acad. Sci. USA 93, 4879-4884.

Adamson, P. C. (1994). Clinical and pharmacokinetic studies of all-trans-retinoic acid in pediatric-patients with cancer. Leukemia 8, S 22-S 25.

Adamson, P. C., Murphy, R. F., Godwin, K. A., Ulm, E. H., & Balis, F. M. (1995). pharmacokinetics of 9-*cis*-retinoic acid in the rhesus-monkey. Cancer Res. 55, 482-485.

Allegretto, E. A., McClurg, M. R., Lazarchik, S. B., Clemm, D. L., Kerner, S. A., Elgort, M. G., Boehm, M. F., White, S. K., Pike, J. W., & Heyman, R. A. (1993). Transactivation properties of retinoic acid and retinoid-X receptors in mammalian-cells and yeast-correlation with hormone-binding and effects of metabolism. J. Biol. Chem. 268, 26625-26633.

Allenby, G., Bocquel, M. T., Saunders, M., Kazmer, S., Speck, J., Rosenberger, M., Lovey, A., Kastner, P., Grippo, J. F., Chambon, P., & Levin, A. A. (1993). Retinoic acid receptors and retinoid X-receptors-interactions with endogenous retinoic acids. Proc. Natl. Acad. Sci. USA 90, 30-34.

Allenby, G., Janocha, R., Kazmer, S., Speck, J., Grippo, J. F., & Levin, A. A. (1994). Binding of 9-*cis*-retinoic acid and all-*trans*-retinoic acid to retinoic acid receptor-alpha, receptor-beta, and receptor-gamma-retinoic acid receptor-gamma binds all-*trans*-retinoic acid preferentially over 9-*cis*-retinoic acid. J. Biol. Chem. 269, 16689-16695.

Almasan, A., Mangelsdorf, D. J., Ong, E. S., Wahl, G. M., & Evans, R. M. (1994). Chromosomal localization of the human retinoid-X receptors. Genomics 20, 397-403.

Angel, P., & Karin, M. (1991). The role of *jun, fos* and the AP-1 complex in cell proliferation and transformation. Biochim. Biophys. Acta 1072, 129-157.

Angulo, A., Suto, C., Heyman, R. A., & Ghazal, P. (1996). Characterization of the sequences of the human cytomegalovirus enhancer that mediate differential regulation by natural and synthetic retinoids. Mol. Endocrinol. 10, 781-793.

Anzano, M. A., Byers, S. W., Smith, J. M., Peer, C. W., Mullen, L. T., Brown, C. C., Roberts, A. B., & Sporn, M. B. (1994). prevention of breast-cancer in the rat with 9-*cis*-retinoic acid as a single-agent and in combination with tamoxifen. Cancer Res. 54, 4614-4617.

Apfel, C., Bauer, F., Crettaz, M., Forni, L., Kamber, M., Kaufmann, F., Lemotte, P., Pirson, W., & Klaus, M. (1992). A retinoic acid receptor-alpha antagonist selectively counteracts retinoic acid effects. Proc. Natl. Acad. Sci. USA 89, 7129-7133.

Arany, Z., Newsome, D., Oldread, E., Livingston, D. M., & Eckner, R. (1995). A family of transcriptional adaptor proteins targeted by the E1a oncoprotein. Nature 374, 81-84.

Astrom, A., Pettersson, U., Chambon, P., & Voorhees, J. J. (1994). Retinoic acid induction of human cellular retinoic acid-binding protein-II gene-transcription is mediated by retinoic acid receptor retinoid-X receptor heterodimers bound to one far upstream retinoic acid-responsive element with 5-base pair spacing. J. Biol. Chem. 269, 22334-22339.

Astrom, A., Pettersson, U., & Voorhees, J. J. (1992). Structure of the human cellular retinoic acid-binding protein-II gene—early transcriptional regulation by retinoic acid. J. Biol. Chem. 267, 25251-25255.

Bailey, J. S., & Siu, C. H. (1988). Purification and partial characterization of a novel binding-protein for retinoic acid from neonatal rat. J. Biol. Chem. 263, 9326-9332.

Bailey, J. S., & Siu, C. H. (1990). Unique tissue distribution of 2 distinct cellular retinoic acid binding-proteins in neonatal and adult-rat. Biochim. Biophys. Acta 1033, 267-272.

Baur, E. V., Zechel, C., Heery, D., Heine, M. J. S., Garnier, J. M., Vivat, V., Ledouarin, B., Gronemeyer, H., Chambon, P., & Losson, R. (1996). Differential ligand-dependent interactions between the AF-2 activating domain of nuclear receptors and the putative transcriptional intermediary factors mSUG1 and TIF1. EMBO J. 15, 110-124.

Bavik, C. O., Busch, C., & Eriksson, U. (1992). Characterization of a plasma retinol-binding protein membrane-receptor expressed in the retinal-pigment epithelium. J. Biol. Chem. 267, 23035-23042.

Bavik, C. O., Eriksson, U., Allen, R. A., & Peterson, P. A. (1991). Identification and partial characterization of a retinal-pigment epithelial membrane-receptor for plasma retinol-binding protein. J. Biol. Chem. 266, 14978-14985.

Bavik, C. O., Levy, F., Hellman, U., Wernstedt, C., & Eriksson, U. (1993). The retinal-pigment epithelial membrane-receptor for plasma retinol-binding protein-isolation and cDNA cloning of the 63-kDa protein. J. Biol. Chem. 268, 20540-20546.

Bavik, C. O., Peterson, P. A., & Eriksson, U. (1995). Retinol-binding protein mediates uptake of retinol to cultured human keratinocytes. Exp. Cell Res. 216, 358-362.

Benbrook, D., Lernhardt, E., & Pfahl, M. (1988). A new retinoic acid receptor identified from a hepatocellular- carcinoma. Nature 333, 669-672.

Berard, J., Gaboury, L., Landers, M., Derepentigny, Y., Houle, B., Kothary, R., & Bradley, W. E. C. (1994). Hyperplasia and tumors in lung, breast and other tissues in mice carrying a RAR-beta-4-like transgene. EMBO J. 13, 5570-5580.

Berkenstam, A., Ruiz, M. D. V., Barettino, D., Horikoshi, M., & Stunnenberg, H. G. (1992). Cooperativity in transactivation between retinoic acid receptor and TFIID requires an activity analogous to E1a. Cell 69, 401-412.

Bernard, B. A., Bernardon, J. M., Delescluse, C., Martin, B., Lenoir, M. C., Maignan, J., Charpentier, B., Pilgrim, W. R., Reichert, U., & Shroot, B. (1992). Identification of synthetic retinoids with selectivity for human nuclear retinoic acid receptor-gamma. Biochem. Biophys. Res. Comm. 186, 977-983.

Block, G., Patterson, B., & Subar, A. (1992). Fruit, vegetables and cancer prevention: a review of the epidemiological evidence. Nutr. Cancer 18, 1-29.

Boehm, M. F., Zhang, L., Badea, B. A., White, S. K., Mais, D. E., Berger, E., Suto, C. M., Goldman, M. E., & Heyman, R. A. (1994). Synthesis and structure-activity-relationships of novel retinoid-X receptor-selective retinoids. J. Med. Chem. 37, 2930-2941.

Boerman, M., & Napoli, J. L. (1991). Cholate-independent retinyl ester hydrolysis-stimulation by apo-cellular retinol-binding protein. J. Biol. Chem. 266, 22273-22278.

Boerman, M., & Napoli, J. L. (1995). Characterization of a microsomal retinol dehydrogenase—a short-chain alcohol-dehydrogenase with integral and peripheral membrane forms that interacts with holo-CRBP (type-I). Biochemistry 34, 7027-7037.

Boerman, M., & Napoli, J. L. (1996). Cellular retinol-binding protein-supported retinoic acid synthesis—relative roles of microsomes and cytosol. J. Biol. Chem. 271, 5610-5616.

Bourguet, W., Ruff, M., Chambon, P., Gronemeyer, H., & Moras, D. (1995). Crystal-structure of the ligand-binding domain of the human nuclear receptor RXR-alpha. Nature 375, 377-382.

Boylan, J. F., & Gudas, L. J. (1991). Overexpression of the cellular retinoic acid binding protein-I (CRABP-I) results in a reduction in differentiation-specific gene-expression in F9 teratocarcinoma cells. J. Cell Biol. 112, 965-979.

Boylan, J. F., & Gudas, L. J. (1992). The level of CRABP-I expression influences the amounts and types of all-*trans*-retinoic acid metabolites in F9 teratocarcinoma stem-cells. J. Biol. Chem. 267, 21486-21491.

Boylan, J. F., Lohnes, D., Taneja, R., Chambon, P., & Gudas, L. J. (1993). Loss of retinoic acid receptor-gamma function in F9 cells by gene disruption results in aberrant hoxa-1 expression and differentiation upon retinoic acid treatment. Proc. Natl. Acad. Sci. USA 90, 9601-9605.

Boylan, J. F., Lufkin, T., Achkar, C. C., Taneja, R., Chambon, P., & Gudas, L. J. (1995). Targeted disruption of retinoic acid receptor-alpha (RAR-alpha) and RAR-gamma results in receptor-specific alterations in retinoic acid-mediated differentiation and retinoic acid metabolism. Mol. Cell. Biol. 15, 843-851.

Brand, N., Petkovich, M., Krust, A., Chambon, P., Dethe, H., Marchio, A., Tiollais, P., & Dejean, A. (1988). Identification of a 2nd human retinoic acid receptor. Nature 332, 850-853.

Brand, N. J., Petkovich, M., & Chambon, P. (1990). Characterization of a functional promoter for the human retinoic acid receptor-alpha (hRAR-alpha). Nucl. Acids Res. 18, 6799-6806.

Buck, J., Derguini, F., Levi, E., Nakanishi, K., & Hammerling, U. (1991). Intracellular signaling by 14-hydroxy-4,14-retro-retinol. Science 254, 1654-1656.

Buck, J., Ritter, G., Dannecker, L., Katta, V., Cohen, S. L., Chait, B. T., & Hammerling, U. (1990). Retinol is essential for growth of activated human B-cells. J. Exp. Med. 171, 1613-1624.

Carughi, A., & Hooper, F. G. (1994). Plasma carotenoid concentrations before and after supplementation with a carotenoid mixture. Am. J. Clin. Nutr. 59, 896-899.

Cavailles, V., Dauvois, S., Danielian, P. S., & Parker, M. G. (1994). Interaction of proteins with transcriptionally-active oestrogen receptors. Proc. Natl. Acad. Sci. USA 91, 10009-10013.

Cavailles, V., Dauvois, S., L'Horset, F., Lopez, G., Hoare, S., Kushner, P. J., & Parker, M. G. (1995). Nuclear factor RIP140 modulates transcriptional activation by the oestrogen receptor. EMBO J. 14, 3741-3751.

Chai, X. Y., Boerman, M., Zhai, Y., & Napoli, J. L. (1995a). Cloning of a cDNA for liver microsomal retinol dehydrogenase—a tissue-specific, short-chain alcohol-dehydrogenase. J. Biol. Chem. 270, 3900-3904.

Chai, X. Y., Zhai, Y., & Napoli, J. L. (1996). Cloning of a rat cDNA-encoding retinol dehydrogenase isozyme type-III. Gene 169, 219-222.

Chai, X. Y., Zhai, Y., Popescu, G., & Napoli, J. L. (1995b). Cloning of a cDNA for a 2nd retinol dehydrogenase type-II—expression of its messenger-RNA relative to type-I. J. Biol. Chem. 270, 28408-28412.

Chakravarti, D., LaMorte, V. J., Nelson, M. C., Nakajima, T., Schulman, I. G., Juguilon, H., Montminy, M., & Evans, R. M. (1996). Role of CBP/P300 in nuclear receptor signalling. Nature 383, 99-103.

Chao, D. M., & Young, R. A. (1996). Activation without a vital ingredient. Nature 383, 119-120.

Charpentier, B., Bernardon, J. M., Eustache, J., Millois, C., Martin, B., Michel, S., & Shroot, B. (1995). Synthesis, structure-affinity relationships, and biological activities of ligands binding to retinoic acid receptor subtypes. J. Med. Chem. 38, 4993-5006.

Chen, H. W., & Privalsky, M. L. (1995). Cooperative formation of high-order oligomers by retinoid-X rceptors—an unexpected mode of DNA recognition. Proc. Natl. Acad. Sci. USA 92, 422-426.

Chen, J. D., & Evans, R. M. (1995). A transcriptional co-repressor that interacts with nuclear hormone receptors. Nature 377, 454-457.

Chen, J. D., Umesono, K., & Evans, R. M. (1996a). SMRT isoforms mediate repression and anti-repression of nuclear receptor heterodimers. Proc. Natl. Acad. Sci. USA 93, 7567-7571.

Chen, J.-Y., Clifford, J., Zusi, C., Starrett, J., Tortolani, D., Ostrowski, J., Reczek, P. R., Chambon, P., & Gronemeyer, H. (1996b). Two distinct actions of retinoid receptor ligands. Nature 382, 819-822.

Chytil, F., & Ong, D. E. (1984). Cellular retinoid binding proteins. In: *The Retinoids*, Vol. 2 (Sporn, M. B., Roberts A. B., & Goodman, D. S., eds.) Academic Press, Orlando.

Clevidence, B. A., & Bieri, J. G. (1993). Association of carotenoids with human plasma lipoproteins. Meths. Enzymol. 214, 33-46.

Cornic, M., Delva, L., Castaigne, S., Lefebvre, P., Balitrand, N., Degos, L., & Chomienne, C. (1994). *In vitro* all-*trans*-retinoic acid (ATRA) sensitivity and cellular retinoic acid-binding protein (CRABP) levels in relapse leukemic-cells after remission induction by ATRA in acute promyelocytic leukemia. Leukemia 8, 914-917.

Covant, H. A., & Hardy, M. H. (1990). Excess retinoid acts through the stroma to produce mucous glands from newborn hamster-cheek pouch *in vitro*. J. Exp. Zool. 253, 271-279.

Cowan, S. W., Newcomer, M. E., & Jones, T. A. (1990). Crystallographic refinement of human serum retinol binding-protein at 2A resolution. Proteins Struct. Funct. Genet. 8, 44-61.

Cowan, S. W., Newcomer, M. E., & Jones, T. A. (1993). Crystallographic studies on a family of cellular lipophilic transport proteins—refinement of P2 myelin protein and the structure determination and refinement of cellular retinol-binding protein in complex with all-*trans*-retinol. J. Mol. Biol. 230, 1225-1246.

Daly, A. K., & Redfern, C. P. F. (1987). Characterization of a retinoic acid binding component from F9 embryonal carcinoma cell nuclei. Eur. J. Biochem. 168, 133-139.

Daly, A. K., & Redfern, C. P. F. (1988). Purification and properties of cellular retinoic acid-binding protein from neonatal rat skin. Biochim. Biophys. Acta 965, 118-126.

Davis, J. T., & Ong, D. E. (1992). Synthesis and secretion of retinol-binding protein by cultured rat sertoli cells. Biol. Reprod. 47, 528-533.

Davis, J. T., & Ong, D. E. (1995). Retinol processing by the peritubular cell from rat testis. Biol. Reprod. 52, 356-364.

Dawson, M. I., Jong, L., Hobbs, P. D., Cameron, J. F., Chao, W. R., Pfahl, M., Lee, M. O., Shroot, B., & Pfahl, M. (1995). Conformational effects on retinoid receptor selectivity .2. effects of retinoid bridging group on retinoid-X-receptor activity and selectivity. J. Med. Chem. 38, 3368-3383.

de The, H., Marchio, A., Tiollais, P., & Dejean, A. (1989). Differential expression and ligand regulation of the retinoic acid receptor-alpha and receptor-beta genes. EMBO J. 8, 429-433.

de The, H., Vivancoruiz, M. D., Tiollais, P., Stunnenberg, H., & Dejean, A. (1990). Identification of a retinoic acid responsive element in the Retinoic Acid Receptor-beta gene. Nature 343, 177-180.

Delescluse, C., Cavey, M. T., Martin, B., Bernard, B. A., Reichert, U., Maignan, J., Darmon, M., & Shroot, B. (1991). Selective high-affinity retinoic acid receptor-alpha or receptor-beta-gamma ligands. Mol. Pharm. 40, 556-562.

Demmer, L. A., Birkenmeier, E. H., Sweetser, D. A., Levin, M. S., Zollman, S., Sparkes, R. S., Mohandas, T., Lusis, A. J., & Gordon, J. I. (1987). The cellular retinol binding protein-II gene-sequence-analysis of the rat gene, chromosomal localization in mice and humans, and documentation of its close linkage to the cellular retinol binding-protein gene. J. Biol. Chem. 262, 2458-2467.

Derguini, F., Nakanishi, K., Hammerling, U., Chua, R., Eppinger, T., Levi, E., & Buck, J. (1995). 13,14-Dihydroxy-retinol, a new bioactive retinol metabolite. J. Biol. Chem. 270, 18875-18880.

Dew, S. E., & Ong, D. E. (1994). Specificity of the retinol transporter of the rat small-intestine brush-border. Biochemistry 33, 12340-12345.

Dollé, P., Fraulob, V., Kastner, P., & Chambon, P. (1994). Developmental expression of murine retinoid-x-receptor (RXR) genes. Mech. Dev. 45, 91-104.

Dollé, P., Ruberte, E., Kastner, P., Petkovich, M., Stoner, C. M., Gudas, L. J., & Chambon, P. (1989). Differential expression of genes encoding alpha,beta and gamma-retinoic acid receptors and CRABP in the developing limbs of the mouse. Nature 342, 702-705.

Dollé, P., Ruberte, E., Leroy, P., Morrisskay, G., & Chambon, P. (1990). Retinoic acid receptors and cellular retinoid binding-proteins .1. A systematic study of their differential pattern of transcription during mouse organogenesis. Development 110, 1133-1151.

Doucas, V., Brockes, J. P., Yaniv, M., Dethe, H., & Dejean, A. (1993). The PML retinoic acid receptor-alpha translocation converts the receptor from an inhibitor to a retinoic acid-dependent activator of transcription factor-AP-1. Proc. Natl. Acad. Sci. USA 90, 9345-9349.

Duell, E. A., Kang, S., & Voorhees, J. J. (1996). Retinoic acid isomers applied to human skin in-vivo each induce a 4-hydroxylase that inactivates only *trans* retinoic acid. J. Invest. Dermatol. 106, 316-320.

El Akawi, Z., & Napoli, J. L. (1994). Rat-liver cytosolic retinal dehydrogenase—comparison of 13-*cis*-, 8-*cis*- , and all-*trans*-retinal as substrates and effects of cellular retinoid-binding proteins and retinoic acid on activity. Biochemistry 33, 1938-1943.

Elder, J. T., Astrom, A., Pettersson, U., Voorhees, J. J., & Trent, J. M. (1992). Assignment of the human CRABP-II gene to chromosome 1q21 by nonisotopic in situ hybridization. Hum. Genet. 89, 487-490.

Eppinger, T. M., Buck, J., & Hammerling, U. (1993). Growth-control or terminal differentiation—endogenous production and differential activities of vitamin A metabolites in Hl-60 cells. J. Exp. Med. 178, 1995-2005.

Fanjul, A., Dawson, M. I., Hobbs, P. D., Jong, L., Cameron, J. F., Harlev, E., Graupner, G., Lu, X. P., & Pfahl, M. (1994). A new class of retinoids with selective-inhibition of AP-1 inhibits proliferation. Nature 372, 107-111.

Fanjul, A. N., Bouterfa, H., Dawson, M., & Pfahl, M. (1996). Potential role for retinoic acid receptor-gamma in the inhibition of breast-cancer cells by selective retinoids and interferons. Cancer Res. 56, 1571-1577.

Fawcett, D., Pasceri, P., Fraser, R., Colbert, M., Rossant, J., & Giguere, V. (1995). Postaxial polydactyly in forelimbs of CRABP-II mutant mice. Development 121, 671-679.

Fell, H. B. (1957). The effects of excess vitamin A on cultures of embryonic chicken skin explanted at different stages of differentiation. Proceedings of the Royal Society B 146, 242-256.

Fell, H. B., & Mellanby, E. (1953). Metaplasia produced in cultures of chick endoderm by high vitamin A. J. Physiol. 119, 470-488.

Fiorella, P. D., Giguere, V., & Napoli, J. L. (1993). Expression of cellular retinoic acid-binding protein (type-II) in Escherichia-coli—characterization and comparison to cellular retinoic acid-binding protein (type-I). J. Biol. Chem. 268, 21545-21552.

Fiorella, P. D., & Napoli, J. L. (1991). Expression of cellular retinoic acid binding-protein (CRABP) in Escherichia-coli—characterization and evidence that holo-CRABP is a substrate in retinoic acid metabolism. J. Biol. Chem. 266, 16572-16579.

Fiorella, P. D., & Napoli, J. L. (1994). Microsomal retinoic acid metabolism—effects of cellular retinoic acid-binding protein (type-I) and C18-hydroxylation as an initial step. J. Biol. Chem. 269, 10538-10544.

Fogh, K., Voorhees, J. J., & Astrom, A. (1993). Expression, purification, and binding-properties of human cellular retinoic acid-binding protein typeI and type-II. Arch. Biochem. Biophys. 300, 751-755.

Folkers, G. E., & Vandersaag, P. T. (1995). Adenovirus E1a functions as a cofactor for retinoic acid receptor-beta (RAR-beta) through direct interaction with RAR-beta. Mol. Cell. Biol. 15, 5868-5878.

Garbe, A., Buck, J., & Hammerling, U. (1992). Retinoids are important cofactors in T-cell activation. J. Exp. Med. 176, 109-117.

Giguere, V., Ong, E. S., Segui, P., & Evans, R. M. (1987). Identification of a receptor for the morphogen retinoic acid. Nature 330, 624-629.

Glass, C. K., Devary, O. V., & Rosenfeld, M. G. (1990). Multiple cell type specific proteins differentially regulate target sequence recognition by the alpha-retinoic acid receptor. Cell 63, 729-738.

Gorry, P., Lufkin, T., Kierich, A., Rochette-Egly, C., Decimo, D., Dolle, P., Mark, M., Durand, B., & Chambon, P. (1994). The cellular retinoic acid binding protein-I is dispensible. Proc. Natl. Acad. Sci. USA 91, 9032-9036.

Han, G. Y., Chang, B. S., Connor, M. J., & Sidell, N. (1995). Enhanced potency of 9-*cis* versus all-*trans* retinoic acid to induce the differentiation of human neuroblastoma cells. Differentiation 59, 61-69.

Hardy, M. H., Dhouailly, D., Torma, H., & Vahlquist, A. (1990). Either chick-embryo dermis or retinoid-treated mouse dermis can initiate glandular morphogenesis from mammalian epidermal tissue. J. Exp. Zool. 256, 279-289.

Hashimoto, Y., Kagechika, H., & Shudo, K. (1990). Expression of retinoic acid receptor genes and the ligand-binding selectivity of retinoic acid receptors (RARs). Biochem. Biophys. Res. Comm. 166, 1300-1307.

Hebuterne, X., Wang, X. D., Johnson, E. J., Krinsky, N. I., & Russell, R. M. (1995). Intestinal-absorption and metabolism of 9-*cis*-beta-carotene *in-vivo*-biosynthesis of 9-*cis*-retinoic acid. J. Lipid Res. 36, 1264-1273.

Herr, F. M., & Ong, D. E. (1992). Differential interaction of lecithin retinol acyltransferase with cellular retinol binding-proteins. Biochemistry 31, 6748-6755.

Heyman, R. A., Mangelsdorf, D. J., Dyck, J. A., Stein, R. B., Eichele, G., Evans, R. M., & Thaller, C. (1992). 9-*cis* retinoic acid is a high-affinity ligand for the retinoid-X receptor. Cell 68, 397-406.

Ho, S. N., Biggar, S. R., Spencer, D. M., Schreiber, S. L., & Crabtree, G. R. (1996). Dimeric ligands define a role for transcriptional activation domains in reinitiation. Nature 382, 822-826.

Hogan, B. L. M., Thaller, C., & Eichele, G. (1992). Evidence that hensen node is a site of retinoic acid synthesis. Nature 359, 237-241.

Horlein, A. J., Naar, A. M., Heinzel, T., Torchia, J., Gloss, B., Kurokawa, R., Ryan, A., Kamel, Y., Soderstrom, M., Glass, C. K., & Rosenfeld, M. G. (1995). Ligand-independent repression by the thyroid-hormone receptor—mediated by a nuclear receptor co-repressor. Nature 377, 397-404.

Hughes, G. S., Ringer, T. V., Francom, S. F., Means, L. K., & DeLoof, M. J. (1994). Lack of effects of ß-carotene on lipids and sex-steroid hormones in hyperlipidemics. Am. J. Med. Sci. 308, 16-22.

Issemann, I., Prince, R. A., Tugwood, J. D., & Green, S. (1993). The retinoid X receptor enhances the function of the peroxisome proliferator activated receptor. Biochimie 75, 251-256.

Jamison, R. S., Newcomer, M. E., & Ong, D. E. (1994). Cellular retinoid-binding proteins—limited proteolysis reveals a conformational change upon ligand-binding. Biochemistry 33, 2873-2879.

Jones, B. B., Ohno, C. K., Allenby, G., Boffa, M. B., Levin, A. A., Grippo, J. F., & Petkovich, M. (1995). New retinoid-X receptor subtypes in zebra fish (Danio-rerio) differentially modulate transcription and do not bind 9-*cis* retinoic acid. Mol. Cell. Biol. 15, 5226-5234.

Jong, L., Lehmann, J. M., Hobbs, P. D., Harlev, E., Huffman, J. C., Pfahl, M., & Dawson, M. I. (1993). Conformational effects on retinoid receptor selectivity .1. Effect of 9-double bond geometry on retinoid-X receptor activity. J. Med. Chem. 36, 2605-2613.

Kaji, E. H., & Lodish, H. F. (1993). Unfolding of newly made retinol-binding protein by dithiothreitol—sensitivity to retinoids. J. Biol. Chem. 268, 22188-22194.

Kamei, Y., Xu, L., Heinzel, T., Torchia, J., Kurokawa, R., Gloss, B., Lin, S. C., Heyman, R. A., Rose, D. W., Glass, C. K., & Rosenfeld, M. G. (1996). A CBP integrator complex mediates transcriptional activation and AP-1 inhibition by nuclear receptors. Cell 85, 403-414.

Kastner, P., Krust, A., Mendelsohn, C., Garnier, J. M., Zelent, A., Leroy, P., Staub, A., & Chambon, P. (1990). Murine isoforms of retinoic acid receptor-gamma with specific patterns of expression. Proc. Natl. Acad. Sci. USA 87, 2700-2704.

Kersten, S., Dawson, M. I., Lewis, B. A., & Noy, N. (1996). Individual subunits of heterodimers comprised of retinoic acid and retinoid X receptors interact with their ligands independently. Biochemistry 35, 3816-3824.

Kersten, S., Kelleher, D., Chambon, P., Gronemeyer, H., & Noy, N. (1995a). Retinoid-X-receptor-alpha forms tetramers in solution. Proc. Natl. Acad. Sci. USA 92, 8645-8649.

Kersten, S., Pan, L., Chambon, P., Gronemeyer, H., & Noy, N. (1995b). Role of ligand in retinoid signaling-9-*cis*-retinoic acid modulates the oligomeric state of the retinoid-X-receptor. Biochemistry 34, 13717-13721.

Kersten, S., Pan, L., & Noy, N. (1995c). On the role of ligand in retinoid signaling—positive cooperativity in the interactions of 9-*cis* retinoic acid with tetramers of the retinoid-X-receptor. Biochemistry 34, 14263-14269.

Kliewer, S. A., Umesono, K., Mangelsdorf, D. J., & Evans, R. M. (1992). Retinoid X-receptor interacts with nuclear receptors in retinoic acid, thyroid-hormone and vitamin-d3 signaling. Nature 355, 446-449.

Kraft, J. C., Schuh, T., Juchau, M., & Kimelman, D. (1994). The retinoid-X receptor-ligand, 9-*cis*-retinoic acid, is a potential regulator of early xenopus development. Proc. Natl. Acad. Sci. USA 91, 3067-3071.

Krust, A., Kastner, P., Petkovich, M., Zelent, A., & Chambon, P. (1989). A 3rd Human retinoic acid receptor, hRAR-gamma. Proc. Natl. Acad. Sci. USA 86, 5310-5314.

Kruyt, F. A. E., Folkers, G. E., Walhout, A. J. M., Vanderleede, B. M., & Vandersaag, P. T. (1993). E1a functions as a coactivator of retinoic acid-dependent retinoic acid receptor-beta-2 promoter activation. Mol. Endocrinol. 7, 604-615.

Kumarendran, M. K., Loughney, A. D., Prentice, A., Thomas, E. J., & Redfern, C. P. F. (1996). Nuclear retinoid receptor expression in normal human endometrium throughout the menstrual cycle. Mol. Hum. Reprod. 2, 123-129.

Kumarendran, M. K., Matthews, C. J., Levasseur, M. D., Prentice, A., Thomas, E. J., & Redfern, C. P. F. (1994). Estrogen and progesterone do not regulate the expression of retinoic acid receptors and retinoid X receptors in human endometrial stromal cells *in vitro*. Hum. Reprod. 9, 229-234.

Kurokawa, R., Direnzo, J., Boehm, M., Sugarman, J., Gloss, B., Rosenfeld, M. G., Heyman, R. A., & Glass, C. K. (1994). Regulation of retinoid signaling by receptor polarity and allosteric control of ligand-binding. Nature 371, 528-531.

Kurokawa, R., Soderstrom, M., Horlein, A., Halachmi, S., Brown, M., Rosenfeld, M. G., & Glass, C. K. (1995). Polarity-specific activities of retinoic acid receptors determined by a co-repressor. Nature 377, 451-454.

Kurokawa, R., Yu, V. C., Naar, A., Kyakumoto, S., Han, Z. H., Silverman, S., Rosenfeld, M. G., & Glass, C. K. (1993). Differential orientations of the DNA-binding domain and carboxy-terminal dimerization interface regulate binding-site selection by nuclear receptor heterodimers. Genes Dev. 7, 1423-1435.

Lampron, C., Rochette-Egly, C., Gorry, P., Dolle, P., Mark, M., Lufkin, T., Lemeur, M., & Chambon, P. (1995). Mice deficient in cellular retinoic acid binding protein-II (CRABPII) or in both CRABPI and CRABPII are essentially normal. Development 121, 539-548.

Le Douarin, B., Zechel, C., Garnier, J. M., Lutz, Y., Tora, L., Pierrat, B., Heery, D., Gronemeyer, H., Chambon, P., & Losson, R. (1995). The N-terminal part of TIF1, a putative mediator of the ligand- dependent activation function (AF-2) of nuclear receptors, is fused to B-Raf in the oncogenic protein T18. EMBO J. 14, 2020-2033.

Lee, M. O., Dawson, M. I., Picard, N., Hobbs, P. D., & Pfahl, M. (1996). A novel class of retinoid antagonists and their mechanism of action. J. Biol. Chem. 271, 11897-11903.

Lee, M. O., Hobbs, P. D., Zhang, X. K., Dawson, M. I., & Pfahl, M. (1994). A synthetic retinoid antagonist inhibits the human-immunodeficiency-virus type-1 promoter. Proc. Natl. Acad. Sci. USA 91, 5632-5636.

Lee, M. S., Kliewer, S. A., Provencal, J., Wright, P. E., & Evans, R. M. (1993). Structure of the retinoid-X receptor-alpha DNA-binding domain—a helix required for homodimeric DNA-binding. Science 260, 1117-1121.

Lehmann, J. M., Dawson, M. I., Hobbs, P. D., Husmann, M., & Pfahl, M. (1991a). Identification of retinoids with nuclear receptor subtype-selective activities. Cancer Res. 51, 4804-4809.

Lehmann, J. M., Hoffmann, B., & Pfahl, M. (1991b). Genomic organization of the retinoic acid receptor gamma-gene. Nucl. Acids Res. 19, 573-578.

Lehmann, J. M., Jong, L., Fanjul, A., Cameron, J. F., Lu, X. P., Haefner, P., Dawson, M. I., & Pfahl, M. (1992a). Retinoids selective for retinoid-X receptor response pathways. Science 258, 1944-1946.

Lehmann, J. M., Zhang, X. K., & Pfahl, M. (1992b). RAR-gamma-2 expression is regulated through a retinoic acid response element embedded in Sp1-sites. Mol. Cell. Biol. 12, 2976-2985.

Leng, X. H., Blanco, J., Tsai, S. Y., Ozato, K., Omalley, B. W., & Tsai, M. J. (1995). Mouse retinoid-X receptor contains a separable ligand-binding and transactivation domain in its E-region. Mol. Cell. Biol. 15, 255-263.

Leroy, P., Krust, A., Zelent, A., Mendelsohn, C., Garnier, J. M., Kastner, P., Dierich, A., & Chambon, P. (1991a). Multiple isoforms of the mouse retinoic acid receptor alpha are generated by alternative splicing and differential induction by retinoic acid. EMBO J. 10, 59-69.

Leroy, P., Nakshatri, H., & Chambon, P. (1991b). Mouse retinoic acid receptor-alpha-2 isoform is transcribed from a promoter that contains a retinoic acid response element. Proc. Natl. Acad. Sci. USA 88, 10138-10142.

Levin, A. A., Sturzenbecker, L. J., Kazmer, S., Bosakowski, T., Huselton, C., Allenby, G., Speck, J., Kratzeisen, C., Rosenberger, M., Lovey, A., & Grippo, J. F. (1992). 9-*cis* retinoic acid stereoisomer binds and activates the nuclear receptor RXR-alpha. Nature 355, 359-361.

Li, E., Qian, S. J., Winter, N. S., Davignon, A., Levin, M. S., & Gordon, J. I. (1991). Fluorine nuclear-magnetic-resonance analysis of the ligand-binding properties of 2 homologous rat cellular retinol-binding proteins expressed in Escherichia-coli. J. Biol. Chem. 266, 3622-3629.

Liau, G., Ong, D. E., & Chytil, F. (1981). Interaction of the retinol-cellular retinol-binding protein complex with isolated-nuclei and nuclear-components. J. Cell Biol. 91, 63-68.

Liau, G., Ong, D. E., & Chytil, F. (1985). Partial characterization of nuclear-binding sites for retinol delivered by cellular retinol binding-protein. Arch. Biochem. Biophys. 237, 354-360.

Liu, Q., & Linney, E. (1993). The mouse retinoid-X receptor-gamma gene—genomic organization and evidence for functional isoforms. Mol. Endocrinol. 7, 651-658.

Lohnes, D., Kastner, P., Dierich, A., Mark, M., Lemeur, M., & Chambon, P. (1993). Function of retinoic acid receptor-gamma in the mouse. Cell 73, 643-658.

Loughney, A. D., Kumarendran, M. K., Thomas, E. J., & Redfern, C. P. F. (1995). Variation in the expression of cellular retinoid-binding proteins in human endometrium throughout the menstrual-cycle. Hum. Reprod. 10, 1297-1304.

Loughney, A. D., & Redfern, C. P. F. (1995). Menstrual cycle related differences in the proliferative responses of cultured human endometrial stromal cells to retinoic acid. J. Reprod. Fertil. 105, 153-159.

Lovat, P. E., Pearson, A. D. J., Malcolm, A., & Redfern, C. P. F. (1993). Retinoic acid receptor expression during the *in vitro* differentiation of human neuroblastoma. Neurosci. Letts. 162, 109-113.

Lu, X. P., Eberhardt, N. L., & Pfahl, M. (1993). DNA bending by retinoid-X receptor-containing retinoid and thyroid-hormone receptor complexes. Mol. Cell. Biol. 13, 6509-6519.

Lufkin, T., Lohnes, D., Mark, M., Dierich, A., Gorry, P., Gaub, M. P., Lemeur, M., & Chambon, P. (1993). High postnatal lethality and testis degeneration in retinoic acid receptor-alpha mutant mice. Proc. Natl. Acad. Sci. USA 90, 7225-7229.

Lundblad, J. R., Kwok, R. P. S., Laurance, M. E., Harter, M. L., & Goodman, R. H. (1995). Adenoviral E1a-associated protein P300 as a functional homolog of the transcriptional coactivator CBP. Nature 374, 85-88.

Luo, J. M., Pasceri, P., Conlon, R. A., Rossant, J., & Giguere, V. (1995). Mice lacking all iso-forms of retinoic acid receptor-beta develop normally and are susceptible to the teratogenic effects of retinoic acid. Mech. Dev. 53, 61-71.

Luo, J. M., Sucov, H. M., Bader, J. A., Evans, R. M., & Giguere, V. (1996). Compound mutants for retinoic acid receptor (RAR)-beta and RAR-alpha-1 reveal developmental functions for multiple RAR-beta-isoform. Mech. Dev. 55, 33-44.

Lyn, S., & Giguere, V. (1994). Localization of CRABP-I and CRABP-II messenger-rna in the early mouse embryo by whole-mount in-situ hybridization—implications for teratogenesis and neural development. Dev. Dyn. 199, 280-291.

Macdonald, P. N., Bok, D., & Ong, D. E. (1990). Localization of cellular retinol-binding protein and retinol-binding protein in cells comprising the blood-brain-barrier of rat and human. Proc. Natl. Acad. Sci. USA 87, 4265-4269.

Macdonald, P. N., & Ong, D. E. (1987). Binding specificities of cellular retinol-binding protein and cellular retinol-binding protein, type-II. J. Biol. Chem. 262, 10550-10556.

Macgregor, T. M., Copeland, N. G., Jenkins, N. A., & Giguere, V. (1992). The murine gene for cellular retinoic acid-binding protein type-II—genomic organization, chromosomal localization, and posttranscriptional regulation by retinoic acid. J. Biol. Chem. 267, 7777-7783.

Maden, M., Ong, D. E., & Chytil, F. (1990). Retinoid-binding protein distribution in the developing mammalian nervous-system. Development 109, 75-80.

Maden, M., Ong, D. E., Summerbell, D., & Chytil, F. (1988). Spatial-distribution of cellular protein-binding to retinoic acid in the chick limb bud. Nature 335, 733-735.

Mangelsdorf, D. J., Borgmeyer, U., Heyman, R. A., Zhou, J. Y., Ong, E. S., Oro, A. E., Kakizuka, A., & Evans, R. M. (1992). Characterization of 3 RXR genes that mediate the action of 9-*cis* retinoic acid. Genes Dev. 6, 329-344.

Mangelsdorf, D. J., Ong, E. S., Dyck, J. A., & Evans, R. M. (1990). Nuclear receptor that identifies a novel retinoic acid response pathway. Nature 345, 224-229.

Mangelsdorf, D. J., Umesono, K., Kliewer, S. A., Borgmeyer, U., Ong, E. S., & Evans, R. M. (1991). A direct repeat in the cellular retinol-binding protein type-II gene confers differential regulation by RXR and RAR. Cell 66, 555-561.

Mattei, M. G., Riviere, M., Krust, A., Ingvarsson, S., Vennstrom, B., Islam, M. Q., Levan, G., Kautner, P., Zelent, A., Chambon, P., Szpirer, J., & Szpirer, C. (1991). Chromosomal assignment of retinoic acid receptor (RAR) genes in the human, mouse, and rat genomes. Genomics 10, 1061-1069.

Meeks, R. G., Zaharevitz, D., & Chen, R. F. (1981). Membrane effects of retinoids: possible correlation with toxicity. Arch. Biochem. Biophys. 207, 141-147.

Mendelsohn, C., Mark, M., Dolle, P., Dierich, A., Gaub, M. P., Krust, A., Lampron, C., & Chambon, P. (1994). Retinoic acid receptor beta-2 (RAR-beta-2) null mutant mice appear normal. Dev. Biol. 166, 246-258.

Muindi, J. F., & Young, C. W. (1993). Lipid hydroperoxides greatly increase the rate of oxidative catabolism of all-*trans*-retinoic acid by human cell-culture microsomes genetically enriched in specified cytochrome-P-450 isoforms. Cancer Res. 53, 1226-1229.

Muindi, J. R. F., Young, C. W., & Warrell, R. P. (1994). Clinical-pharmacology of all-*trans*-retinoic acid. Leukemia 8, 1807-1812.

Nagao, A., & Olson, J. A. (1994). Enzymatic formation of 9-*cis*, 13-*cis*, and all-*trans* retinals from isomers of beta-carotene. FASEB J. 8, 968-973.

Nagata, T., Kanno, Y., Ozato, K., & Taketo, M. (1994). The mouse RXRb gene encoding RXR-beta—genomic organization and 2 messenger-RNA isoforms generated by alternative splicing of transcripts initiated from CPG island promoters. Gene 142, 183-189.

Nagpal, S., Friant, S., Nakshatri, H., & Chambon, P. (1993). RARs and RXRs—evidence for 2 autonomous transactivation functions (AF-1 and AF-2) and heterodimerization in vivo. EMBO J. 12, 2349-2360.

Nagpal, S., Saunders, M., Kastner, P., Durand, B., Nakshatri, H., & Chambon, P. (1992a). Promoter context-dependent and response element-dependent specificity of the transcriptional activation and modulating functions of retinoic acid receptors. Cell 70, 1007-1019.

Nagpal, S., Zelent, A., & Chambon, P. (1992b). RAR-beta-4, a retinoic acid receptor isoform is generated from RAR- beta-2 by alternative splicing and usage of a CUG initiator codon. Proc. Natl. Acad. Sci. USA 89, 2718-2722.

Napoli, J. L., Posch, K. C., & Burns, R. D. (1992). Microsomal retinal synthesis—retinol vs holo-CRBP as substrate and evaluation of NADP, NAD and NADPH as cofactors. Biochim. Biophys. Acta 1120, 183-186.

Napoli, J. L., Posch, K. P., Fiorella, P. D., & Boerman, M. (1991). Physiological occurrence, biosynthesis and metabolism of retinoic acid—evidence for roles of cellular retinol-binding protein (CRBP) and cellular retinoic acid-binding protein (CRABP) in the pathway of retinoic acid homeostasis. Biomed. Pharmacoth. 45, 131-143.

Napoli, J. L., & Race, K. R. (1987). The biosynthesis of retinoic acid from retinol by rat-tissues *in vitro*. Arch. Biochem. Biophys. 255, 95-101.

Napoli, J. L., & Race, K. R. (1988). Biogenesis of retinoic acid from beta-carotene—differences between the metabolism of beta-carotene and retinal. J. Biol. Chem. 263, 17372-17377.

Newcomer, M. E. (1993). Structure of the epididymal retinoic acid-binding protein at 2.1 angstrom resolution. Structure 1, 7-18.

Newcomer, M. E., Jones, T. A., Aqvist, J., Sundelin, J., Eriksson, U., Rask, L., & Peterson, P. A. (1984). The 3-dimensional structure of retinol-binding protein. EMBO J. 3, 1451-1454.

Newcomer, M. E., Pappas, R. S., & Ong, D. E. (1993). X-Ray crystallographic identification of a protein-binding site for both all-*trans*-retinoic and 9-*cis*-retinoic acid. Proc. Natl. Acad. Sci. USA 90, 9223-9227.

Nilsson, M. H. L., Spurr, N. K., Lundvall, J., Rask, L., & Peterson, P. A. (1988). Human cellular retinol-binding protein gene organization and chromosomal location. Eur. J. Biochem. 173, 35-44.

Noy, N., & Xu, Z.-J. (1990). Interactions of retinol with binding proteins: implications for the mechanism of uptake by cells. Biochemistry 29, 3878-3883.

Ong, D. E. (1984). A novel retinol-binding protein from rat—purification and partial characterization. J. Biol. Chem. 259, 1476-1482.

Ong, D. E., Davis, J. T., Oday, W. T., & Bok, D. (1994). Synthesis and secretion of retinol-binding protein and transthyretin by cultured retinal-pigment epithelium. Biochemistry 33, 1835-1842.

Ong, D. E., Kakkad, B., & Macdonald, P. N. (1987). Acyl-Coa-independent esterification of retinol bound to cellular retinol-binding protein (type-II) by microsomes from rat small-intestine. J. Biol. Chem. 262, 2729-2736.

Ong, D. E., Macdonald, P. N., & Gubitosi, A. M. (1988). Esterification of retinol in rat-liver—possible participation by cellular retinol-binding protein and cellular retinol-binding protein-II. J. Biol. Chem. 263, 5789-5796.

Perez-Castro, A. V., Tothrogler, L. E., Wei, L. N., & Nguyenhuu, M. C. (1989). Spatial and temporal pattern of expression of the cellular retinoic acid-binding protein and the cellular retinol-binding protein during mouse embryogenesis. Proc. Natl. Acad. Sci. USA 86, 8813-8817.

Perez-Castro, A. V., Tran, V. T., & Nguyenhuu, M. C. (1993). Defective lens fiber differentiation and pancreatic tumorigenesis caused by ectopic expression of the cellular retinoic acid-binding protein-I. Development 119, 363-375.

Perlmann, T., Rangarajan, P. N., Umesono, K., & Evans, R. M. (1993). Determinants for selective RAR and TR recognition of direct repeat HREs. Genes Dev. 7, 1411-1422.

Perlmann, T., Umesono, K., Rangarajan, P. N., Forman, B. M., & Evans, R. M. (1996). 2 distinct dimerization interfaces differentially modulate target gene specificity of nuclear hormone receptors. Mol. Endocrinol. 10, 958-966.

Petkovich, M., Brand, N. J., Krust, A., & Chambon, P. (1987). A human retinoic acid receptor which belongs to the family of nuclear receptors. Nature 330, 444-450.

Pijnappel, W. W. M., Hendriks, H. F. J., Folkers, G. E., Vandenbrink, C. E., Dekker, E. J., Edelenbosch, C., Vandersaag, P. T., & Durston, A. J. (1993). The retinoid ligand 4-oxo-retinoic acid is a highly active modulator of positional specification. Nature 366, 340-344.

Plet, A., Gerbaud, P., Anderson, W. B., & Brion, D. E. (1985). Changes in the activity and subcellular-distribution of cyclic-AMP-dependent protein-kinases with retinoic-acid-induced differentiation of embryonal carcinoma-cells. Differentiation 30, 159-164.

Plet, A., Gerbaud, P., Sherman, M. I., Anderson, W. B., & Brion, D. E. (1986). Retinoic acid effect on cyclic AMP-dependent protein-kinases in embryonal carcinoma-cells—studies with differentiation-defective sublines. J. Cell. Physiol. 127, 341-347.

Posch, K. C., Boerman, M., Burns, R. D., & Napoli, J. L. (1991). Holocellular retinol binding-protein as a substrate for microsomal retinal synthesis. Biochemistry 30, 6224-6230.

Predki, P. F., Zamble, D., Sarkar, B., & Giguere, V. (1994). Ordered binding of retinoic acid and retinoid-X receptors to asymmetric response elements involves determinants adjacent to the DNA-binding domain. Mol. Endocrinol. 8, 31-39.

Prentice, A., Matthews, C. J., Thomas, E. J., & Redfern, C. P. F. (1992). The expression of retinoic acid receptors in cultured human endometrial stromal cells and the effects of retinoic acid. Hum. Reprod. 7, 692-700.

Raner, G. M., Vaz, A. D. N., & Coon, M. J. (1996). Metabolism of all-*trans*, 9-*cis*, and 13-*cis* isomers of retinal by purified isozymes of microsomal cytochrome-P450 and mechanism-based inhibition of retinoid oxidation by citral. Mol. Pharm. 49, 515-522.

Rask, L., Arundi, H., & Peterson, P. A. (1979). The primary structure of the human retinol binding protein. FEBS Letts. 104, 55-58.

Rask, L., & Peterson, P. A. (1976). *In vitro* uptake of vitamin A from the retinol-binding plasma protein to mucosal epithelial cells from the monkeys small intestine. J. Biol. Chem. 251, 6360-6366.

Rastinejad, F., Perlmann, T., Evans, R. M., & Sigler, P. B. (1995). Structural determinants of nuclear receptor assembly on DNA direct repeats. Nature 375, 203-211.

Redfern, C. P. F., Daly, A. K., Latham, J. A. E., & Todd, C. (1990). The biological-activity of retinoids in melanoma-cells—induction of expression of retinoic acid receptor-beta by retinoic acid in S91 melanoma-cells. FEBS Letts. 273, 19-22.

Redfern, C. P. F., Lovat, P. E., Malcolm, A. J., & Pearson, A. D. J. (1994). Differential-effects of 9-*cis* and all-*trans*-retinoic acid on the induction of retinoic acid receptor-beta and cellular retinoic acid- binding protein-II in human neuroblastoma-cells. Biochem. J. 304, 147-154.

Redfern, C. P. F., Lovat, P. E., Malcolm, A. J., & Pearson, A. D. J. (1995). Gene-expression and neuroblastoma cell differentiation in response to retinoic acid—differential effects of 9-*cis* and all-*trans* retinoic acid. Eur. J. Cancer 31A, 486-494.

Redfern, C. P. F., & Todd, C. (1992). Retinoic acid receptor expression in human skin keratinocytes and dermal fibroblasts *in vitro*. J. Cell Sci. 102, 113-121.

Redfern, C. P. F., & Wilson, K. E. (1993). Ligand-binding properties of human cellular retinoic acid-binding protein-II expressed in Escherichia-coli as a glutathione-s- transferase fusion protein. FEBS Letts. 321, 163-168.

Rees, J. L., Daly, A. K., & Redfern, C. P. F. (1989). Differential expression of the α and β receptors in tissues of the rat. Biochem. J. 259, 917-919.

Repa, J. J., Hanson, K. K., & Clagettdame, M. (1993). All-*trans*-retinoyl is a ligand for the retinoic acid receptors. Proc. Natl. Acad. Sci. USA 90, 7293-7297.

Rigtrup, K. M., Kakkad, B., & Ong, D. E. (1994). Purification and partial characterization of a retinyl ester hydrolase from the brush-border of rat small-intestine mucosa—probable identity with brush-border phospholipase-B. Biochemistry 33, 2661-2666.

Roberts, E. S., Vaz, A. D. N., & Coon, M. J. (1992). Role of isozymes of rabbit microsomal P-450 in the metabolism of retinoic acid, retinol and retinal. Mol. Pharm. 41, 427-433.

Rocchi, M., Covone, A., Romeo, G., Faraonio, R., & Colantuoni, V. (1989). Regional mapping of RBP4 to 10q23-q24 and RBP1 to 3q21-q22 in Man. Som. Cell Mol. Genet. 15, 185-190.

Roy, D. Y., Taneja, R., & Chambon, P. (1995). Synergistic activation of retinoic acid (RA)-responsive genes and induction of embryonal carcinoma cell-differentiation by an RA receptor-alpha (RAR-alpha)-selective, RAR-beta-selective, or RAR-gamma-selective ligand in combination with a retinoid X receptor-specific ligand. Mol. Cell. Biol. 15, 6481-6487.

Ruberte, E., Dollé, P., Chambon, P., & Morrisskay, G. (1991). Retinoic acid receptors and cellular retinoid binding-proteins .2. Their differential pattern of transcription during early morphogenesis in mouse embryos. Development 111, 45-60.

Ruberte, E., Dollé, P., Krust, A., Zelent, A., Morrisskay, G., & Chambon, P. (1990). Specific spatial and temporal distribution of retinoic acid receptor-gamma transcripts during mouse embryogenesis. Development 108, 213-222.

Ruberte, E., Friederich, V., Chambon, P., & Morrisskay, G. (1993). Retinoic acid receptors and cellular retinoid-binding proteins .3. Their differential transcript distribution during mouse nervous-system development. Development 118, 267-282.

Salbert, G., Fanjul, A., Piedrafita, F. J., Lu, X. P., Kim, S. J., Tran, P., & Pfahl, M. (1993). Retinoic acid receptors and retinoid-X receptor-alpha down-regulate the transforming growth factor-beta (1) promoter by antagonizing AP-1 activity. Mol. Endocrinol. 7, 1347-1356.

Samokyszyn, V. M., Chen, T., Maddipati, K. R., Franz, T. J., Lehman, P. A., & Lloyd, R. V. (1995). Free-radical oxidation of (E)-retinoic acid by Prostaglandin-H Synthase. Chem. Res. Toxicol. 8, 807-815.

Samokyszyn, V. M., & Marnett, L. J. (1987). Hydroperoxide-dependent cooxidation of 13-*cis* retinoic acid by Prostaglandin H Synthase. J. Biol. Chem. 262, 14119-14133.

Sande, S., & Privalsky, M. L. (1996). Identification of TRACs (T-3 receptor-associating cofactors), a family of cofactors that associate with, and modulate the activity of, nuclear hormone receptors. Mol. Endocrinol. 10, 813-825.

Schaefer, W. H., Kakkad, B., Crow, J. A., Blair, I. A., & Ong, D. E. (1989). Purification, primary structure characterization, and cellular-distribution of 2 forms of cellular retinol-binding protein, type-II from adult-rat small-intestine. J. Biol. Chem. 264, 4212-4221.

Schallreuter, K. U., & Woods, J. M. (1989). The stereospecific suicide inhibition of human melanoma thioredoxin reductase by 13-*cis* retinoic acid. Biochem. Biophys. Res. Comm. 160, 573-579.

Schwabe, J. W. R., Neuhaus, D., & Rhodes, D. (1990). Solution structure of the DNA-binding domain of the oestrogen receptor. Nature 348, 458-461.

Schwartz, E. L., Hallam, S., Gallagher, R. E., & Wiernik, P. H. (1995). Inhibition of all-*trans*-retinoic acid metabolism by fluconazole *in vitro* and in patients with acute promyelocytic leukemia. Biochem. Pharm. 50, 923-928.

Shingleton, J. L., Skinner, M. K., & Ong, D. E. (1989). Characteristics of retinol accumulation from serum retinol-binding protein by cultured sertoli cells. Biochemistry 28, 9641-9647.

Shirley, M. A., Bennani, Y. L., Boehm, M. F., Breau, A. P., Pathirana, C., & Ulm, E. H. (1996). Oxidative and reductive metabolism of 9-*cis*-retinoic acid in the rat—identification of 13,14-dihydro-9-*cis*-retinoic acid and its taurine conjugate. Drug Met. Disp. 24, 293-302.

Siddiqui, N. A., Loughney, A., Thomas, E. J., Dunlop, W., & Redfern, C. P. F. (1994). Cellular retinoid-binding proteins and nuclear retinoic acid receptors in endometrial epithelial-cells. Hum. Reprod. 9, 1410-1416.

Siddiqui, N. A., Thomas, E. J., Dunlop, W., & Redfern, C. P. F. (1995). Retinoic acid receptors and retinoid-binding proteins in endometrial adenocarcinoma—differential expression of cellular retinoid-binding proteins in endometrioid tumors. Int. J. Cancer 64, 253-263.

Sivaprasadarao, A., & Findlay, J. B. C. (1988). The interaction of retinol-binding protein with its plasma-membrane receptor. Biochem. J. 255, 561-569.

Smith, W. C., Nakshatri, H., Leroy, P., Rees, J., & Chambon, P. (1991). A retinoic acid response element is present in the mouse cellular retinol binding protein-i (mCRBP-I) promoter. EMBO J. 10, 2223-2230.

Soprano, D. R., Chen, L. X., Wu, S., Donigan, A. M., Borghaei, R. C., & Soprano, K. J. (1996). Overexpression of both RAR and RXR restores AP-1 repression in ovarian adenocarcinoma cells resistant to retinoic acid-dependent growth-inhibition. Oncogene 12, 577-584.

Stahl, W., Sundquist, A. R., Hanusch, M., Schwarz, W., & Sies, H. (1993). Separation of beta-carotene and lycopene geometrical-isomers in biological samples. Clin. Chem. 39, 810-814.

Stillwell, W., Ricketts, M., Hudson, H., & Nahmias, S. (1982). Effect of retinol and retinoic acid on permeability, electrical resistance and phase transition of lipid bilayers. Biochim. Biophys. Acta 688, 653-659.

Stunnenberg, H. G. (1993). Mechanisms of transactivation by retinoic acid receptors. BioEssays 15, 309-315.

Sucov, H. M., Dyson, E., Gumeringer, C. L., Price, J., Chien, K. R., & Evans, R. M. (1994). RXR-alpha mutant mice establish a genetic-basis for vitamin-A signaling in heart morphogenesis. Genes Dev. 8, 1007-1018.

Sucov, H. M., Izpisuabelmonte, J. C., Ganan, Y., & Evans, R. M. (1995). Mouse embryos lacking RXR-alpha are resistant to retinoic-acid-induced limb defects. Development 121, 3997-4003.

Takahashi, N., & Breitman, T. R. (1990a). Retinoic acid acylation-retinoylation. Meths. Enzymol. 189, 233-238.

Takahashi, N., & Breitman, T. R. (1990b). Retinoylation of HL-60 proteins—comparison to labeling by palmitic and myristic acids. J. Biol. Chem. 265, 19158-19162.

Takahashi, N., & Breitman, T. R. (1991). Retinoylation of proteins in leukemia, embryonal carcinoma, and normal kidney-cell lines—differences associated with differential responses to retinoic acid. Arch. Biochem. Biophys. 285, 105-110.

Takahashi, N., & Breitman, T. R. (1994). Retinoylation of vimentin in the human myeloid-leukemia cell-line HL- 60. J. Biol. Chem. 269, 5913-5917.

Takahashi, N., Jetten, A. M., & Breitman, T. R. (1991a). Retinoylation of cytokeratins in normal human epidermal-keratinocytes. Biochem. Biophys. Res. Comm. 180, 393-400.

Takahashi, N., Liapi, C., Anderson, W. B., & Breitman, T. R. (1991b). Retinoylation of the cAMP-binding regulatory subunits of type-I and type-II cAMP-dependent protein-kinases in HL-60-cells. Arch. Biochem. Biophys. 290, 293-302.

Takase, S., Ong, D. E., & Chytil, F. (1986). Transfer of retinoic acid from its complex with cellular retinoic acid-binding protein to the nucleus. Arch. Biochem. Biophys. 247, 328-334.

Taneja, R., Bouillet, P., Boylan, J. F., Gaub, M. P., Roy, B., Gudas, L. J., & Chambon, P. (1995). Reexpression of retinoic acid receptor (RAR)gamma or overexpression of RAR-alpha or RAR-beta in RAR-gamma-null f9 cells reveals a partial functional redundancy between the 3 RAR types. Proc. Natl. Acad. Sci. USA 92, 7854-7858.

Taneja, R., Roy, B., Plassat, J. L., Zusi, C. F., Ostrowski, J., Reczek, P. R., & Chambon, P. (1996). Cell-type and promoter-context dependent retinoic acid receptor (RAR) redundancies for RAR-beta-2 and HoxA-1 activation in F9 and P19 cells can be artifactually generated by gene knockouts. Proc. Natl. Acad. Sci. USA 93, 6197-6202.

Tate, B. F., Allenby, G., Janocha, R., Kazmer, S., Speck, J., Stürzenbecker, L. J., Abarzua, P., Levin, A. A., & Grippo, J. F. (1994). Distinct binding determinants for 9-*cis* retinoic acid are located within AF-2 of retinoic acid receptor-alpha. Mol. Cell. Biol. 14, 2323-2330.

Tate, B. F., & Grippo, J. F. (1995). Mutagenesis of the ligand-binding domain of the human retinoic acid receptor-alpha identifies critical residues for 9-*cis*-retinoic acid-binding. J. Biol. Chem. 270, 20258-20263.

Thaller, C., & Eichele, G. (1987). Identification and spatial-distribution of retinoids in the developing chick limb bud. Nature 327, 625-628.

Tickle, C., Alberts, B., Wolpert, L., & Lee, J. (1982). Local application of retinoic acid to the limb bond mimics the action of the polarizing region. Nature 296, 564-566.

Tini, M., Otulakowski, G., Breitman, M. L., Tsui, L. C., & Giguere, V. (1993). An everted repeat mediates retinoic acid induction of the gamma-F- crystallin gene—evidence of a direct role for retinoids in lens development. Genes Dev. 7, 295-307.

Tomita, S., Tsujita, M., Matsuo, Y., Yubisui, T., & Ichikawa, Y. (1993). Identification of a microsomal retinoic acid synthase as a microsomal cytochrome P-450-linked monooxygenase system. Int. J. Biochem. 25, 1775-1784.

Tournier, S., Raynaud, F., Gerbaud, P., Lohmann, S. M., Anderson, W. B., & Evainbrion, D. (1996). Retinoylation of the type-II cAMP-binding regulatory subunit of cAMP-dependent protein-kinase is increased in psoriatic human fibroblasts. J. Cell. Physiol. 167, 196-203.

Tsujita, M., Tomita, S., Miura, S., & Ichikawa, Y. (1994). Characteristic properties of retinal oxidase (retinoic acid synthase) from rabbit hepatocytes. Biochim. Biophys. Acta 1204, 108-116.

Umesono, K., Murakami, K. K., Thompson, C. C., & Evans, R. M. (1991). Direct repeats as selective response elements for the thyroid-hormone, retinoic acid, and vitamin-D3 receptors. Cell 65, 1255-1266.

Urbach, J., & Rando, R. R. (1994). Isomerisation of all-*trans* to 9-*cis* retinoic acid. Biochem. J. 299, 459-465.

van Kessel, A. G., Deleeuw, H., Dekker, E. J., Rijks, L., Spurr, N., Ledbetter, D., Kootwijk, E., & Vaessen, M. J. (1991). Localization of the cellular retinoic acid binding-protein (CRABP) gene relative to the acute promyelocytic leukemia-associated breakpoint on human chromosome-15. Hum. Genet. 87, 201-204.

Varani, J., Burmeister, W., Bleavins, M. R., & Johnson, K. (1996). All-trans-retinoic acid reduces membrane fluidity of human dermal fibroblasts—assessment by fluorescence redistribution after photobleaching. Am. J. Pathol. 148, 1307-1312.

Viallet, J. P., Ruberte, E., Dumanoir, S., Krust, A., Zelent, A., & Dhouailly, D. (1991). Retinoic acid-induced glandular metaplasia in mouse skin is linked to the dermal expression of retinoic acid receptor-beta messenger-RNA. Dev. Biol. 144, 424-428.

Vudac, N., Schoonjans, K., Kosykh, V., Dallongeville, J., Heyman, R. A., Staels, B., & Auwerx, J. (1996). Retinoids increase human apolipoprotein-a-ii expression through activation of the retinoid-X receptor but not the retinoic acid receptor. Mol. Cell. Biol. 16, 3350-3360.

Wagner, M., Han, B., & Jessell, T. M. (1992). Regional differences in retinoid release from embryonic neural tissue detected by an *in vitro* reporter assay. Development 116, 55.

Walker, S. S., Reese, J. R., Apone, L. M., & Green, M. R. (1996). Transcription activation in cells lacking TAFIIs. Nature 383, 185-188.

Wang, X. D., Krinsky, N. I., Benotti, P. N., & Russell, R. M. (1994). Biosynthesis of 9-*cis*-retinoic acid from 9-*cis*-beta-carotene in human intestinal-mucosa *in vitro*. Arch. Biochem. Biophys. 313, 150-155.

Wardlaw, S. A., & Ong, D. E. (1996). Characterization of the microsomal and partially purified retinal reductase of rat small-intestine. J. Nutr. Biochem. 7, 222-229.

Wei, L. N., Blaner, W. S., Goodman, D. S., & Nguyenhuu, M. C. (1989). Regulation of the cellular retinoid-binding proteins and their messenger ribonucleic-acids during P-19 embryonal carcinoma cell-differentiation induced by retinoic acid. Mol. Endocrinol. 3, 454-463.

Wei, L. N., & Lee, C. H. (1994). Demethylation in the 5'-flanking region of mouse cellular retinoic acid-binding protein-I gene is associated with its high-level of expression in mouse embryos and facilitates its induction by retinoic acid in P19 embryonal carcinoma-cells. Dev. Dyn. 201, 1-10.

Wei, L. N., Tsao, J. L., Chu, Y. S., Jeannotte, L., & Nguyenhuu, M. C. (1990). Molecular-cloning and transcriptional mapping of the mouse cellular retinoic acid-binding protein gene. DNA Cell Biol. 9, 471-478.

Williams, J. B., & Napoli, J. L. (1985). Metabolism of retinoic acid and retinol during differentiation of F9 embryonal carcinoma-cells. Proc. Natl. Acad. Sci. USA 82, 4658-4662.

Wilson, J. G., Roth, C. B., & Warkany, J. (1953). An analysis of the syndrome of malformations induced by maternal vitamin A deficiency. Effects of restoration of vitamin A at various times during gestation. Am. J. Anat. 92, 189-217.

Winter, N. S., Bratt, J. M., & Banaszak, L. J. (1993). Crystal-structures of holo- and apo-cellular retinol-binding protein- II. J. Mol. Biol. 230, 1247-1259.

Wolbach, S. B., & Howe, P. R. (1925). Tissue changes following deprivation of fat-soluble vitamin A. J. Exp. Med. 43, 753-777.

Wu, Y. N., Gadina, M., Taocheng, J. H., & Youle, R. J. (1994). Retinoic acid disrupts the golgi-apparatus and increases the cytosolic routing of specific protein toxins. J. Cell Biol. 125, 743-753.

Xiao, J. H., Durand, B., Chambon, P., & Voorhees, J. J. (1995). Endogenous retinoic-acid receptor (RAR) retinoid-x receptor (RXR) heterodimers are the major functional forms regulating retinoid- responsive elements in adult human keratinocytes—binding of ligands to RAR only is sufficient for RAR-RXR heterodimers to confer ligand-dependent activation of hRAR-beta-2/RARe (DR5). J. Biol. Chem. 270, 3001-3011.

Xu, X. C., Ro, J. Y., Lee, J. S., Shin, D. M., Hong, W. K., & Lotan, R. (1994). Differential expression of nuclear retinoid receptors in normal, premalignant, and malignant head and neck tissues. Cancer Res. 54, 3580-3587.

Yu, V. C., Delsert, C., Andersen, B., Holloway, J. M., Devary, O. V., Naar, A. M., Kim, S. Y., Boutin, J. M., Glass, C. K., & Rosenfeld, M. G. (1991). RXR-beta—a coregulator that enhances binding of retinoic acid, thyroid-hormone, and vitamin-D receptors to their cognate response elements. Cell 67, 1251-1266.

Zawel, L., & Reinberg, D. (1995). Common themes in assembly and function of eukaryotic transcription complexes. Annual Reviews in Biochemistry 64, 533-562.

Zechel, C., Shen, X. Q., Chen, J. Y., Chen, Z. P., Chambon, P., & Gronemeyer, H. (1994). The dimerization interfaces formed between the DNA-binding domains of RXR, RAR and TR determine the binding-specificity and polarity of the full-length receptors to direct repeats. EMBO J. 13, 1425-1433.

Zelent, A., Krust, A., Petkovich, M., Kastner, P., & Chambon, P. (1989). Cloning of murine alpha-retinoic and beta-retinoic acid receptors and a novel receptor-gamma predominantly expressed in skin. Nature 339, 714-717.

Zelent, A., Mendelsohn, C., Kastner, P., Krust, A., Garnier, J. M., Ruffenach, F., Leroy, P., & Chambon, P. (1991). Differentially expressed isoforms of the mouse retinoic acid receptor-beta are generated by usage of 2 promoters and alternative splicing. EMBO J. 10, 71-81.

Zhang, X. K., Hoffmann, B., Tran, P. B. V., Graupner, G., & Pfahl, M. (1992a). Retinoid X-receptor is an auxiliary protein for thyroid-hormone and retinoic acid receptors. Nature 355, 441-446.

Zhang, X. K., Lehmann, J., Hoffmann, B., Dawson, M. I., Cameron, J., Graupner, G., Hermann, T., Tran, P., & Pfahl, M. (1992b). Homodimer formation of retinoid X-receptor induced by 9-*cis* retinoic acid. Nature 358, 587-591.

RETINOIDS IN MAMMALIAN EMBRYONIC DEVELOPMENT

Gillian M. Morriss-Kay

Abstract 80
I. Introduction 80
II. Maternal Factors, Embryonic Uptakc, and Rctinoic Acid Synthesis 81
A. Maternal Retinoid Intake, Storage, and Homeostasis 81
B. Uptake of Retinol by the Embryo 81
C. Embryonic Retinoic Acid Synthesis 82
III. The Retinoic Acid Signaling Pathway 84
A. Access of Retinoic Acid to the Nuclear Receptors 84
B. Nuclear Retinoid Receptors 85
C. Retinoic Acid-Induced Alterations of Gene Expression and Morphogenesis 86
IV. Developmental Consequences of RA Signaling Pathway Defects 86
A. Vitamin A Deficiency 86
B. Retinoid Receptor Defects 87
C. Gene-Environment Interactions 88
Acknowledgments 89
References 90

Advances in Organ Biology
Volume 3, pages 79-92.

ISBN: 0-7623-0285-2

ABSTRACT

Studies on retinoic acid signaling in relation to mouse embryogenesis have made major contributions to our understanding of mechanisms of mammalian development. Sufficiency, but not excess, of retinoic acid within the embryonic tissues is required for normal patterns of cell proliferation, morphogenesis, and differentiation. The mammalian embryo is protected against maternal vitamin A deficiency by maternal retinoid storage and homeostatic mechanisms until the stores reach exceptionally low levels. Retinol (vitamin A) is bound to a retinol binding protien (RBP) in the maternal blood and transferred to embryonic RBP in the placenta. Retinoic acid is synthesized from retinol in the embryonic tissues, possibly facilitated by cytoplasmic retinol binding protein. Its access to the nucleus may also be controlled by mechanisms involving cytoplasmic binding, although this is now a point of controversy. In the nucleus, retinoic acid activates nuclear retinoid receptors (RARs and RXRs), which, as heterodimers, initiate transcription of genes with a retinoic acid response element. Loss of one retinoid receptor, through targeted gene disruption, may have no effect or only minor effects on development; loss of two receptors causes multiple congenital abnormalities similar to those caused by severe vitamin A deficiency. Although such catastrophic (double null) mutations are unlikely to occur in real life, mutations of any retinoic acid pathway gene may be of unrecognized importance as an underlying cause of human congenital abnormality, particularly through gene-environment interactions in conditions of mild vitamin A deficiency.

I. INTRODUCTION

Achievement of an understanding of the roles of retinoids in human pregnancy, and of their significance in congenital disease, is complicated by a number of factors. Chief among these is the sheer invisibility of the embryo and fetus; ultrasound scanning only brings the most serious abnormalities to light, whereas the most common retinoid-related developmental defects concern differentiation, and hence affect the microstructure and function of the affected systems rather than their morphology. On the other hand, the sporadic and unpredictable incidence of congenital defects means that the involvement of retinoid-related mechanisms in causing morphological abnormalities is hard to detect and may be underestimated. The embryo is genetically different from both parents, and quite apart from the possibility of new retinoid-related mutations occuring during gametogenesis, susceptibility to minor defects in maternal retinoid intake, metabolism, and transport may differ between siblings. Familial patterns of human abnormality with a retinoid-related basis may hence be masked by genetic variability and small family size.

Studies on experimental animals, especially rodents, can provide important insights into the possible roles of retinoids in human congenital disease. Inbred strains minimize genetic variability, while large litter sizes and short generation times enable assessments to be made at the population level. Mice have contributed enormously to our understanding of retinoid signaling in recent years through genetic manipulation. It is more difficult to achieve viable vitamin A-deficient pregnancy in mice than in rats, but here, too, there have been significant advances, opening up opportunities for investigating the effects of interaction between moderate levels of deficiency and genetic susceptibility. The availability of mouse-specific molecular probes, the variety of different strains of normal mice, and the vast number of mutations with specific phenotypes (many of which resemble recognized patterns of human abnormality) make the mouse the species of choice as a model for elucidating the importance of retinoids in prenatal development of humans and domestic animals. This chapter will review the contribution of mouse developmental studies to our understanding of the roles of retinoids in mammalian development.

II. MATERNAL FACTORS, EMBRYONIC UPTAKE, AND RETINOIC ACID SYNTHESIS

A. Maternal Retinoid Intake, Storage, and Homeostasis

Dietary sources of retinoids are mainly in the form of retinyl esters from animal products, especially liver, and β-carotene from plant foods. Retinoids, mainly in the form of esters, are stored in a number of sites including lung, kidney, gonads, and eyes, but the most important site is the liver (Wake, 1980). Retinyl esters are converted to retinol (vitamin A) for circulation in the blood, in which retinol is transported bound to a retinol binding protein (RBP)-transthyretin complex (Soprano et al., 1986). Retinol homeostasis is maintained by a mechanism ensuring that when retinol levels fall, hydrolysis of retinyl esters restores normal cellular and blood levels (Napoli et al., 1995, and references therein). The maternal retinol-RBP complex, from which the embryos are supplied, is therefore relatively unaffected by variations in the maternal dietary intake until the retinoid stores fall to very low levels (Ross and Gardner, 1994). The transcriptionally active retinoids are mainly all-*trans*-retinoic acid and 9-*cis*-retinoic acid; these are not transferred to the embryo, but synthesized within the embryonic tissues. (See Table 1.)

B. Uptake of Retinol by the Embryo

Normal development of mammalian embryos relies on adequate transfer of retinol from the maternal blood via the placenta. During early postimplantation development in rodents, the yolk sac (Figure 1) is the functional placenta. It is also the earliest site of synthesis of embryonic RBP (Soprano et al., 1986). The essential nature of yolk sac RBP has recently been demonstrated by means of injecting antisense RBP oligodeoxynucleotides into the exocoelom of mouse embryos, which were then cultured for up to 24 hours (Bavik et al., 1996) (the exocoelomic

Table 1. Key Events in Maternal-Embryonic Retinoid Metabolism

retinyl esters (animal) β carotene (plant)	dietary sources
⇘ ⇙	
retinyl esters	stores in maternal tissues (liver and extra-hepatic sites)
↓ ↑	
retinol	
⇩	retinoid homeostasis and transport in maternal blood
retinol-RBP-transthyretin	
↓	
(?) RBP receptor	placental uptake (visceral yolk sac endoderm)
↓	
retinol - RBP (embryonic)	transport to embryonic tissues
↓	
all-trans-retinol	
⇩	
all-trans retinaldehyde	synthesis of retinoic acids in embryonic tissues
⇩	
all-trans-retinoic acid	
↑ ↓	
9-cis-retinoic acid	
retinoic acid	
⇙ ⇘	metabolism of retinoic acid in embryonic tissues
other active retinoids? inactive metabolites	

cavity lies between the visceral yolk sac and the amnion; see Figure 1). Yolk sac RBP synthesis was rapidly inhibited; within six hours embryonic retinoic acid (RA) levels had fallen sufficiently to affect expression of an RA-sensitive LacZ-coupled promoter, and by 12 hours RA-induced promoter activity was undetectable. The cultured embryos showed developmental defects including failure of cranial neurulation and lens induction, and altered gene expression (down-regulation of *TGF-β1* and *Shh*). These effects could be prevented by supplying the embryos directly with RA, indicating that the embryonic RBP-retinol complex is essential for the provision of retinol to the embryo as the raw material for synthesis of transcriptionally active RA. The same group has now presented strong evidence indicating that uptake of maternal retinol into the yolk sac placenta involves an RBP receptor (Ward et al., 1997).

C. Embryonic Retinoic Acid Synthesis

The mechanism by which retinol is converted to retinoic acid in the cytoplasm of cells of the embryo and elsewhere is thought to involve a cytoplasmic retinol

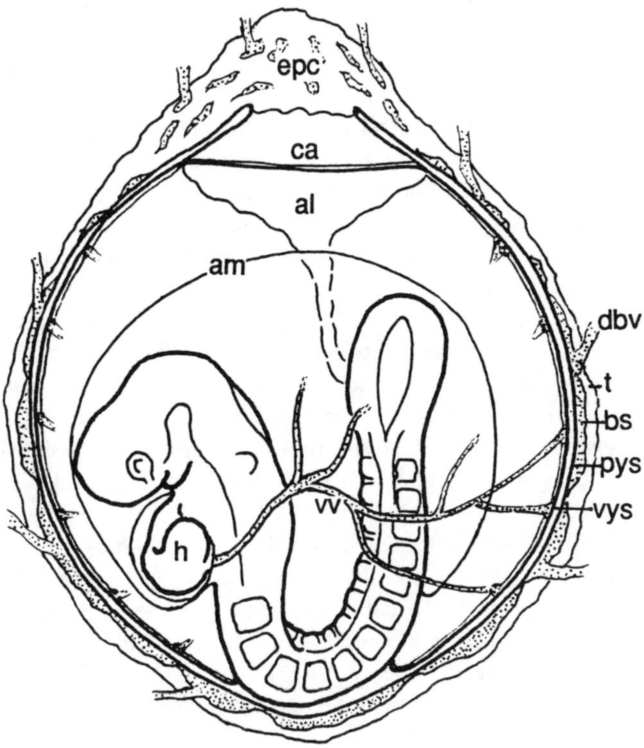

Figure 1. Diagrammatic representation of a day 8.5 mouse embryo, showing the maternal-embryonic relationship. Maternal blood traverses the decidual tissue in small vessels (dbv) that empty into (and drain from) a blood sinus (bs) between the trophoblast (t) and the parietal yolk sac (pys), a thick basement membrane-like structure. Retinol is taken up by the outer (endodermal) layer of the visceral yolk sac (vys) and passes into the vitelline veins (vv) in the inner (mesodermal) layer of the visceral yolk sac, for transport to the embryonic heart (h), and thence to the embryonic tissues. al, allantois; am, amnion; ca, chorioallantoic membrane; epc, ectoplacental cone trophoblast.

binding protein, CRBP I. Although there is no direct evidence that this protein is essential for retinoic acid synthesis in embryonic cells, its involvement is supported by evidence from metabolic studies (Napoli 1996; Means and Gudas, 1995, and references therein), and by its tissue-specific pattern of transcription and protein localization (Dollé et al., 1990; Ruberte et al., 1991; Gustafson et al., 1993). CRBP I is expressed in concentrations that exceed retinol availability; it interacts with an NADP-dependent microsomal retinol dehydrogenase, which catalyzes the conversion of retinol to retinaldehyde; RA synthesis is then mediated by an NAD-dependent retinal dehydrogenase (Napoli, 1996).

CRBP I transcripts are present in abundance in embryonic tissues that are particularly susceptible to vitamin A deficiency, such as the respiratory system and the face (Dollé et al., 1990). This pattern of localization supports the metabolic evidence that CRBP I is involved in creating differential tissue-specific levels of RA. It is also present in the visceral yolk sac (Ruberte et al., 1991), together with CRBP II (Sapin et al., 1996). However, this function cannot be essential, since in vitamin A deficiency (VAD) induced either by dietary means (Howell et al., 1964) or by antisense RBP oligonucleotides (Bavik et al., 1996), abnormal embryogenesis can be prevented by administration of RA, thereby bypassing the retinol-sourced RA synthetic machinery.

Ang and colleagues (1996) found that the onset of RA synthesis in mouse embryos, which they were first able to detect at the primitive streak stage, coincided with the onset of class IV alcohol dehygrogenase (ADH) activity. RA and ADH were colocalized in the developing craniofacial region, including the neural crest, and in mesenchyme of the trunk and limb buds. An aldehyde dehydrogenase, isoform AHD-2, known to oxidize retinaldehyde to retinoic acid, has similarly been identified in sites of retinoic acid synthesis in the developing eye (McCaffery et al., 1992). This study also indicates that the biochemistry of RA synthesis is different in the dorsal and ventral halves of the retina, reflecting the fact that shortening of the ventral retina is a feature of vitamin A deficiency that appears to act specifically through RXRα (see below). The complexity of RA signaling may include many as yet undetected tissue-specific features of RA synthesis as well as tissue-specific aspects of nuclear access of RA and of retinoid receptor function.

III. THE RETINOIC ACID SIGNALING PATHWAY

A. Access of Retinoic Acid to the Nuclear Receptors

Once synthesized within the cytoplasm of embryonic cells, RA must reach the nucleus in order to activate the nuclear retinoid receptors. The quantity of RA reaching the nucleus must in some way be tissue-specific, because different regions of the embryo show differential susceptibilities to RA excess and deficiency. Control of access of RA to the nucleus is thought to involve one or both of the cytoplasmic RA binding proteins, CRABP I and CRABP II. The rate of metabolism of RA and other retinoids in the cytoplasm is dependent on whether they are present in unbound or CRABP-bound forms (Napoli, 1996). However, this is now a controversial area, since although there is good evidence to support the roles of CRABPs in controlling the rate of RA catabolism, there are also data indicating that they are completely dispensable. The supporting evidence includes *in vitro* observations indicating that overexpression of CRABP I inhibits access of RA to the nuclear receptors (Means and Gudas, 1995, and references therein). Furthermore, the spatiotemporal patterns of distribution of CRABP I transcripts show high levels in embryonic tissues that are specifically vulnerable to retinoid excess, including the early hindbrain and neural crest (Ruberte et al., 1991). Expression of CRABP I and II also shows specific patterns of expression in the developing placenta and decidual tissue (Sapin et al., 1996).

The evidence against CRABP-mediated regulation of access of RA to the nucleus comes from null mutation studies: targeted disruption of both CRABP I and CRABP II has very little effect on embryonic development (the exception is a partial supernumerary digit on the forelimbs of some embryos), nor does lack of both RA binding proteins alter the sensitivity of the embryos to maternally administered RA excess (Lampron et al., 1995).

The existence of a mechanism controlling acess of RA to the nuclear receptors differentially in different embryonic (and adult) tissues makes logical sense as part of the global pattern by which tissue differences are generated; clearly, more information is required.

B. Nuclear Retinoid Receptors

Embryonic development is governed by a sequence of gene expression events that specify patterns of cell proliferation, morphogenesis, and differentiation. These events are controlled by a large number of transcription factors, many of which interact with each other's and their own promoters. The developmental program thus involves a complex meshwork of molecular interactions within the nucleus, as well as transcription of mRNAs that are exported to the cytoplasm for translation into proteins. Only this second group is directly involved in developmental change. The products of retinoid receptor-activated transcription include both the intra-nuclear and the cytoplasmic components of the developmental program.

There are two families of retinoic acid receptors, the RARs and RXRs, each of which has three isotypes designated α, β, and γ, and a numer of isoforms that are transcribed by alternative splicing from two promoters in each receptor sequence (see Mangelsdorf et al., 1994 and Chambon, 1996 for recent reviews). RARs and RXRs form heterodimers, activated by all-*trans*-RA (which binds specifically to RARs) and 9-*cis*-RA (which binds to both RARs and RXRs). RARs are highly conserved within mammals; they are present in all vertebrates but have no equivalent in invertebrates. RXRs are also highly conserved, and appear to be older in evolutionary terms, being related to the *Drosophila* homologue *ultraspiracle*. The ligand-activated RAR-RXR heterodimers bind to specific DNA sequences, retinoic acid response elements (RAREs), in the promoters of target genes, thereby initiating transcription. There are RAREs in the second promoters of each of the RARs, so that the RAR isoforms transcribed from these promoters are themselves RA-inducible. Other genes induced directly by RA include the 3′ *HOX* genes (Marshall et al., 1996), whose protein products are themselves transcription factors that are involved in embryonic pattern formation. Examples of RA-induced genes whose transcripts are exported to the cytoplasm include the extracellular matrix genes laminin B1 and collagen type IV (Vasios et al., 1989; Wang and Gudas, 1983).

In contrast, retinoid receptors can also mediate inhibition of gene expression: activity of the promoters of TGFβ1, collagenase, and cFos, all of which are controlled by transcription factor AP-1, can be inhibited by RARα, RARβ, or RXRα by a mechanism involving direct retinoid receptor interaction with AP-1 (Salbert et al., 1993).

C. Retinoic Acid-Induced Alterations of Gene Expression and Morphogenesis in Mouse Embryos

Exposure of gastrulation-stage mouse embryos to excess RA leads to alterations of forebrain pattern at early stages and of hindbrain pattern at slightly later stages. These patterns of altered morphogenesis are correlated with altered expression of many genes that are induced either directly or secondarily by RA, including *Hoxb-1, Hoxb-2, Krox-20, Pax-2, Emx1, Emx2, Dlx1, Wnt-1, Wnt-8, En-1,* and *En-2* (Morriss-Kay et al., 1991; Wood et al., 1994; Avantaggiato et al., 1996; Bouillet et al., 1996; Marshall et al., 1996). In general, mammalian embryos respond in similar ways to other vertebrate embryos (see chapter by Maden, this volume) when exposed to excess RA. This similarity includes human embryos, in which craniofacial abnormalities first described in rat embryos (Morriss, 1972) were induced by pharmacological levels of 13-*cis*-retinoic acid, taken as a treatment for severe cystic acne (Lammer et al., 1985).

Perhaps the most bizarre pattern alteration observed in the mouse to date is the induction of supernumerary hindlimbs by excess RA administered at pregastrulation stages of development (Rutledge et al., 1994; Niederreither et al., 1996). This is the nearest a mouse study has come to achieving the digital duplications observed in avian embryos after implantation of RA-soaked beads (see Maden, this volume, for details). Exposure of mouse embryos to RA excess during stages of limb bud outgrowth and patterning results in loss of digits and retardation of osteogenesis, possibly involving effects on *HoxD* gene expression (Wood et al., 1996).

IV. DEVELOPMENTAL CONSEQUENCES OF RA SIGNALING PATHWAY DEFECTS

A. Vitamin A Deficiency

Normal RA signaling depends on adequate levels of retinol reaching the embryo, on retinol being converted to RA, and on RA reaching the nuclear receptors for activation of signal transduction. Vitamin A homeostasis in the maternal blood (see above) provides a protective mechanism against inadequate supply of retinol to the embryo. The system fails when the maternal stores are severely depleted by long periods of dietary deprivation, although genetic defects affecting the storage or homeostatic mechanisms may be an unrecognized cause of congenital vitamin A deficiency. Below a certain level of maternal vitamin A, vitamin A deficiency is incompatible with pregnancy (Howell et al., 1964). However, when maternal stores are around the minimal level required for the maintenance of pregnancy, the spectrum of abnormalities indicates that the most vulnerable developing systems are the eyes, urogenital tract, cardiovascular system, and lungs (Wilson et al., 1953).

Severe vitamin A deficiency in human pregnancy is hard to detect because it is usually complicated by protein malnutrition and occurs in parts of the world in which facilities for chemical analysis are not available. One case reported by Sarma (1959) appears convincing: a mother suffering from night blindness gave birth prematurely to an infant with microcephaly and anophthalmia. Functional deficiencies, especially of the lungs, are a more common outcome of human vitamin A deficiency (Chytil, 1996), and have also been described in detail in the mouse (Sokolova, 1996). Other abnormalities observed in mouse and in other experimental animal studies include microphthalmia, umbilical hernia, edema, and spongy tissue structure of the thymus, liver, and heart ventricular muscle (Bauernfiend, 1986; Sokolova, 1996, and references therein). Only in the mouse have major morphological defects of the face and limbs been observed: these are midline facial clefting and severe forelimb deficiencies, and minor forelimb defects including persistent webbing and forelimb-specific supernumerary postaxial digit formation (Morriss-Kay and Sokolova, 1996). The supernumerary digit of these mice is identical to that of CRABP II-deficient mice (Lampron et al., 1995); all of the other morphological and histological abnormalities characteristic of experimental vitamin A deficiency have now been observed in animals lacking RA signaling pathway genes. Restriction of effects to the forelimbs is interesting, and suggests that higher RA levels are required for normal forelimb than for hindlimb development. The forelimbs/upper limbs may be more vulnerable to abnormal development in general: a recent large survey of human limb malformations reported that upper limb defects are approximately twice as common as those of the lower limbs (Evans et al., 1994).

B. Retinoid Receptor Defects

Targeted disruption (null mutation) has now been carried out for all of the RARs, for RXRα and RXRβ, and for some specific RAR isoforms; many of the mutant mice have been studied in paired RAR/RAR or RAR/RXR combinations (Kastner et al., 1997, and references therein). Some single null mutants are apparently normal (RARβ, RARα1, RARγ2). These results are surprising, given the specificity of the spatiotemporal expression patterns of RARs during embryogenesis (Dollé et al., 1990; Ruberte et al., 1991). It is too easy to dismiss the apparently normal mutants as demonstrating functional redundancy of the missing receptors; these studies are carried out in mice that are otherwise genetically normal and live in ideal conditions protected against variations of temperature and nutrition. Other mutants show some minor morphological abnormalities, together with problems of growth, defects of tissue maturation, and male sterility (RARα and RARγ). RXRβ null mutants appear to be normal apart from male sterility, and RXRα mutants show vitamin A deficiency-like abnormalities of the eye and heart. One of the RXRα-associated eye defects, shortening of the ventral retina, has not been observed in RAR mutants, suggesting that this aspect of normal eye development depends on an RXRα-dependent RA transduction event.

With the exception of the ventral retina anomaly, RAR double mutants show the whole spectrum of congenital abnormalities associated with vitamin A deficiency, and some that are not (Lohnes et al., 1994; Mendelsohn et al., 1994). Two of the abnormalities not associated with VAD are exencephaly and lens agenesis; however, these have been induced in cultured embryos that were unable to take up retinol because of RBP deficiency (Bavik et al., 1996). Other abnormalities observed in RAR double null mutants but not in any VAD studies may be VAD-inducible in theory, but require levels of deficiency that are incompatible with pregnancy. Double null mutant embryos have a survival advantage over severely vitamin A-deficient embryos in that they are developing within single null mutant mothers maintained on an adequate diet.

The results of the null mutant studies have emphasized that the unit of RA signal transduction is the RAR/RXR heterodimer. It is clear that although binding to specific types of RA reponse elements is normally heterodimer type specific (see Chambon, 1996, for details), in the absence of one RAR or RXR, a substitution may be made. This can only work if a suitable substitute is present, and the fact that minor anomalies do occur in most of the single null mutants suggests that the ability to make the substitution involves tissue-specific factors.

The seriousness of the morphogenetic effects of receptor loss is directly related to the number of missing alleles. For instance, RARγ null mutants show homeotic transformations and malformations of vertebrae (Lohnes et al., 1993), confirming that RA signal transduction plays an important role in patterning of the axial structures, as originally suggested on the basis of the effects of RA excess (Kessel and Gruss, 1991). Loss of further RAR alleles against the RARγ null background, that is, loss of RARα1 and then RAR α2, progressively increases the severity of the axial defects (Lohnes et al., 1994). A cumulative effect is also seen with respect to eye defects: the severity of eye defects seen in RXRα null mutants increases progressively with the further loss of one then both of RARβ2 and RARγ. Similarly, loss of one RXRα allele from an RXRγ null mutant reproduces the eye anomalies seen in RXRα null animals. Equivalent cumulative developmental effects associated with incremental loss of alleles occur with respect to heart defects, and syndactyly (Kastner et al., 1997).

C. Gene-Environment Interactions

Loss of function mutations involving retinoic acid receptor genes can occur in humans, as shown by the example of acute promyelocytic leukemia (Chomienne et al., 1990). However, the chances of losing two retinoid receptor alleles in a single individual must be vanishingly small. Interactions between minor mutations of retinoic acid signaling pathway genes and mild levels of VAD may be a significant unrecognized underlying cause of human congenital defects. Mutations affecting maternal retinoid metabolism could similarly affect the supply of retinol to the embryo even when dietary intake is apparently sufficient and the embryo is genetically normal.

One interesting model of gene-environment interactions showing RA-induced modulation of the phenotype is the *curly tail (ct/ct)* mouse. This mutant shows an approximately 54 percent incidence of spinal neural tube defects, including open spina bifida and tail flexion defects (Seller et al., 1979). Administration of 5mg/kg RA on day 10 prevents up to 50 percent of the tail defects, and up to 36 percent of the spina bifida, depending on the precise time of administration (Chen et al., 1994). (This level of exogenous RA is one twentieth of the level required for teratogenesis in genetically normal mice at the same developmental stage). The *ct/ct* embryos show a defect of RARγ expression. Normally, RARγ is expressed in the open caudal neuropore throughout the period of primary neurulation, and in the tail bud during secondary neurulation (Ruberte et al., 1991). In pre-neural tube defect *ct/ct* embryos, the caudal neuropore is enlarged and RARγ expression within it and in the tail bud is deficient (Chen et al., 1995). RA administration results in decreased caudal neurpore size and up-regulation of RARγ expression; although neither is restored to normal, there is a direct correlation between levels of RARγ transcripts and caudal neuropore size.

These observations on interactions between the *ct* gene and RA, in relation to the expression of the neural tube defect phenotype, provide an interesting contrast with the axial defects in RARγ null mutants. These are minor and entirely confined to the cervical and thoracic region, and morphogenesis of the vertebral column is unaffected by the administration of 100 mg/kg RA (Lohnes et al., 1993), a treatment that induces major homeotic repatterning in wild type mice (Kessel and Gruss, 1991). RARγ is clearly required for transduction of the RA signal in teratogenesis, and expression of this gene or function of the gene product is altered by the *ct* mutation.

The *curly tail* mouse is not an ideal model for gene-environment interaction, but it does serve to indicate that the phenotypic effects of mutations affecting retinoic acid signaling genes can be modulated by the addition of low levels of RA at the appropriate stage of development. This may also be true of a number of human abnormalities, although it will be difficult to verify this suggestion for the reasons set out in the introduction to this review.

ACKNOWLEDGMENTS

I thank the Medical Research Council and the Human Frontier Science Organisation for financial support; Simon Ward and Andrea Brown for helpful comments on the manuscript.

REFERENCES

Ang, H.L., Deltour, L., Hayamizu, T.F., Zgombic-Knight, M., & Duester, G. (1996). Retinoic acid synthesis in mouse embryos during gastrulation and craniofacial development linked to class IV alcohol dehydrogenase gene expression. J. Biol. Chem. 271, 9526-9534.

Avantaggiato, V., Acampora, D., Tuorto, F., & Simeone, A. (1996). Retinoic acid induces stage-specific repatterning of the rostral central nervous system. Dev. Biol. 175, 347-357.

Bauernfiend, C.J. (1986). *Vitamin A Deficiency and its Control.* Academic Press, New York.

Bavik, C., Ward, S.J., & Chambon, P. (1996). Developmental abnormalities in cultured mouse embryos depived of retinoic acid by inhibition of yolk-sac retinol binding protein synthesis. Proc. Natl. Acad. Sci. USA 93, 3110-3114.

Bouillet, P., Oulad-Abdelghani, M., Ward, S.J., Bronner, S., Chambon, P., & Dollé, P. (1996). A new member of the *Wnt* gene family, *mWnt-8*, is expressed during early embryogenesis and is ectopically induced by retinoic acid. Mech. Dev. 58, 141-152.

Chambon, P. (1996). A decade of molecular biology of retinoic acid receptors. FASEB J. 10940-954.

Chen, W-H., Morriss-Kay, G.M., & Copp, A.J. (1994). Prevention of spinal neural tube defects in the curly tail mouse mutant by a specific effect of retinoic acid. Developmental Dynamics 199, 93-102.

Chen, W-H., Morriss-Kay, G.M., & Copp, A.J. (1995). Genesis and prevention of spinal neural tube defects in the curly tail mouse mutant: involvement of retinoic acid and its nuclear receptors RAR-β and RAR-γ. Development 121, 681-691.

Chomienne, C., Ballerini, P., Balitrand, N., Huang, M.E., Krawice, I., Castaigne, S., Fenaux, P., Tiollais, P., Dejean, A., & Degos, L. (1990). The retinoic acid receptor α gene is rearranged in retinoic acid sensitive promyelocytic leukemia. Leukemia 4, 802-807.

Chytil, F. (1996). Retinoids in lung develpment. FASEB J. 10, 986-992.

Dollé, P., Ruberte, E., LeRoy, P., Morriss-Kay, G., & Chambon, P. (1990). Retinoic acid receptors and cellular binding proteins. I. A systematic study of their differential pattern of transcription during mouse organogenesis. Development 110, 1133-1151.

Evans, J.A., Vitez, M., & Czeizel, A. (1994). Congenital abnormalities associated with limb deficiency defects: A population study based on case from the Hungarian Congenital Malformation Registry (1975-1984). Am. J. Med. Genet. 49, 52-66.

Gustafson, A.-L., Dencker, L., & Eriksson, U. (1993). Non-overlapping expression of CRBP I and CRABP I during pattern formation of limbs and craniofacial structures in the early mouse embryo. Development 117, 451-460.

Howell, J.McC, Thompson, J.N., & Pitt, G.A.J. (1964). Histology of the lesions produced in the reproductive tract of animals fed a diet deficient in vitamin A alcohol but containing vitamin A acid. J.Reprod. Fertil. 7, 251-258.

Kastner, P., Mark, M., Ghyselinck, N., Krezel, W., Dupé, V., Grondona, J.M., & Chambon, P. (1997). Genetic evidence that the retinoid signal is transduced by heterodimeric RXR/RAR functional units during mouse development. Development 124, 313-326.

Kessel, M., & Gruss, P. (1991). Homeotic transformations of murine vertebrae and concomitant alteration of *Hox* codes induced by retinoic acid. Cell 67, 89-104.

Lammer, E.J., Chen, D.T., Hoar, R.M., Agnish, N.D., Benke, P.J., Braun, J.T., Curry, C.J., Fernhoff, P.M., Grix, A.W., Lott, I.T., Richard, J.M., & Sun, S.C. (1985). Retinoic acid embryopathy. N. Eng. J. Med. 313, 837-841.

Lampron, C., Rochette-Egly, C., Gorry, P.,Dollé, P., Mark, M., Lufkin, T., LeMeur, M., & Chambon, P. (1995). Mice deficient in cellular retinoic acid binding protein II (CRABP II) or in both CRABP I and CRABP II are esentially normal. Development 121, 539-548.

Lohnes, D., Kastner, P., Dierich, A., Mark M., LeMeur, M., & Chambon, P. (1993). Function of retinoic acid receptor γ (RARγ) in the mouse. Cell 73, 643-658.

Lohnes, D., Mark, M., Mendelsohn, C., Dollé, P., Dierich, A., Gorry, P., Gansmuller, A., & Chambon, P. (1994). Function of the retinoic acid receptors (RARs) during development. I. Craniofacial and skeletal abnormalities in RAR double mutants. Development 120, 2723-2748.

McCaffery, P., Lee, M-O., Wagner, M.A., Sladek, N.E., & Dräger, U.C. (1992). Asymmetrical retinoic acid synthesis in the dorsoventral axis of the retina. Development 115, 371-382.

Mangelsdorf, D., Umesono, K., & Evans, R. (1994). The retinoid receptors. In: *The Retinoids*, 2nd edition (Sporn, M.B., Roberts, A.B., & Goodman, D.S., eds), pp. 319-349, Raven Press, New York.

Marshall, H., Morrison, A., Studer, M., Pöpperl, H., & Krumlauf, R. (1996). Retinoids and *Hox* genes. FASEB J. 10 968-978.

Means, A.L., & Gudas, L.J. (1995). The roles of retinoids in vertebrate development. Annu. Rev. Biochem. 64, 201-233.

Mendelsohn, C., Lohnes, D., Décimo, D., Lufkin, T., LeMeur, M., Chambon, P., & Mark, M. (1994). Function of the retinoic acid receptors (RARs) during development. II. Multiple abnormalities at various stages of organogenesis in RAR double mutants. Development 120, 2749-2771.

Morriss, G.M. (1972). Morphogenesis of the malformations induced in rate embryos by maternal hypervitaminosis A. J. Anat. 113, 241-250.

Morriss-Kay, G.M., Murphy, P., Hill, R.E., & Davidson, D.R. (1991). Effects of retinoic acid excess on expression of *Hox-2.9* and *Krox-20* and on morphological segmentation of the hindbrain of mouse embryos. EMBO J. 10, 2985-2995.

Morriss-Kay, G.M., & Sokolova, N.V. (1996). Embryonic development and pattern formation. FASEB J. 10 961-968.

Napoli, J.L. (1996). Regulation of the biosynthesis and catabolism of retinoids. FASEB. J. 993-1001.

Ross, A.C., & Gardner, E.M. (1994). The function of vitamin A in cellular growth and differentiation, and its roles during pregnancy and lactation. Adv. Exp. Med. Biol. 352, 187- 200.

Ruberte, E., Dollé, P., Chambon, P., & Morriss-Kay, G. (1991). Retinoic acid receptors and cellular binding proteins. II. Their differential pattern of transcription during early morphogenesis in mouse embryos. Development 111, 45-60.

Salbert, G., Fanjul, A., Piedrafita, F.J., Lu, X.P., Kim, S-J., Tran, P., & Pfahl, M. (1993). Retinoic acid receptors and retinoid X-receptor-α down-regulate the transforming growth factor-β1 promoter by antagonizing AP-1 activity. Mol. Endocrinol. 7, 1347-1356.

Sapin, V., Ward, S.J., Bronner, S., Chambon, P., & Dollé, P. (1996). Differential expression of transcripts encoding retinoic binding proteins and retinoic acid receptors during placentation of the mouse. Dev. Dynam. 208, 199-210.

Sarma, V. (1959). Maternal vitamin A deficiency and fetal microcephaly and anophthalmia. Obstet. Gynecol. 13, 299-301.

Seller, M.J., Embury, S., Polani, P.E., & Adinolfi, M. (1979). Neural tube defects in curly-tail mice. II. Effects of maternal administration of vitamin A. Proc. R. Soc. Lond. Biol. 206, 95-107.

Sokolova, N.V. (1996). The roles of vitamin A in embryonic lung development. D.Phil Thesis, University of Oxford, U.K.

Soprano, R.D., Soprano, K.J., & Goodman, D.S. (1986). Retinol-binding protein and transthyretin mRNA levels in visceral yolk sac and liver during fetal development in the rat. Proc. Natl. Acad. Sci. USA. 83, 7330-7334.

Vasios, G.W., Gold, J.D., Petkovich, M., Chambon, P., & Gudas, L.J. (1989). A retinoic acid-responsive element is present in the 5' flanking region of the laminin B1 gene. Proc. Natl. Acad. Sci. USA 86, 9099-9103.

Wake, K. (1980). Perisinusoidal stellate cells (fat-storing cells, interstitial cells, lipocytes), their related structure in and around the liver sinusoids, and vitamin A storing cells in the liver and extra-hepatic organs. Int. Rev. Cytol. 66, 303-353.

Wang, S.-Y., & Gudas, L.J. (1983). Isolation of cDNA clones specific for collagen IV and laminin from mouse teratocarcinoma cells. Proc. Natl. Acad. Sci. USA 80, 5880-5884.

Ward, S.J., Chambon, P., Ong, D.E., & Bavik, C. (1997). A mechanism for embryonic uptake of vitamin A. Biol. Reprod. (in press).

Wilson, J.C., Roth, C.B., & Warkany, J. (1953). An analysis of the syndrome of malformations induced by maternal vitamin A deficiency. Effects of restoration of vitamin A at various times during gestation. Am. J. Anat. 92, 189-217.

Wood, H.B., Pall, G.S., & Morriss-Kay, G.M. (1994). Exposure to retinoic acid before or after the onset of somitogenesis reveals separate effects on rhombomeric segmentation and 3′ *HoxB* gene expression domains. Development 120, 2279-2285.

Wood, H.B., Ward, S.J., & Morriss-Kay, G.M. (1996). Effects of all-*trans*-retinoic acid on skeletal pattern, 5′*HoxD* gene expression, and RARβ2/β4 promoter activity in embryonic mouse limbs. Devel. Genet. 18 (in press).

THE ROLE OF RETINOIDS IN PATTERNING FISH, AMPHIBIAN, AND CHICK EMBRYOS

Malcolm Maden and John Pizzey

Abstract 94

I. Introduction 94

II. Zebrafish 94

A. Excess RA 94

B. Endogenous Retinoids and RA Synthesis 98

C. Retinoic Acid Receptors 100

III. *Xenopus* 101

A. Excess RA 101

B. Endogenous Retinoids and Their Distribution 105

C. Retinoid Binding Proteins and Nuclear Receptors 108

IV. Regenerating Amphibian Appendages 110

A. RA and Limb Regeneration 110

B. RA and Tail Regeneration 112

C. The Cellular Effects of Retinoids 113

D. Endogenous Retinoids in Limb Regeneration 115

E. Retinoid Binding Proteins and Receptors and Their Function 116

Advances in Organ Biology
Volume 3, pages 93-139.

ISBN: 0-7623-0285-2

V. The Early Chick Embryo ... 117
A. Excess RA ... 117
B. Endogenous Retinoids in the Early Chick Embryo and Development in their Absence ... 121
C. Retinoid Binding Proteins and Receptors in the Early Chick Embryo ... 123
VI. The Chick Limb Bud ... 125
A. Excess RA ... 125
B. Endogenous RA and the Retinoid Pathway ... 127
C. Retinoid Binding Proteins and Receptors ... 128
References ... 129

ABSTRACT

We describe here the role that retinoids play in the development of non-mammalian vertebrates, namely, fish (zebrarish development), amphibians (*Xenopus* development, newt and axolotl limb regeneration), and birds (chick and quail development, chick limb bud development). In each system the effects of excess administration of retinoids are discussed with regard to the genes that are affected and the resultant anatomical structures that are altered. The measurements and techniques used to determine the presence of endogenous retinoids are then described for each system and finally the patterns of expression of the cytosolic retinoid binding proteins and the nuclear retinoic acid receptors are considered. Even though the systems are diverse, excess retinoids have a surprising similarity of effect both on the phenotype of these embryos and on the genes that are affected. Similar studies on mammalian embryos are described in the following chapter.

I. INTRODUCTION

This chapter describes the role of retinoids in the developing embryos of non-mammalian vertebrates: zebrafish, *Xenopus*, the regenerating amphibian limb and tail, the chick embryo, and limb bud. We describe each system in terms of the effects of excess retinoids on the morphology of the embryos, experiments on the determination of endogenous retinoids in the system, and, finally, the presence and spatial localization of the various retinoid receptors (binding proteins, RARs and RXRs). The subsequent chapter deals with aspects of the mammalian embryo, so we have only referred to mammalian studies where relevant.

II. ZEBRAFISH

A. Excess RA

Treatment of the zebrafish embryo with 10^{-7}M tRA at gastrulation stages has a specific effect on the anteroposterior organization of the CNS. The effect is precisely localized to the anterior hindbrain and the midbrain/hindbrain junction, which includes the precursor of the cerebellum (Holder and Hill, 1991). Part of the affected tissue is identified in the normal embryo as the expression domain of the *engrailed* protein and this domain is absent in treated embryos, which results in the shortening of the head region and the anterior shift in the position of the otic vesicle (Figures 1A & B). The cranial ganglia, which are associated with this region, the Vth, and the VIIIth ganglia had far fewer cells within them, suggesting an effect on the neural crest cells of this region, but the remainder of the CNS, that is the rest of the midbrain, the forebrain, and the spinal cord were completely unaffected by RA treatment.

Further analysis of these effects of RA on the zebrafish CNS by Hill and colleagues (Hill et al., 1995) has shown that the deleted region corresponds to a posterior midbrain stripe at the midbrain/hindbrain border, a region that expressed the *pax2* gene, and rhombomeres 1-3. The effects of excess RA are incredibly specific—only the engrailed expressing cells at the midbrain/hindbrain border are affected, not those cells expressing engrailed in the somites; only the *pax2* expressing cells in the midbrain are affected, not those cells expressing *pax2* in the eye stalk and otic vesicle; and only the *Krox20* expressing cells in rhombomere 3 are affected, not those expressing *Krox20* in rhombomere 5.

In addition, when the dose of tRA is slightly lowered then a completely different phenotype is generated. Instead of a deletion of tissue in the hindbrain, a respecification of one rhombomere is revealed, a result similar to the effects on the mouse embryo (Marshall et al., 1992). By analyzing the reticulospinal neurons of the hindbrain that are individually identifiable and together form a reproducible, segmentally arranged column of cells in the rhombomeres, Hill and colleagues (Hill et al., 1995) demonstrated that rhombomere 2 takes on a hybrid phenotype containing its normal cells plus a neuron normally only found in rhombomere 4, the Mauthner neuron.

These effects on the CNS were produced by treating the embryos during gastrulation when pattern within the mesoderm and the CNS is being specified. There is a rise in sensitivity to treatment at 50 percent epiboly (Zhang et al., 1996). Interestingly, treatment of embryos with 9-*cis*-RA instead of tRA induces a different type of CNS malformation—in this case the rostral CNS (forebrain, eyes, and midbrain) were absent (Zhang et al., 1996) instead of the missing segment of midbrain/hindbrain as is the case with tRA (Figure 1C). This result has a bearing on an explanation for the two completely different effects of RA on the CNS of the embryo: the loss of forebrain in combination with a normal hindbrain, a result typical of *Xenopus* (see below), but also seen in the mouse (Simeone et al., 1995) on the one hand, and the deletion/respecification of anterior hindbrain in combination with a normal forebrain, seen in the mouse and zebrafish, on the other hand. These contrasting effects may be related to stage or dose, or they may be ligand-specific, and this may also depend on the relative abundance of endogenous ligands present at the time of administration. Interestingly, the RXR-specific ligands SR

11237 and Ro25-6603 do not apparently produce any teratogenic effects on zebrafish embryos (Minucci et al., 1996). However, another RXR-specific ligand SR 11217 does act with similar characteristics to 9-*cis*-RA, but at a 10 fold higher concentration (Zhang et al., 1996).

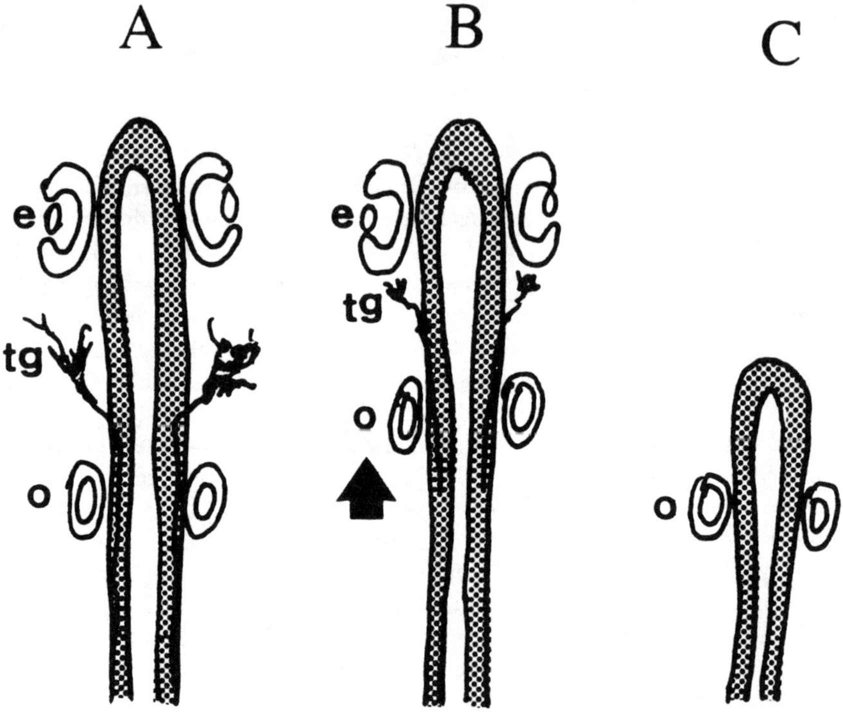

***Figure 1*.** (A.-C.) Comparison of the effects of all-*trans*-RA and 9-*cis*-RA on the zebrafish embryo. (A) Drawing of the head structures of a normal 24hr embryo showing the relationship between the eye (e), trigeminal ganglion (tg), and otic vesicle (o). The neuroepithelium of the forbrain, midbrain, and hindbrain is shown dotted. (B) 24hr embryo treated with a pulse of 10-7M all-*trans*-RA at gastrulation stages showing normal anterior structures (the eye and forebrain), but the trigeminal ganglion is reduced in size and is moved anteriorly towards the eye, as is the otic vesicle. This anterior shift of these structures is shown by the arrow and is caused by a loss of CNS tissue in the posterior midbrain/ anterior hindbrain region. (C) 24hr embryo treated with a pulse of 10-7M 9-*cis*-RA at gastrulation stages. In contrast to tRA, 9-*cis* causes the loss of forebrain, eyes, and anterior structures. (D) The endogenous distribution of a RARE-TKpr-βgal transgene in a 18-20hr embryo. This is the earliest stage at which the transgene was activated and reveals intense activation in the anterior trunk posterior to the hindbrain, with activation levels declining towards the tail. Drawings taken from Holder and Hill (1991), Zhang et al. (1996), and Marsh-Armstrong et al. (1995).

Another effect that tRA has on the hindbrain of zebrafish embryos is to alter the domains of expression of *svp* [40] (Fjose et al., 1995). This gene is a COUP/ARP-1 type of nuclear receptor that may function as a negative regulator of transcription and play a role in the establishment of the segmental nature of the hindbrain. In some exciting recent experiments, the mechanism of action of tRA in altering the structure of the CNS and branchial arches has been approached. It is well known that one of the earliest effects of tRA administration to embryos is to up-regulate the expression of many 3′ *Hox* genes, usually in an anterior direction. *Hoxa*-1 is one such gene in particular and to test its role Alexandre and colleagues (1996) overexpressed zebrafish *Hoxa*-1 by RNA injection into the fertilized egg. This resulted in the appearance of many of the effects of RA—an alteration of the *Krox*-20 expression in rhombomere 3; a duplication of the Mauthner neurons characteristic of rhombomere 4 in more anterior rhombomeres; an abnormal development of the jaw apparatus; missing cartilagenous derivatives of the first branchial arch; abnormally shaped cartilagenous derivatives of the second branchial arch; fusion of the Vth and VIIth cranial ganglia; and a missing stream of anterior neural crest from the rhombomere 2 region.

A good deal of research on zebrafish embryos has been concentrated upon CNS development, as described earlier. However, three other systems have been studied in these embryos.

The first is heart development. When low doses (10^{-8} - 10^{-7}M) of tRA are administered at gastrulation stages there is a progressive truncation of the heart tube

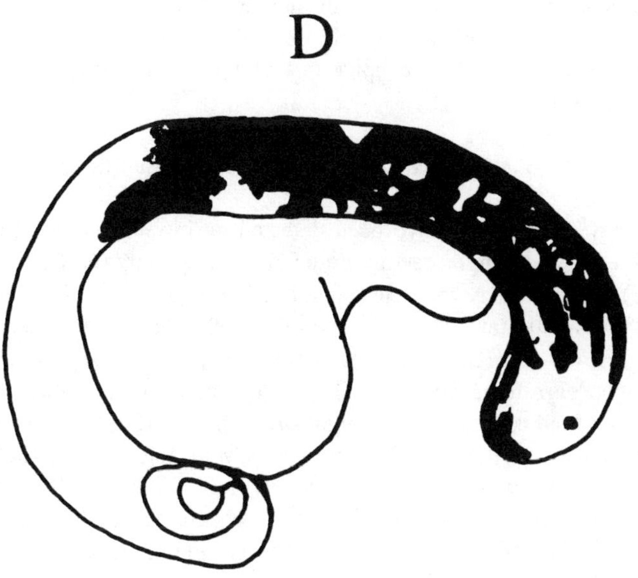

Figure 1. Continued

(Stainier and Fishman, 1992). This increasing loss with increasing dose progresses from the arterial end of the heart to the veinous end of the heart. Thus increasing RA exposure first causes a loss of the bulbus arteriosus, then the ventricle, then the atrium, and finally the sinus venosus. This is a similar situation to that seen in the chick embryo (see below) where atrial tissue is induced at the expense of ventricular tissue (Yutzey et al., 1995).

The second is retina development. tRA administration during the period of formation of the eye primordia (early somite stages) results in the dupliction of the ventral retina (Hyatt et al., 1992). This result fits very neatly with the converse experiment: in the absence of RA synthesis the ventral retina fails to develop (Marsh-Armstrong et al., 1994).

The third system is fin development and regeneration. The developing fins express the gene *shh/vhh-1* on the posterior margin exactly as the limb buds of chick and mouse embryos do (see below) (Akimenko and Ekker, 1995). When embryos are treated with RA this *shh* expressing region is duplicated on the anterior margin. Again, in chick limb buds this same phenomenon occurs and it leads to the duplication of the limb. Unfortunately, zebrafish embryos treated systemically with this dose of RA (10^{-6}M) do not survive long enough to see whether the fins become duplicated.

The pectoral and caudal fins of the adult zebrafish can regenerate with slight differences in the branching patterns of the rays between unamputated and regenerated fins. RA treatment initially slows down fin regeneration, but the final length of the regenerated fin is the same as the controls. However, internally, defects occur with fusion of rays and an increase in the number of segments in the rays (Geraudie et al. 1994; 1995). The explanation for this effect of RA may be due to increased cell death, specifically in the epidermis of the regenerating fin (Ferretti and Geraudie, 1995).

B. Endogenous Retinoids and RA Synthesis

A variety of retinoids have been detected during early development of the zebrafish embryo by high pressure liquid chromatography (HPLC) (Costaridis et al., 1996). All-*trans*-RA was detected at low levels (11nM) and the levels did not change consistently throughout development. Other active retinoic acids such as 9-*cis*-RA and 4-*oxo*-RA could not be detected, but ddRA was detectable at increasing levels after gastrulation. The unusual feature of the zebrafish egg is that the major retinoid is retinal at very high concentrations (2μM), although the levels of the retinoid esters, the normal storage forms, were not reported. The same is true in *Xenopus* (see below) and in tunicates (Kawamura et al., 1993) and in clear contrast to higher vertebrates where in the yolk of the chicken's egg, retinal is virtually undetectable and it is retinol that is present at high levels in addition to the esters. Thus embryos of lower vertebrates must have a different type of retinoid metabolism from those of higher vertebrates.

Data on the spatial localization of RA in the zebrafish embryo has been provided by injecting a reporter construct containing a retinoic acid response element (RARE) linked to a minimal promoter and the β-galactosidase gene (Marsh-Armstrong et al., 1995). In the presence of RA this construct is activated and after histochemical processing for *lac*Z, blue cells can be seen in the embryo. Such constructs have been used to good effect in transgenic mouse embryo studies (Rossant et al., 1991; Balkan et al., 1992). In the zebrafish embryo the construct was injected into the cytoplasm of 1 and 2 cell embryos and the first localization of expression was found at the 18-21 somite stage when transgene activity was concentrated in the anterior trunk in a variety of cell types including spinal cord, floor plate, notochord, muscle, and epidermis (Figure 1D).

In addition, RA was also detected in the zebrafish by dissecting pieces of the embryo and assaying for RA release with a reporter cell line (Wagner et al., 1992). This is a mouse embryonal carcinoma cell line that has been transfected with these RARE-tk-βgal constructs and the cells respond to the release of RA by turning blue after histochemical staining for *lac*Z. Either pieces of tissue dissected from embryos can be placed on lawns of these cells or homogenates of tissue can be added to the medium. In the zebrafish embryo, using this alterative system (Marsh-Armstrong et al., 1995), a similar picture emerged to the transgenic studies. There was a gradient of RA along the trunk with levels the highest in the anterior trunk region (at the level of somites 7-9). The hindbrain region showed only low levels. Finally, this thorough analysis demonstrated that thc localization of RA in the embryo was due to the localized presence of the enzyme retinaldehyde dehydrogenase (which converts retinal to retinoic acid).

It is surprising that localization of RA was only demonstrated at relatively late stages (15 somite stage onwards) in this study of Marsh-Armstrong and coworkers (Marsh-Armstrong et al., 1995) when similar studies in *Xenopus*, chick, and mouse embryos (see below) have shown that gastrulation is the stage when RA can first be detected and RA is certainly present in the zebrafish embryo from fertilization onwards (Costaridis et al. 1996). Furthermore, the inhibition of RA synthesis at gastrulation stages with enzyme inhibitors (see below) results in abnormal antero-posterior development. Presumably this rather late detection reflects the limited sensitivity of this transgene construct in the zebrafish embryo.

As further proof of the role of RA synthesis in the development of the antero-posterior axis of the zebrafish embryo, inhibition of RA synthesis with the sulphydryl reagent, disulphiram, causes abnormal development of the trunk:tail axis such that the CNS is shorter than normal and the notochord is longer than normal. This results in the most obvious feature of a wavy, undulating notochord (Marsh-Armstrong et al., 1995; Costaridis et al., 1996) and can be induced by treatment as early as gastrulation stages, confirming the importance of RA synthesis early in development. It is intriguing that a similar phenotype can be induced by overexpression of two genes, sonic hedgehog and moonrat (the latter is closely related to the murine Indian hedgehog gene) (Currie et al., 1995) as the level of sonic hedgehog has been shown to be regulated by RA in the chick limb bud (Stratford et al., 1996, and see below).

Furthermore, after disulphiram treatment of embryos the pectoral fins fail to develop (Marsh-Armstrong et al., 1995), which ties in well with studies on the effects of excess RA on fin development.

Finally, RA synthesis has also been detected in a specific region of the late zebrafish embryo, the retina (Marsh-Armstrong et al., 1994). In the developing zebrafish retina there are two different retinaldehyde dehydrogenases, as is the case in the mouse embryo, one found dorsally and the other found ventrally. The ventral enzyme appears at around the 15 somite stage of embryogenesis, several hours before the dorsal enzyme and the inhibition of this RA synthesis in the ventral retina by the administration of citral results in the failure of the ventral retina to develop.

Thus RA and RA synthesis in the zebrafish is implicated in the development of the anteroposterior axis of the embryo, the eye, and the fins.

C. Retinoic Acid Receptors

In light of the previously mentioned studies, we would expect the retinoic acid receptors (RARs) to be present in the zebrafish embryo. cDNAs corresponding to RARα, RARβ, and RARγ have indeed been cloned from a zebrafish embryo library (White et al., 1994). The α and β isoforms respond differently to RA treatment of embryos and individually the effects vary depending on the developmental stage of treatment (Joore et al., 1994). Generally, RARα expression is either induced or not affected, whereas RARγ expression is either decreased or not affected. RARα is strongly expressed in the neural epithelium of the hindbrain and the tail bud. RARγ is expressed in anterior lateral mesenchyme located on both sides of the future brain and in the tail bud. In adults, both the normal and regenerating caudal fin express RARα and RARγ whereas RARβ is undetectable (White et al., 1994). RARγ is expressed at far higher levels than RARα in the fin regenerate and it is in the mesenchymal cells at the distal ends of the bony rays and not in the epidermis of the regenerate.

Surprisingly, the RXRs do not follow the conventional mammalian situation. Instead of three genes there are five in zebrafish, a homologue of each of the mouse RXRα, β, and γ, and two novel ones, RARδ and RXRε (Jones et al., 1995). These two novel receptors each have a 14-amino acid sequence insert in the ligand-binding domain and do not bind 9-*cis*-RA as the normal RXRs do. One of them (RXRε) forms heterodimers with the thyroid hormone receptors (in the absence of 9-*cis* binding) and the other acts to repress transcriptional activity in a dominant negative manner. This phenomenon clearly has important implications for the control of gene activity during development and it will be interesting to see whether counterparts of RARδ and RXRε exist in other vertebrates.

Finally, there have been no reports of the cloning of any CRABPs from zebrafish.

III. *XENOPUS*

A. Excess RA

The effects of retinoids on embryonic development, and *Xenopus* embryos in particular, are frequently interpreted in terms of RA perturbing an existing gradient of retinoids across a developing system (such as the primary body axis) and many of the experiments described ahead are consistent with the existence of such a gradient. It has also been suggested that a normal gradient of retinoids could differentially induce homeobox genes according to their particular RA thresholds and by perturbing this gradient with RA or synthetic retinoids, the pattern of homeobox expression may result in some embryonic regions assuming an inappropriate identity. Indeed, RA enhances the expression of many homeobox genes in *Xenopus* embryos (Sive et al., 1990); Papalopulu et al., 1991b; Sive and Cheng, 1991; Leroy and De Robertis, 1992; Dekker et al., 1992; Lopez et al., 1995). The first evidence for exogenous RA respecifying tissues in *Xenopus* was the demonstration by Durston and colleagues (Durston et al., 1989) that administration of all-*trans* RA caused microcephaly. The effect was dose dependent and clearly graded (Figure 2) over the 0.1-10 μM range applied to blastula or gastrula embryos and nonspecific, toxic effects of RA did not account for these changes (Durston et al., 1989). The reduction in the forebrain and midbrain in RA-treated embryos was accompanied by an increase in hindbrain and spinal cord (but see below) and it was suggested that RA may be transforming the neuroectoderm along the AP axis such that anterior elements took on the identity of more posterior parts.

It has also been shown that treating stage 10 embryos with 1 μM RA for varying lengths of time also elicits a graded response (Sive et al, 1990); the eyes being affected after only 10 minutes of treatment, the cement gland being lost after 40-80 minutes exposure, and, after 160 minutes, no beating heart forms. These results show that anterior non-neural ectoderm and mesoderm are also sensitive to RA treatment. The possibility therefore exists that rather than acting directly on the ectoderm, RA may be affecting the signaling properties of the underlying mesoderm. To test this, Sive and coworkers (Sive et al., 1990; Sive and Cheng, 1991) conducted a series of experiments where mesoderm was incubated with dorsal ectoderm derived from from RA-treated embryos and taken before (stage 10) and after (stage 11.5) the dorsal ectoderm is thought to receive neuralizing signals. Using a panel of ectodermal markers, she showed that it appears that both mechanisms are operating, that is, RA can both act directly on the dorsal anterior ectoderm *and* affect anterior mesoderm/ectoderm interactions.

The possibility that RA may act both directly on ectoderm and indirectly via the underlying mesoderm is supported by the studies of Ruiz i Altaba and Jessell (1991a; 1991b). Using isolated animal caps exposed to RA and mesoderm inducers, they demonstrated that 0.1-100 nM RA appears to affect the normal induction of mesoderm from the animal cap ectoderm in a dose-dependent manner. It was found

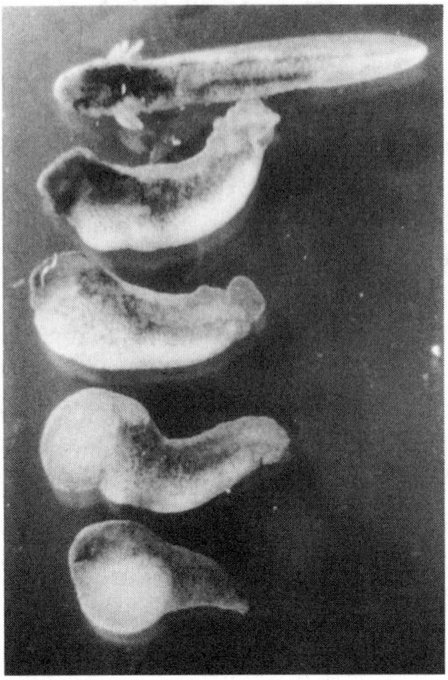

Figure 2. The effects of tRA on early amphibian development using the axolotl as an example. Embryos were treated continuously with varying concentrations from early gastrulae stages and a graded truncation of the anterior structures can be seen as described by Durston et al. (1989) in *Xenopus*. Top, control embryo. Underneath, embryos treated with 8 x 10-7, 4 x 10-6, 2 x 10-5 and 10-4 M respectively.

that the formation of anterior dorsal mesoderm was suppressed, while more posterior or ventral mesoderm appeared more abundant (Ruiz i Altaba and Jessell, 1991a). Thus while RA does not affect initial mesoderm induction, it may act endogenously to modify its character. In this case, the observed effects of RA on the CNS may be secondary (or additional) to these earlier, mesodermal changes. However, the same authors have also shown that the same concentrations of RA used on animal caps can also perturb CNS development when applied to neurula (stage 14-16) embryos (Ruiz i Altaba and Jessell, 1991b). The same general effects on anterior neural structures was seen as described by other reports treating at blastula or gastrula stages (Durston et al., 1989; Cho and De Robertis, 1990; Sive et al, 1990; Ruiz i Altaba and Jessell, 1991a), although much less severe. Since development of the neuroectoderm is largely autonomous at the neurula stage, this result strongly suggests that RA is exerting a direct effect on this tissue although later action through axial mesoderm cannot be excluded. The effects on the CNS of RA treatment at neurula stages also demonstrates that the observed phenotype is not simply a consequence of affecting cell movement during gastrulation.

Down-regulation of dorsal markers such as actin, MyoD, and goosecoid has been reported in RA-treated *Xenopus* embryos (Cho et al., 1991; Sive and Cheng, 1991; Brennan, 1992). These observations, together with those mentioned earlier, suggest that RA may be having a ventralizing as well as posteriorizing effect on *Xenopus* development. Although expression of the dorsally and laterally expressed LIM class homeobox gene, Xlim-1, is up-regulated by RA (Taira et al., 1994). This suppression, like that of the fork head-related gene, Pintallavis (Ruiz i Altaba and Jessell, 1992), is cycloheximide-sensitive and therefore RA is probably acting indirectly in at least some elements of its control.

Under the conditions used by Sive and Cheng (embryos treated with 1 μM RA for 4 hours over the period of gastrulation), the amount of ventral mesoderm (using globin as a marker) was also substantially decreased at stage 30, although no difference in globin levels between control and RA-treated embryos was seen in earlier embryos. The difference between these results and those of Ruiz i Albata and Jessell (1991a) may be due to the very different RA dose regimes employed, although it is also possible that globin expression could be reduced even if ventral mesoderm formation is not suppressed. Indeed, Bertwistle and colleagues (Bertwistle et al., 1996) have shown that while globin and GATA-1 (a transcription factor which precedes globin expression during erythropoiesis) are suppressed by treating gastrulating *Xenopus* embryos with 1 μM RA, GATA-2 (which appears earlier in erythropoiesis) is still expressed in the ventral blood islands. Thus the RA-induced loss of globin is probably related to a later effect on erthryopoiesis rather than simply due to transformation of the ventral mesoderm such that blood precursors fail to appear.

Clearly, a large number of structures are affected by RA treatment of *Xenopus* embryos. In addition to anterior ectoderm (neural and non-neural derivatives), heart and blood are also affected, as mentioned earlier. Furthermore, tail defects have been reported in several of the studies also documenting anterior defects and RA-induced expansion of the pronephros (one of the major expression domains of Xlim-1) has also been seen both in whole embryos (Taira et al., 1994) and in animal caps (Moriya et al., 1993). However, most work has concentrated on perturbation of CNS development by RA, although it remains equivocal whether the RA effects on the anterior CNS reflect a transformation to a more posterior identity or a selective loss of specific domains. Up-regulation of a mid-trunk/posterior neural maker, Hox-B9 (XlHbox6), has been observed in RA-treated embryos both *in vivo* (Sive et al, 1990) and *in vitro* (Sharpe, 1991) while expression of an anterior neural marked, *engrailed*, is suppressed (Brivanlou and Harland, 1989; Sive et al., 1990). Furthermore, a concentration-dependent ectopic expression of Evx-1 (Xhox3) and serotonin in more anterior domains of the CNS, including the forebrain, has also been seen in RA-treated embryos (Ruiz i Albata and Jessell, 1991b). Cell lineage studies have also indicated that RA may be changing cell fate within the CNS since treatment of late gastrula/early neurula embryos with 100 nM all-*trans* RA converts cells of the A1 lineage from a mainly neuronal phenotype to epidermal cells

(Agarwal and Sato, 1993). Such a transformation may be related to at least some of the subsequent major defects seen in the CNS.

Expansion of the hindbrain has been cited in support of the posteriorizing effects of RA on the CNS (Durston et al., 1989). However, in a detailed study of the hindbrain in RA-treated embryos utilizing expression of Krox-20 and identification of individual cranial nerves, Papalopulu and colleagues (Papalopulu et al., 1991a) concluded that this was not due to an increase in neurons. Rather, they found that in addition to microcephaly, the anterior hindbrain was also a target for RA and concluded that the hindbrain was compressed due to a loss of specific anterior rhombomeres. Lopez and coworkers (Lopez et al., 1995) also noted loss of rhombomeres (but with a corresponding enlargement of rhombomere 7) and an anterior shift in the boundary of HoxB-7 following treatment with all-*trans* RA. In contrast, HoxB-7 was down-regulated and there was a caudal shift of its expression when embryos were treated with an RARα/RARβ-specific antagonist. These findings are generally consistent with observations in other species such as zebrafish (Holder and Hill, 1991, and see above) and mouse (Morriss-Kay et al., 1991; Leonard et al., 1995), although RA treatment of these embryos also induces a transformation such that some anterior rhombomeres acquire the characteristics of more posterior ones in addition to the simple loss of specific anterior rhombomeres (see above). Thus the original proposal of Durston and colleagues (Durston et al., 1989) that RA is changing the identity of anterior compartments of the CNS to a more posterior fate may need to be refined, for example by determining whether RA is exerting separate effects on the spinal cord, hindbrain, and forebrain.

Using RA-saturated oil droplets injected between the two ectodermal layers of the presumptive head region of stage 10-11 *Xenopus* embryos, Drysdale and Crawford (1994) directly addressed the possibility of a retinoid gradient having a patterning role by creating a focal artificial source compared to global application in most other *Xenopus* studies. They demonstrated the loss of anterior structures (cement gland and eye) with no evidence for either a posterior conversion nor a transposition to more caudal sites. Equivalent injections into the presumptive tail or trunk produced no discernible defects, but many of the problems in reaching a consensus of the way RA may be acting from studies involving immersion in RA also apply here. For example, to what temporal extent is the embryo sensitive to this type of manipulation and how relevant is the dose of RA used (the injections were given at estimated doses between 8.3 μM and 3.2 mM although the authors acknowledge that the poor solubility of RA in oil make these values only estimates)? However, the observations are consistent with RA having a suppressive effect on these anterior structures while subtler changes may be occuring within other sites such as the hindbrain.

Most of the above studies have involved treatment of *Xenopus* embryos with all-*trans* RA. However, *Xenopus* embryos are extremely active in retinoid metabolism (Creech Kraft and Juchau, 1995; Creech Kraft et al., 1995a; 1995b) and some of the other retinoids that have been detected have also been shown to be biologically active dysmorphogens. For example, both 9-*cis* RA and 4-*oxo*-RA mimic all-*trans* RA in terms of anterior truncations and/or respecifications when administered to stage 9-10 *Xenopus* embryos (Pijnappel et al., 1993; Creech Kraft et al., 1994b). Indeed, 4-*oxo*-RA is approximately 2-3 times more potent than all-*trans* RA in activating HoxB-4 and HoxB-9 in *Xenopus* (Pijnappel et al., 1993), while 9-cis RA induces morphological changes at 2-6 fold lower concentrations than all-*trans* RA (Creech Kraft et al., 1994b; Creech Kraft and Juchau,1995).

Some of the binding properties of 4-*oxo*-RA are discussed ahead, but 9-*cis* RA could be exerting its effects in *Xenopus* through RARs or RXRs since it binds both classes of receptors with similar affinities, in contrast to all-*trans* RA, which preferentially binds to RARs (Heyman et al., 1992). However, evidence for a nonpermissive role for RXRs comes from observations that administration of RXR-specific ligands to stage 9.5 embryos yields essentially the same defects in AP axis formation as seen by exposure to 9-*cis* RA, all-*trans* RA, or RAR-selective ligands (Minucci et al., 1996). In all of these studies, whether using different RA isoforms or receptor ligands, subtle differences may exist in the malformations induced although none have been detected to date.

Finally, very little information is available on the role of retinoids in *Xenopus* embryogenesis by specifically perturbing the expression of genes that mediate RA signaling. However, it is noteworthy that over-expression of xCRABP by injection of its mRNA into fertilized eggs mimics RA treatment of later embryos both in terms of up-regulation of HoxB-4 and HoxB-9 and in producing AP defects (Dekker et al., 1994). This finding suggests that xCRABP may act by enhancing the action of endogenous retinoids and is in contrast to the observed reduced sensitivity to RA of F9 cells transfected with mouse CRABP-I (Boylan and Gudas, 1991). These differences may reflect varied functions of different CRABP family members or simply be a consequence of the complexity of assessing the biological action of CRABPs *in vivo* compared to analyses in a cell line.

The role of RARs in retinoid signaling has been directly addressed by Smith and coworkers (Smith et al., 1994) who injected *Xenopus* embryos with mRNAs encoding a wild-type or dominant negative RARγ. It was found that while over-expression of the wild-type receptor increased the sensitivity of the embryos to RA treatment, injection of the dominant negative form gave partial resistance to RA. The mutant receptor had no effect on normal development in the absence of exogenous RA and strongly suggests that retinoid receptors are playing a role in RA-mediated teratogenesis. Clearly, a fuller understanding of the action of retinoids during *Xenopus* embryogenesis is going to necessitate further studies manipulating the expression of genes involved in RA signaling.

B. Endogenous Retinoids and Their Distribution

The first report of retinoids in *Xenopus* embryos was in 1989 when Durston and colleagues (Durston et al., 1989) demonstrated the presence of all-*trans* RA in stages 10-15 (early gastrula to mid-neural fold) *Xenopus* embryos at a mean concentration of 150 nM using HPLC analyses. They also detected low levels of 13-*cis* RA and very high levels of retinal. Over the period of gastrulation (stages 10-13), the same group has also detected 4-oxo RA (Pijnappel et al, 1993) at a concentration of 20 nM and this RA metabolite is still present in tailbud (stage 28) embryos (Creech Kraft et al., 1994a). All-*trans* retinal, 3,4-didehydroretinal (dd-retinal), and the sugar-conjugated RA metabolites, 13-*cis* retinoyl β glucuronide (13-*cis* RAG), and all-*trans* retinoyl β glucuronide (all-*trans* RAG), have also been found to be abundant in unfertilized *Xenopus* eggs (Azuma et al, 1990; Creech Kraft et al., 1994a). More detailed developmental studies have identified further retinoids together with providing information on their spatial and temporal distribution although unfortunately, and as described ahead, no consistent pattern has emerged.

Creech Kraft and coworkers (Creech Kraft et al., 1994b), also using HPLC, detected all-*trans* RA at a concentration of 2.5 μM in stage 12 (mid-gastrula) *Xenopus* embryos, vastly higher than that reported by Durston and colleagues (Durston et al., 1989). 9-*cis* RA was found to be almost twice as abundant as all-*trans* RA and the temporal and spatial distribution of both appear to be highly dynamic during *Xenopus* embryogenesis. In whole embryos, the levels of all-*trans* RA remained relatively constant from stage 8 (mid-blastula and the onset of zygotic transcription) to stage 18 (neural groove stage, marking the end of neurulation). However, over this same period the levels of 9-*cis* RA increased while those of *trans*-RAG fell substantially. At the start of gastrulation (stage 10), 9-*cis* RA is preferentially concentrated in the dorsal half of the embryo while all-*trans* RA is mainly found ventrally. In both cases, there is a difference of approximately 3.5-fold across the embryo. Although the levels of all-*trans* RA appear equal in dorsal and ventral halves of stage 13 (late gastrulation) embryos, by stage 18 the dorsal and ventral distribution of both RA species is essentially the same as in stage 10 embryos. Over the same time period, *trans*-RAG accumulates in the ventral halves of the embryos.

The distribution of these retinoids has also been studied along the anteroposterior (AP) axis of the embryos between stage 13 (the time when the AP axis is first clearly established) and stage 18 (Creech Kraft et al.,1994b). At stage 13, 9-*cis* RA was essentially absent from the middle of the embryo, rather it was confined (and in almost equal amounts) to the anterior and posterior thirds of the embryo. By stage 18, 9-*cis* RA could be detected in the middle region although it was still more abundant at the anterior and posterior ends. In contrast, all-*trans* RA was found in a shallow gradient along the AP axis, with approximately 2-fold higher levels found anteriorly, at both embryonic stages analysed. Distribution of *trans*-RAG was relatively even throughout the AP axis of stage 13 embryos although by stage 18 the middle section contained nearly 20-fold higher levels than that found in either the anterior or posterior thirds of the embryo.

The distribution of the likely precursors for all-*trans* RA and and 9-*cis* RA have also been determined by HPLC (Creech Kraft et al., 1994a). 3,4-didehydroretinol (dd-retinol), all-*trans* retinol and all-*trans* retinal were all detectable in stage 14 (neural plate stage) embryos. The levels of all three retinoids fell from stage 8 to stage 18 despite previous observations (Creech Kraft et al., 1994b) that levels of their apparent immediate product, all-*trans* RA, remain relatively constant throughout this period. The spatial distribution of the three RA precursors during early embryogenesis is complex and elements of it are difficult to interpret. The distribution across the dorsoventral axis is relatively clear; all three retinoids are preferentially found in the ventral halves of embryos taken at stages 10, 13, and 18. In this respect they broadly follow the all-*trans* RA distribution at these stages (Creech Kraft et al., 1994b). However, the distribution along the AP axis appears, in general, to be the *opposite* of that seen for all-*trans* RA (Creech Kraft et al., 1994b). Thus levels of dd-retinol, all-*trans* retinol, and all-*trans* retinal were all found to be higher in the posterior, rather than anterior, third of stage 13 and stage 18 embryos. However, this does not represent a simple gradient since by stage 18 maximal levels of all three retinoids were in the middle third of the embryo. This illustrates one of the major limitations of such experiments using dissected embryos. For although one or more retinoids may be distributed in a gradient across the embryo in some tissues, this may be masked by additional foci of exceptionally high (or low) levels in other domains. For example, a potential gradient of a retinoid along the AP axis of the developing CNS may be missed by the possible presence of high levels of the same retinoid in the developing heart mesoderm or blood islands, which are present in the middle third of the embryo.

The indirect method for quantitating retinoids involving mouse embryonal carcinoma cells transfected with a luciferase reporter gene construct containing two RAREs from the mouse RARβ gene has also been used in *Xenopus* embryos. In this way, Chen and colleagues (Chen et al., 1994) measured the RA equivalents (in which terms activity must be measured since the identity of the retinoid[s] responsible for driving reporter expression is unknown) in retinoid extracts from stage 2 (2 cell stage) to stage 14 (neural plate stage) *Xenopus* embryos. They found the concentration of RA equivalents in stage 14 embryos to be 76 nM, 3 times higher than that detected at the 2-cell stage. They also noted that activity was concentrated at the posterior end of the older embryos, being approximately 12 times greater than at the anterior end, with the middle section yielding an intermediate value. It was also found that the dorsal marginal zone (DMZ), which contains the organizer, of stage 12 embryos had approximately 10 times the retinoid activity of the equivalent ventral marginal zone. Thus this is consistent with the high levels of retinoids found in Hensen's node of mouse (Hogan et al., 1992) and chick (Chen et al., 1992; Wagner et al., 1992) embryos, suggesting a role for retinoids in primary axis formation.

As mentioned earlier, the major limitation of this approach is the uncertainty of the nature of the retinoids mediating reporter expression. The RARE used by Chen and colleagues (Chen et al., 1994) is a DR5 response element that represents the most potent and abundant class of RARE and can bind to RAR/RXR heterodimers

(Minucci et al., 1994). It has also been shown to be activated by all-*trans* RA, 9-*cis* RA, and synthetic ligands selective for RARs (but not RXRs) in *Xenopus* embryos (Minucci et at., 1996). Furthermore, 3,4-didehydro RA has also been shown to be biologically active (Thaller and Eichele, 1990) and 4-oxo RA binds to and activates RARs but not RXRs (Pijnappel et al., 1993). Together, these observations suggest that activation of this reporter is mediated by endogenous RARs, although whether binding is due to 9-*cis* RA, all-*trans* RA, or another, perhaps novel, retinoid is not clear at present.

C. Retinoid Binding Proteins and Nuclear Receptors

Two CRABPs have been identified in *Xenopus*, xCRABP (Dekker et al., 1994) and xCRABP-b (Ho et al., 1994). On the basis of homology, it is not possible to allocate either as being the *Xenopus* homologue of CRABP-I or CRABP-II. xCRABP shows 92 percent amino acid homology with both mouse CRABPs, whereas the predicted xCRABP-b protein has 76 percent and 85 percent homologies to mouse CRABP-I and CRABP-II respectively. Whereas xCRABP-b has the same predicted size as other CRABP proteins, xCRABP contains an additional 9 amino acid sequence. xCRABP and xCRABP-b transcript sizes are different but both are much larger than those of previously identified CRABP cDNAs.

The expression pattern of xCRABP generally resembles both CRABP species identified so far, although some notable differences also exist. For example, xCRABP is expressed in stage 32 (tailbud stage) embryos in the central nervous system (among other sites) and is prominent in the hindbrain (resembling both murine CRABPs). However, xCRABP is expressed in rhombomeres 4, 6, 7, and 8 whereas at equivalent mouse stages, hindbrain expression of CRABP-I is confined to rhombomeres 2, 4, 5, and 6 (Leonard et al, 1995) and mouse CRABP-II is preferentially expressed in rhombomere 4, with no or low expression elsewhere in the hindbrain (Ruberte et al., 1992). Some of the other expression domains also make it difficult to define xCRABP as either a CRABP-I or CRABP-II homologue. Thus xCRABP is expressed in the heart and neural crest thereby resembling CRABP-I distribution but also in the telencephalon, diencephalon, and spinal cord, characteristic of CRABP-II. Perhaps the main difference between xCRABP and other CRABPs is the very early expression of the former; xCRABP transcripts are present as a maternal message at low levels but their abundance increases dramatically at the start of gastrulation. After mid-late gastrulation, the levels fall once more but remain at a relatively high level to at least stage 35 (tailbud stage). At gastrulation, xCRABP RNA is mainly found in the DMZ and the dorsal part of the animal cap. By mid-gastrula, expression is detected in an anterior domain in the mid-neural plate and a posterior domain in the dorsal blastopore lip. Subsequently, the posterior domain disappears while the anterior domain resolves into the hindbrain profile described earlier.

In contrast to xCRABP, xCRABP-b is not present as a maternal message, rather it appears at the start of zygotic transcription, its levels increase, and then remain high until at least stage 23. The expression profile of xCRABP-b is initially similar to that of xCRABP in that discrete anterior and posterior domains can be identified in mid-gastrula embryos. However, a number of differences in the expression between the two genes later become apparent. For example, although xCRABP-b is also expressed in the CNS and neural crest, it is additionally present in the paraxial mesoderm and rhombomeric expression is different from that of xCRABP—a gradient of expression being seen along the AP axis of the hindbrain, with highest levels cranially. Although there are still some differences, xCRABP-b generally resembles the pattern of expression of mouse CRABP-II. Consistent with this, RA-treated *Xenopus* embryos show no significant change in xCRABP-b expression (Ho et al., 1994) similar to our own (unpublished) observations on CRABP-II expression in RA-treated mouse embryos.

Thus the identity of *Xenopus* CRABPs is unclear at present and the pseudotetraploid nature of *Xenopus* means that xCRABP and xCRABP-b may represent pseudoalleles. However, the difference of 26 percent at the amino acid level probably suggests that they represent different genes. The question remains as to whether either (or both) *Xenopus* CRABPs are novel members of the CRABP family. If more than two vertebrate CRABPs exist, it may account for the relatively benign phenotype of mice bearing null mutations in both their CRABP-I and CRABP-II genes (Lampron et al., 1995). In support of this possibility, Southern blotting has indicated the presence of at least 3 CRABP genes in a range of vertebrates, including *Xenopus* (Giguere et al., 1990; Pizzey, unpublished data).

Although the identity of *Xenopus* CRABPs is equivocal, the nature of the RAR and RXRs cloned to date from *Xenopus* is less complex. RARα, RARγ, RXRα, RXRβ, and RXRγ have all been cloned from *Xenopus* and, furthermore, have also been shown to be present as maternal messages (Ellinger-Ziegelbauer and Dreyer, 1991; Blumberg et al., 1992; Marklew et al., 1994). However, of these, only RXRβ and RARγ are expressed at relatively high levels throughout early development. RARα and RXRγ are degraded just before gastrulation and RXRα mRNA drops sharply before gastrulation but remains expressed at a low level until the tailbud stage (stage 26) when it accumulates until at least stage 45.

Using a riboprobe, which would not discriminate between the RARγ1 and RARγ2 isoforms (Ellinger-Ziegelbauer and Dreyer, 1991), and an anti-RARγ antiserum (Ellinger-Ziegelbauer and Dreyer, 1993), high levels of mRNA expression and protein could be detected at the anterior and posterior ends of late gastrulation *Xenopus* embryos. These expression domains remain essentially the same up to the tadpole stage and all three germ layer derivatives (but not all tissues) were represented in both domains. Subsequently, it became clear that in stage 24 embryos, RARγ1 was expressed in the anterior region while RARγ2 was found in both (Pfeffer and De Robertis, 1994). The anterior domain of neurula (stage 19) embryos included the pharyngeal endoderm, the prechordal plate, and brain while the posterior domain included all the tissues just posterior to the site of somitogenesis. Of particular interest is the expression of very high levels of RARγ in the chordoneural hinge in the posterior domain. The chordoneural hinge is the anterior limit of the neurenteric canal, which connects the post-anal gut with the central canal, of the spinal cord. It is derived from the dorsal lip and probably represents the amphibian homologue of the regressing Hensen's node containing a mixing of cells from all three primary germ layers. Significantly, it also retains inducing activity in grafting experiments (Gont et al., 1993) and the high levels of RARγ may reflect high levels of RA as found in other vertebrate structures with organizing capacity. Analyses have not been performed specifically on the chordoneural hinge, but this may be the site of the high levels of retinoids found in the posterior region of *Xenopus* embryos in some of the studies described earlier. Recently, the complexity of RARγ expression in *Xenopus* has been increased by the identification of two isoforms of RARγ2 (Crawford et al., 1995). RARγ2.2 represents the transcript described earlier (Elinger-Ziegelbauer and Dreyer, 1991,1993; Pfeffer and De Robertis, 1994), but RARγ2.1 first appears during gastrulation and is strongly expressed in the branchial arches. The expression of these two RARγ2 isoforms also differ in their sensitivities to RA in that head and tail expression of RARγ2.2 is refractory to RA treatment, whereas RARγ2.1 is suppressed in the presumptive head structures of treated embryos. Furthermore, microinjection of RA indicates that this suppression is both local and rapid (Crawford et al., 1995).

IV. REGENERATING AMPHIBIAN APPENDAGES

A. RA and Limb Regeneration

In 1978, Niazi and Saxena (1978) made the amazing discovery that when the hindlimb buds of the toad, *Bufo andersonii*, were amputated through the shank and treated with retinyl palmitate, the regenerates seemed to contain *extra* elements in the proximodistal axis and often two regenerates appeared instead of one. This was the first example of the violation of the "law of distal transformation" because normally only those elements that have been removed (i.e., distal to the amputation plane) are replaced (Figure 3A). Investigating this phenomenon in more detail in the axolotl it was shown that the effects of retinoids were to produce extra elements in the proximodistal axis in a very precise and concentration-dependent fashion (Maden, 1982; 1983a). Thus after amputating limbs through the mid-forearm level, at low concentrations of retinyl palmitate the regenerated radius and ulna were too long or an extra complete radius and ulna was regenerated (Figure 3B). At a higher dose an extra elbow joint appeared (Figure 3C) and at a higher dose still a complete limb could be regenerated from the amputation plane (Figure 3D). The same effect occurs from other levels of amputation, for example, after amputation through the carpals and treating with a high dose of retinoid, a complete limb could be regenerated from the hand and the same phenomenon can be seen in hindlimbs. Thus what is happening is that retinoids are respecifying the positional information of the cells at the amputation plane in a gradual, concentration-dependent manner and in a proximal (towards the shoulder) direction. That is, retinoids are proximalizing the cells of the regenerating limb.

The effect of retinoids is thus concentration-dependent: the higher the concentration the more proximalized the blastemal tissue. The effects are also time-dependent and stage-dependent: longer treatment times induce greater degrees of proximalization and if treatment is delayed beyond a certain stage, instead of stimulating extra growth of elements, the regenerates are phocomelic (Maden, 1982; Niazi et al., 1985). Of the naturally occurring retinoids, all-*trans*-retinoic acid is more potent than retinol, retinyl palmitate, and retinyl acetate (Maden, 1983a), but 9-*cis*-RA seems to be more potent than tRA (Tsonis et al., 1994). Several synthetic retinoids are far more potent than tRA (Kim & Stocum, 1986a; Keeble and Maden, 1989; Maden et al., 1991). This phenomenon of proximalization has also been reported in a variety of other amphibian species such as *Notophthalmus, Pleurodeles, Triturus,* and *Xenopus* (Thoms and Stocum, 1984; Niazi et al., 1985; Lheureux et al., 1986; Scadding & Maden, 1986).

When these experiments were extended to the frog, *Rana temporaria*, a further axial effect of retinyl palmitate was observed, as had originally been seen by Niazi (Niazi and Saxena, 1978). Not only were the hindlimbs proximalized, but at higher concentrations the anteroposterior axis (thumb to little finger) was also respecified (Maden, 1983b). Thus after amputation of the foot, instead of just replacing the foot as controls do (Figure 3E), treated animals produced limbs that were complete in the proximodistal axis (Figure 3F). At higher concentrations, regenerates were not only complete in the proximodistal axis, but also duplicated in the anteroposterior (AP) axis (Figures 3G & H). These regenerates were always double posterior in structure, that is, the sequence of digits was 5, 4, 3, 2, 1, 1, 2, 3, 4, 5. This showed the cells of the regenerate had not only been proximalized, but were also posteriorized. It is worth remembering (in relation to the results on the chick limb bud described ahead) that these effects appear after systemic treatment of the animals: they are dosed by placing them in solutions of retinoids.

This effect on the AP axis of the limb has subsequently been seen in Urodeles (Stocum and Thoms, 1984; Kim and Stocum, 1986b; Wigmore, 1990) and it is also a concentration-dependent phenomenon (Monkmeyer et al., 1992). To complete the picture, it is now apparent that retinoids can also ventralize the cells of the regenerating limb, that is, it affects the third axis of the limb, the dorsoventral (DV) axis. When axolotl limbs are cut in half in the DV axis to produce half-ventral or half-dorsal limbs, then half-ventral limbs do not regenerate at all, whereas half-dorsal limbs regenerate a complete limb from the amputation plane (Ludolph et al., 1990). Thus in the regenerating amphibian limb retinoids have specific effects on each of the cardinal axes of the limb: the proximodistal axis is proximalized, the anteroposterior axis is posteriorized, and the dorsoventral axis is ventralized.

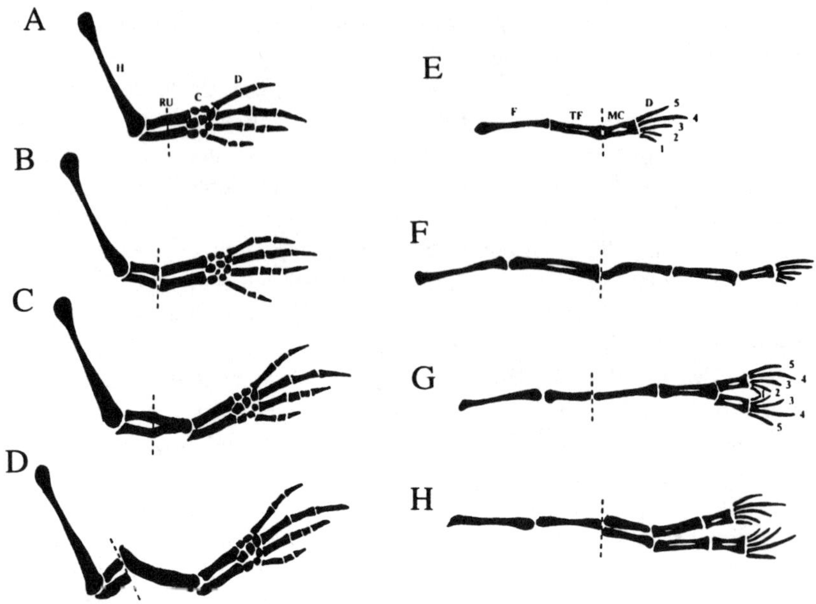

Figure 3. The dose-dependent effects of retinoids on the regenerating axolotl forelimb (A-D) and the regenerating *Rana temporaria* hindlimb (E-H). Broken lines mark the amputation plane. (A) A control limb amputated through the mid radius and ulna that regenerated exactly what was removed. H = humerus, RU = radius and ulna, C = carpals, D = digits. B-D After increasing doses of retinoids a greater amount of limb tissue regenerates, resulting in a complete limb regenerating from high doses (D). The tissue at the amputation plane is gradually proximalised. (E) A control limb amputated through the foot which regenerated exactly what was removed. F = femur, TF = tibia and fibula, MC = metatarsals. D = digits. (F) A complete limb regenerated from the foot level after retinoid treatment showing the same phenomenon of proximalization as in B-D. G-H at higher doses not only is the frog limb proximalized, but also duplicated in the anteroposterior axis, producing double posterior limbs (see sequence of digits marked in G). (H) A complete pair of limbs has regenerated from an amputation though the foot.

B. RA and Tail Regeneration

Mohanty-Hejmadi and colleagues (Mohanty-Hejmadi et al., 1991) made the dramatic discovery that when the amputated tails of the marbled balloon frog *Uperodon systoma* were treated with retinyl palmitate, instead of regenerating tails as would be expected, several hindlimbs were regenerated. This demonstration was all the more remarkable because it had earlier been reported that retinoids have *no* effect on the regeneration of tails in a variety of other species (Niazi and Saxena, 1979; Scadding, 1987). This homeotic transformation also occurs in *Rana temporaria* (Maden, 1993; Figure 4) and *Polypedates maculatus, Bufo melanosticus,* and *Microhyla ornata* (Mahapatra and Mohanty-Hejamdi, 1994). The induction of limbs from tails demonstrates characteristics typical of the retinoid effects described above, that is, it is concentration-dependent, time-dependent, and stage-dependent (Maden, 1993). Furthermore, up to 9 limbs can be produced and there is a strong tendency for them to appear in even numbers (presumably in pairs) rather than in odd numbers. They are always hindlimbs and when their anatomy was examined in detail, a range of abnormalities were found. In addition to normal limbs, double posterior limbs were common (never double anterior limbs) and in the dorsoventral axis double dorsal and double ventral limbs were observed.

This transformation takes place as the animals approach metamorphosis and the levels of circulating thyroid hormone (TH) begin to rise, suggesting a role for TH. A recent test of this hypothesis in which TH levels were suppressed with prolactin has shown that this treatment prevents the homeotic transformation (Maden and Corcoran, 1996). Since the homeosis is both TH dependent and RA dependent it is likely that the thyroid hormone receptors and the retinoid X receptors are involved, perhaps via heterodimerization as both are present in the regenerating tail blastema (Yaoita and Brown, 1990; Maden and Corcoran, 1996). Continuing molecular analyses of this amazing phenomenon, for example, using subtractive hybridization, should provide important insights into the genes and molecules involved.

C. The Cellular Effects of Retinoids

In an attempt to understand the cellular processes involved in this positional respecification induced by RA various histological, immunological, and molecular studies have been performed. By grafting between RA-treated and untreated mesenchyme amd epidermis of the blastema, it was shown that it is the mesoderm and not the epidermis that is responsible for respecification (Maden, 1984). Nevertheless, epidermal changes induced by RA are readily apparent, such as the appearance of ciliated cells (Scadding, 1989), the appearance of Alcian blue staining deposits (Maden, 1983a), the displacement of the apical cap (Maden, 1983a: Scadding, 1990), and the induction of a secretory phenotype characterized by the immunoreactivity to an antibody known as WE3 (Tassava, 1992).

Thus the responsibility for positional respecification lies in changes in the mesenchymal cells. Excessive dedifferentiation of the tissues at the cut surface has been observed (Kim and Stocum, 1986b) after RA treatment. Indeed, the extent of dedifferentiation correlates well with the extent of positional respecification (Ju and Kim, 1994). Cell surface changes in those blastemal cells liberated by excessive dedifferentiation result in the clumping of blastemal cells (Maden, 1983a), possibly due to the retention of fibronectin at the cell surface (Maden and Keeble, 1987). The cell surface must be the ultimate sensor of positional information because

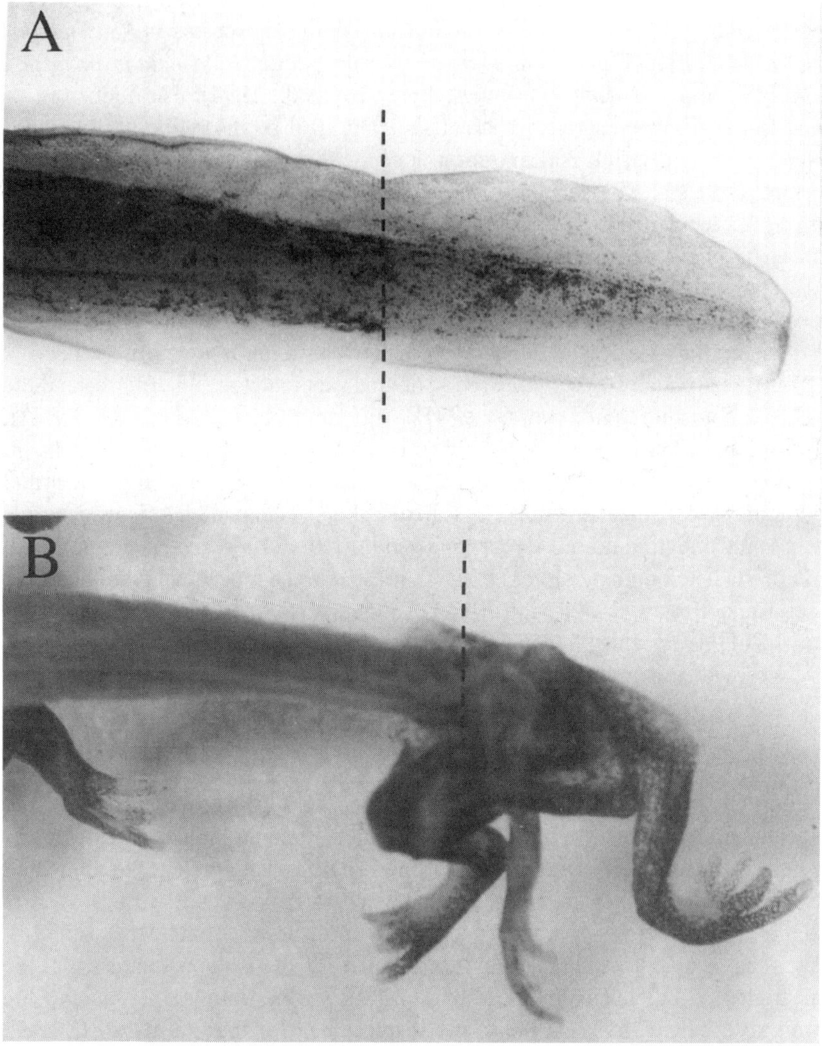

Figure 4. The effect of retinoids on tail regeneration in *Rana temporaria*. (A) A control tail regenerate replacing what was amputated. Broken lines mark the amputation plane. (B) After retinoid treatment 4 hindlimbs have regrown from the amputation plane rather than tail tissue.

whole blastemas will move up or down a regenerating limb to find their correct level, a phenomenon termed "affinophoresis" (Crawford and Stocum, 1988). Thus a distal blastema grafted to the dorsal surface of a proximal blastema will become displaced distally during regeneration to its correct proximodistal level, but a RA treated distal blastema will not because it has been proximally respecified. In support of this concept of cell surface sensing, proximodistal respecification is inhibited by tunicamycin which inhibits the glycosylation of proteins (Johnson and Scadding, 1992).

Other molecular changes reported in blastemal cells include the inhibition of cell division (Maden, 1983a; Pietsch, 1987), the down-regulation of a newt type II keratin (Ferretti et al., 1991), and the down-regulation of a homeobox gene, HoxA13 (Gardiner et al., 1995).

D. Endogenous Retinoids in Limb Regeneration

Three studies have demonstrated that RA is present in the regenerating amphibian limb. In one, the advantages of the large size of the blastema were exploited and by HPLC the retinoids present in the wound epidermis versus the blastemal mesenchyme and in four different quadrants of the blastema could be analysed (Scadding and Maden, 1994). Three retinoids were readily identifiable, all-*trans*-retinol, 3,4-didehydroretinol and all-*trans*-RA. Retinol and tRA were found at similar levels in the epidermis and the mesenchyme, but ddretinol was 12 times higher in the epidermis than the mesenchyme. When the blastemas were divided into four pieces across the anteroposterior axis, it was observed that there was a 5-fold gradient in tRA concentration with a high point on the posterior side. There was no such consistent variation in the other two retinoids. Suggestively, in the regenerating limbs of *Xenopus*, in which only an unpatterned spike of cartilage develops, there is no gradient of tRA across the anteroposterior axis, implying that the tRA gradient may be responsible for the generation of pattern within the regenerate. Comparisons of retinoid levels between blastemas obtained after proximal amputations (through the middle of the upper arm) versus those from distal amputations (through the middle of the lower arm) revealed a 2.7 times higher concentration of tRA in distal blastemas.

In the second study a reporter gene construct was used (Brockes, 1992). Cultured mesenchymal cells were microinjected with a RARE from the RARβ promoter, a plasmid containing the newt RARα and a luciferase or β-galactosidase contruct. When pellets of these transfected cells were implanted under the wound epidermis of a regenerating limb and left for four to six days it was shown that such reporter cells would respond as expected to systemic injections of RA. The activation of the constructs was then compared in cells implanted into proximal versus distal blastemas and it was found that in proximal blastemas there was a 3.5-fold higher level of activation than in distal blastemas. Since excess RA can proximalize distal blastemas this result is exactly what one might expect: proximal blastemas would be expected to have higher endogenous levels of RA than distal blastemas on a simplistic gradient model. However, for reasons that are not clear, this is the opposite result to that described above for HPLC measurements.

The third study used a combination of reporter genes and HPLC and concentrated on the wound epidermis (Viviano et al., 1995). Blastemal mesenchymal cells served as reporter cells that were transfected with a CAT reporter linked to two glucocorticoid response elements and a chimeric reporter in which the ligand binding domain of the glucocorticoid receptor has been replaced with the corresponding domain of RARα_1. In the presence of RA the chimeric receptor activates CAT expression through the glucocorticoid response pathway. When the wound epidermis was cocultured with these transfected blastemal cells then their reporter plasmids were activated to a level that fell into the range of 0.1 - 1nM RA as determined in a parallel dose response analysis. When proximal and distal wound epidermis were compared, the proximal epidermis gave a consistently higher response than distal epidermis in agreement with the experiments just mentioned (Brockes, 1992). Finally, the metabolism of [^{3}H] all-*trans*-retinol by the wound epidermis was examined by HPLC. One major metabolite peak coeluted with 3,4-didehydroretinol, in accordance with the endogenous HPLC measurements described earlier (Scadding & Maden, 1994) and the acid metabolite was shown to be 9-*cis*-RA. 9-*cis*-RA was both produced by the epidermis and released into the medium.

E. Retinoid Binding Proteins and Receptors and Their Function

The capacity of cytoplasmic extracts of blastemas to bind radiolabelled RA at the appropriate molecular weight suggested that cytoplasmic retinoic acid binding protein (CRABP) was present in blastemal cells (Keeble and Maden, 1986; McCormick et al., 1988) and that its level rose 4-fold during the period of regeneration that is sensitive to respecification by RA. However, the identity of this component (CRABP I or CRABP II) was not revealed until axolotl CRABP I was cloned (Ludolph et al., 1993). Unexpectedly, CRABP I was not present during limb regeneration stages and so the binding capacity of blastemas must be due to CRABP II. CRBP was detected at low levels in the blastema in binding studies, although, again, this level was considerably higher than in the unamputated limb (Maden et al., 1989).

The RARs expressed in the regenerating newt blastema have been characterised. There are five in all—RARα_1, RARα_2, RARδ_{1a}, RARδ_{1b}, and RARδ_2 (Ragsdale et al., 1989; 1992a; 1992b; Giguere et al., 1989). Antibodies have been made to the newt δ_1 isoform and immuncytochemical study (Hill et al., 1993) has shown that in the normal unamputated limb about half of the nuclei in both epidermal and mesenchymal tissues are immunoreactive and this proportion rises in the blastemal tissues. There was no detectable variation in staining of cells along the proximodistal or anteroposterior axis of the blastema and there was no modification of staining after RA treatment of the animals. This is the only use of receptor antibodies for developmental studies. All other distributional analyses of receptors, for example, in the mouse embryo, have relied upon mRNA localization, so only in this one case do we know anything about the distribution of the receptor proteins. Apart from the tail regeneration blastema (Maden and Corcoran, 1996) the other family of receptors, the RXRs, have not been characterized in regenerating systems.

In a very ingenious series of studies from Brockes and colleagues; the function of three of these RAR isoforms has been determined. As described earlier, RA has many different effects on the cells of the regenerate in addition to the effect on pattern—epidermal ciliation, induction of WE3 immunoreactivity and glycoprotein synthesis in the epidermis, excessive dedifferrentiation, inhibition of cell division, blastemal cell clumping, keratin and homeobox gene down-regulation. Simplistically, therefore, we might suggest that each RAR and RXR isoform has a single function in translating the RA signal into an altered pattern of gene activity. In order to test their function, these researchers have made a series of chimeric receptors in which the ligand-binding domain of the RAR isoform is swapped with that taken from the *Xenopus* thyroid hormone receptor, thereby making a chimeric receptor that now becomes responsive to thyroid hormone rather than to retinoic acid, but that still activates its original retinoic acid responsive genes. When these constructs are transfected into blastemal cells one can now ask which of the many retinoic acid effects on cells described earlier now becomes induced by thyroid hormone rather than RA.

Using blastemal mesenchymal cells in culture, either the α_1 or the δ_1 chimera was transfected and it was shown that only after α_1 transfection do the cells now become growth inhibited by thyroid hormone (Schilthius et al., 1993). In contrast, in a subsequent experiment these two constructs were transfected into the epidermis of the regenerate and the phenotypic effect assayed was the induction of WE3 immunoreactivity. In this case only when the wound epidermis cells were transfected with the δ_1 chimera was WE3 induced by thyroid hormone and not with the α_1 (Pecorino et al., 1994). Thus we may conclude that RARα_1 is responsible for the inhibition of cell division by RA, and RARδ_1 is responsible for WE3 induction in the epidermis. Finally, the receptor responsible for the proximalization of blastemal cells has been identified by exactly the same strategy. The proximal movement of cells can be visualized with fluorescence by colabelling chimera transfected cells with alkaline phosphatase. Only when distal blastemal cells have been transfected with the RARδ_2 chimera does proximalization occur after thyroid hormone treatment (Pecorino et al., 1996). Thus three of the RAR isoforms have been identified as having specific functions to perform when the regenerating limb is treated with excess RA.

V. THE EARLY CHICK EMBRYO

A. Excess RA

The Node and Gastrulation

Experiments on the effects of RA on early chick embryos have been performed with a view to testing the possibility that RA is involved in setting up the primary body axis. This is the earliest patterning event in the embryo and it is established during gastrulation by an organizing center called Hensen's node. It was demonstrated that Hensen's node, when transplanted to the anterior margin of the chick limb bud, induced mirror-imaged duplications in the same way that RA does (Wagner et al., 1990). This suggested that RA might be a component of the organizer of the AP axis, which would explain why excess RA has a posteriorizing effect on the AP axis, particularly in Xenopus (see above). Chen and colleagues (Chen et al., 1992), using F9 cells transfected with a RARE as an assay, demonstrated that the stage 4 Hensen's node does indeed contain RA, and at a concentration of 33nM compared to the rest of the embryo, which contained only 1.6nM RA. The stage 6 node contained even more RA.

Furthermore, RA can induce the development of a secondary axis in the early chick embryo in the same way that grafting an additional Hensen's node does (Chen and Solursh, 1992). The RA-induced secondary axes never contained head structures and only consisted of trunk-tails, as indeed do Hensen's node-induced secondary axes. Curiously, this induction only took place when the RA bead was placed on the left side or the anterior of the embryo; it did not occur when the beads were placed on the right or posterior. The equivalent structure in the mouse embryo also induced extra digits when grafted to the chick limb bud and the mouse node can synthesize RA from its biosynthetic precursor, retinol (Hogan et al., 1992). These results have led to a model whereby the AP axis is established according to the total concentration of RA seen by gastrulating mesodermal cells with the least amount resulting in an anterior specification and the greatest amount resulting in a posterior specification (Hogan et al., 1992). It is suggested that RA acts via *Hox* genes to establish this axis and this idea fits well with the concept of RA being a posteriorizing factor during neural specification in *Xenopus* (see above).

The CNS

At the same time as RA-induced secondary axes in the experiments of Chen and Solursh (1992) described earlier, RA also acted as a teratogen on the host embryonic axis that was being established by the node. Defects included missing forebrain, open neural tube and the hindbrain was enlarged, a familiar pattern of defects from *Xenopus* and mouse studies. In support of this view of posteriorization and/or loss of anterior structures after RA treatment, a gene expressed in anterior neurectoderm, *otx2*, has been shown to be down-regulated by brief RA treatment at stage 2 or 3 (Bally-Cuif et al., 1995). However, RA treatment can also induce the loss of midbrain/anterior hindbrain characteristic of zebrafish. Treatment with a pulse of RA at stage 4 produced this phenotype in the CNS (Sundin and Eichele, 1992). This was accompanied by an ectopic induction of the chick homologue of the *labial* gene anterior to its normal boundary of expression.

The Heart

In the experiments of Chen and Solursh (1992) there was also an interesting phenomenon regarding the heart: the host heart looped towards whichever side of the embryo the RA bead was placed. This emphasizes the effect of RA on heart development, which is another area of the embryo that has been studied in the chick. Whole embryo exposure to RA caused the complete absence of the heart, or a lesser effect was for the anterior end (ventricle) to be stunted and the posterior end (atrium) to be enlarged (Osmond et al., 1991). Beads soaked in RA solutions and placed into the precardiac mesoderm of stages 3 to 7 embryos resulted in various degrees of cardia bifida, perhaps due to changes in the extracellular matrix environment through which precardiogenic mesenchymal cells migrate. RA treatment of similarly staged embryos analyzed with a specific marker, atrial-specific myosin heavy chain, showed that RA respecified the anterior regions of the heart (ventricle) by switching on AMHC in these regions and thus expanding the atrial domain (Yutzey et al., 1994; 1995), a similar result to that described earlier. Treatment of embryos between days 3 and 6 of development, with 13-*cis*-RA, resulted in classical heart defects such as ventricular septal defects and conotruncal defects (Hart et al., 1990; 1992). These late effects are thought to be mediated by an inhibition of neural crest migration as this is another area of the embryo on which RA has an effect.

The Neural Crest

It was shown several years ago using cultured quail neural crest cells that retinol inhibits their migration by affecting cell-to-substratum adhesiveness (Thorogood et al., 1982). More recently the same effect was seen with 13-*cis*-RA (Smith-Thomas et al., 1987). Whole embryo retinol or RA treatment at stages 8-10 has the same effect on the neural crest, causing the neural crest cells to remain clustered to the dorsal surface of the neural tube rather than migrating away, which resulted in hypoplasia of the branchial arches (Hassell et al., 1977; Moro Balbas et al., 1993). Changes in the composition of the glycosaminoglycans within the cephalic extracellular matrix could be detected. A more precise analysis of neural crest migration has recently been performed by Gale and colleagues (Gale et al., 1996). Rather than treating the whole embryo with RA, it was injected into the individual units of the stage 10 hindbrain, the rhombomeres, and this study revealed that the migration pathway of specific neural crest populations can be altered rather than a wholesale inhibition as described earlier. Only when RA was injected into rhombomere 4 was the crest migration altered; injection into rhombomere 2 or rhombomere 6 left the streams of crest unaffected. The mismigrating crest from rhombomere 4 took an abrupt right angle turn to join the adjacent stream. This resulted in the misrouting of the rhombomere 4-derived cranial sensory nerves and ganglia and could be correlated with alterations in the domains of expression of specific pattern forming genes such as *Krox*-20 and *Hoxb*-1. In contrast, the motor neuron projections from rhombomere 4 were normal. Thus RA can have very precise genetic effects on very specific populations of cells during chick development.

The Face

As a byproduct of implanting beads soaked in RA into the anterior margin of the limb bud (see next section) it was found that the RA diffused out into the amniotic cavity and affected the development of the beak (Tamarin et al., 1984). Curiously, only the upper beak was affected—it was entirely missing whereas the lower beak was unaffected. Further studies administering radiolabelled retinoids followed by HPLC (Wedden et al., 1987) showed that both the frontonasal mass (from which the upper beak develops) incorporates the same amount of retinoid as the mandibles (from which the lower beak develops), so the explanation cannot be due to differential uptake. Recombination experiments showed that the target of retinoid action was the mesenchyme and not the ectoderm (Wedden, 1987), just like the chick limb bud and the regenerating amphibian limb, but unlike the skin, as described next.

Skin Patterning

The anterior surface of the feet of chickens are covered in scales, unlike most of the rest of the body, which is covered in feathers. Injection of a single dose of RA into the amniotic sac of chick embryos at 10, 11, or 12 days of incubation causes these scales to transform into feathers (Dhouailly and Hardy, 1978). Normally, the dermis is responsible for determining the type of differentiation of the epidermal structure (Sengel, 1976), but in this case dermal/epidermal recombination experiments have shown that RA acts directly on the epidermis and not the dermis (Cadi et al., 1983). Twenty-four hours after injection of RA the scale placodes appear subdivided into several smaller rudiments closely resembling feather primordia (Dhouailly, 1983). Structure-activity studies using various retinoids have shown similar results to studies in the limb, that is, only vitamin A analogues containing a terminal carboxyl end group are active (Cadi et al., 1984).

The effect of RA on dorsal skin development and feather buds has also been studied from another viewpoint. This is because various *Hox* genes have been found to be differentially expressed in interesting ways in feather buds. There is both a gradient of expression of a particular *Hox* gene within a feather bud (the so-called microgradient) and a global, position-specific expression of the whole range of *Hox* genes (the so-called macrogradient) (Chuong, 1993). Chuong has suggested that such a *Hox* code for skin appendages could individually specify each feather bud. To test this theory RA was employed to alter *Hox* gene expression. When a RA bead was placed onto a cultured piece of skin there was a radial zone of inhibition close to the bead and then a zone of disorientation such that feather buds turned centrifugally away from the RA source (Chuong et al., 1992). There was also a reorientation of the *Hox* microgradient, thereby supporting the theory. At higher doses RA transformed feather buds into a scale-like morphology, the opposite of the effects described above on foot skin.

The Tail Bud

Finally, the tail bud has been used as a system for examining the effects of RA (Griffith and Wiley, 1989; 1991). Localized injection of RA at stages 11 to 16 of chick development resulted in some fascinating anatomical effects ranging from total dysplasia to the induction of an accessory neural tube and notochord in mirror symmetry. It was suggested that this was caused by an interference with cell-cell interactions because similar effects could be observed when lectins were administered to the tail bud.

B. Endogenous Retinoids in the Chick Embryo and Development in Their Absence

Confirmation of the role of RA in early chick development has been obtained in two ways. In the first, endogenous retinoids have been quantified and in the second, the developmental malformations that appear when the embryo develops in the absence of retinoids have been determined.

Endogenous RA has been detected in the early avian embryo by two methods. Direct measurement by HPLC has been used with quail embryos, and to overcome the large amount of tissue needed, 2000 embryos were pooled for this analysis (Dong and Zile, 1995). 4-oxo-RA, ddRA, tRA, all-*trans* retinal, all-*trans* retinol, ddretinol, retinyl palmitate, and retinyl sterate were readily detectable whereas 9-*cis*-RA and/or 13-*cis*-RA were below detectable limits. A second method involves the transfected murine embryonal carcinoma cell system. Using this technique Chen and colleagues (Chen et al., 1992) showed that the stage 4 Hensen's node contains 33nM RA compared to only 1.6nM in the rest of the embryo. Similarly, the whole stage 6-7 quail embryo contains RA (Chen et al., 1996), although this measurement has been superceded by the complete analysis of retinoids by HPLC described above (Dong and Zile, 1995). The unique value of the F9 cell system is that individual pieces of the embryo can be explanted onto the reporter cells and thus one can obtain very precise localisation data. Wagner and coworkers (Wagner et al., 1992) showed that both the dorsal neural tube and floor plate from the stage 16 chick embryo induced the appearance of *lac*Z positive cells in this assay. Interestingly, similar numbers of reporter cells were induced by these two tissues, in contrast to the commonly held view that the floor plate is a source of retinoids. The stage 4 Hensen's node also induced *lac*Z positive cells.

Secondly, the effect that vitamin A deficiency has on the avian embryos was first examined in 1969 by Thompson and colleagues (Thompson et al., 1969) and the cardiovascular system was most profoundly affected. Subsequent investigations have shown that these embryos are indeed retinoid deficient as no endogenous retinoids can be detected in their tissues either directly by HPLC (Dong and Zile, 1995) or with the embryonal carcinoma cell system (Chen et al., 1995). The cardiovascular abnormalities consist of the failure of the blood islands in the area vasculosa to connect up with the embryo because the omphalomesenteric veins and vitelline arteries do not form (Heine et al., 1985; Dersch and Zile, 1993). The heart is sometimes bifid, but if the two heart tubes do fuse then they form a single closed distended ventricle and situs inversus (Figure 5A), an interesting corollary to the effect of RA on heart looping described earlier. It would be interesting to know what the expression of AMHC was in these embryos as the developmental retardation of these vitamin A-deficient hearts is most severe in the posterior region. Recent studies on the nervous system of these embryos has identified three CNS defects

A

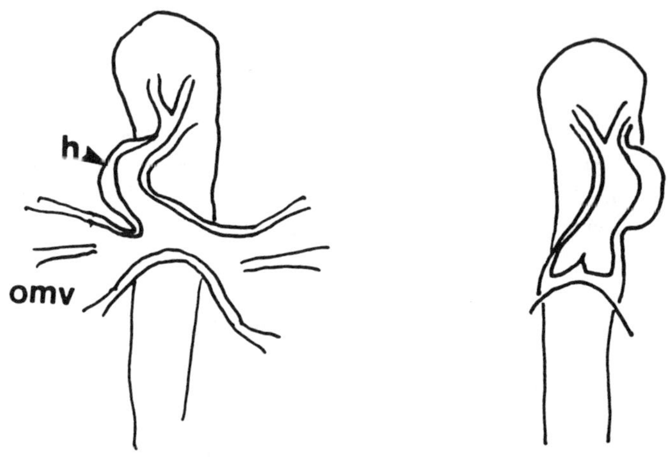

Figure 5. The effect of vitamin A deprivation on the development of the heart (A) and the development of the hindbrain (B). (A) The ventral view of two early somite stage quail embryos to show the developing heart. On the left is a normal embryo showing the heart (h) having fused and looped to the right and formed large omphalomesenteric veins (omv) connecting with the extraembryonic blood vessels and blood islands of the area vasculosa. On the right is the vitamin A-deficient embryo showing a rather swollen heart tube that loops to the left and forms no omvs and thus no connections to the extraembryonic blood vessels. Drawings adapted from Heine et al. (1985). (B) Drawings of the developing CNS of 10-14 somite quail embryos to show the brain and anterior spinal cord structure. On the left is the normal embryo showing the early CNS divided up into segments—forebrain (f), midbrain (mb), the 8 rhombomeres of the hindbrain numbered 1-8, and the spinal cord (sc). On the right is the vitamin A-deficient embryo showing a normal forebrain (f), a normal midbrain (mb), but a hindbrain that consists of only 3 rhombomeres (1-3) instead of the normal 8. A segment of the CNS (rhombomeres 4-8) fails to develop and the spinal cord is moved anteriorly.

(Maden et al., 1996). First, the posterior hindbrain known as the myelencephalon completely fails to develop and instead of having eight rhombomeres there are only three (Figure 5B). This phenomenon is most likely caused by the incorrect patterning of certain genes such as the Hox genes soon after gastrulation. Second, the neural crest cells die by apoptosis having migrated away from the neural tube. This confirms the idea that the neural crest needs RA for survival and proliferation (Henion and Weston, 1994). And third, the neurones in the neural tube never extend neurites into the periphery and those that are extended within the confines of the neural tube seem to follow a chaotic trajectory. This suggests that RA is involved in neurite guidance in some way.

C. Retinoid Binding Proteins and Receptors in the Early Chick Embryo

Measured by radiolabelled binding studies there is a 10-fold excess of CRABP over CBRP in the chick embryo (Maden et al., 1989a) and in an immunoblotting study that discriminated between CRABP I and II it was found that both these proteins were widely expressed in various tissues of the embryo but at different levels (Maden, 1994a). CRABP II was expressed in all tissues studied, with the exception of the heart and somites, but at moderate levels. CRABP I showed more variation from high levels of expression in the frontonasal mass, hindbrain, and branchial arches to low levels in the somites and absent in the heart and epidermis.

B

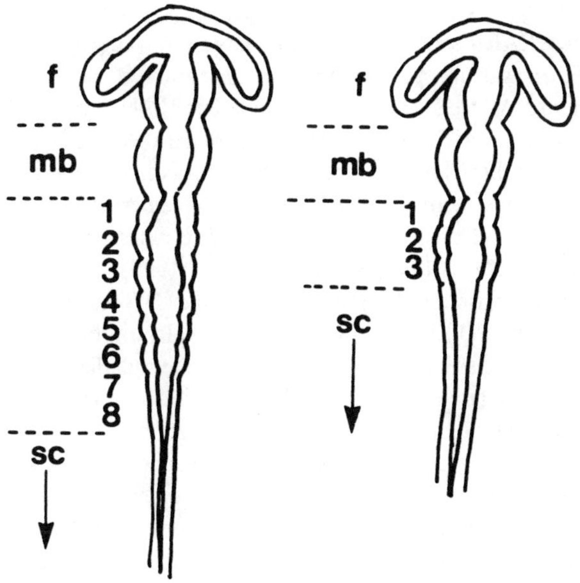

Figure 5. Continued

When the distribution of the protein or mRNA for these genes was studied a fascinating picture emerged. CRBP seems to be localized solely in the ventral floor plate of the developing spinal cord (Maden et al., 1989a). CRABP I, on the other hand, is strongly expressed in the migrating neural crest and its derivatives, namely, branchial arches and facial primordia, sympathetic ganglia, enteric ganglia, dorsal root ganglia, and Schwann cells (Maden et al., 1989b; 1991b; Vaessen et al., 1990). Different branchial arches expressed CRABP I at different levels with the most intense being in arch 1, the next in the frontonasal mesenchyme and arches 3 and 4, and the weakest expression was in arch 2 (Maden et al., 1991b). Within the CNS, CRABP I expression is first detected in individual neuroblasts in the stage 9 chick embryo in the region of the hindbrain, which soon resolves into a striped pattern consisting of neuroblasts in rhombomere 4 and posterior from rhombomere 6 through to the developing spinal cord. Eventually all the commissural neurons and cell bodies of the developing spinal cord show intense expression of CRABP I forming a peripheral region of immunoreactivity. The coincidence of these commissural neurons, which send their axons down towards the ventral floor plate, expressing CRABP I and the ventral floor plate expressing CRBP led to the suggestion that RA was involved in the guidance of neurite outgrowth in the developing CNS (Maden and Holder, 1991; 1992). Recent evidence to suggest that this may be the case was revealed in studies of the vitamin A-deficient quail neural tube, which displays chaotic neurite outgrowth during its development (Maden et al., 1996).

Various RARs and RXRs have been cloned from the chick embryo, although avian homologues of the mammalian RARγ and RXRβ genes have yet to be discovered. RARα seems to be ubiquitously expressed (Noji et al., 1991; Kostetskii et al., 1996), as is the case in the mammalian embryo, and it does not respond to a variation in RA levels. On the other hand, RARβ (Noji et al., 1991; Nohno et al., 1991; Rowe et al., 1991a; 1992; Smith and Eichele, 1991) is expressed in interesting areas of the embryo and is RA responsive. In both the chick and the quail embryo early RARβ expression (stage 6/7) has a sharp anterior boundary across all germ layers at the level of the first somite which crosses the future hindbrain region (Smith and Eichele, 1991; Kostetskii et al., 1996). In the stage 20 embryo it maintains an anterior border of expression in the CNS that is between rhombomeres 5 and 6 (Smith and Eichele, 1991). In the developing face RARβ is expressed in the frontonasal mass and the anterior part of the maxillary process but not in the other branchial arches (Rowe et al., 1991a), and earlier in development it is expressed in the neural crest cells that give rise to these primordia (Rowe et al., 1992). RARβ expression is, as expected, upregulated by RA administration in the facial primordia (Rowe et al., 1991a). It is completely absent in the vitamin A-deficient quail embryo, but expression can be restored by administration of RA (Kostetskii et al., 1996).

Although avian RARγ has not been cloned, probing a Northern blot with the human RARγ probe produced a result of uniform level of expression thoughout quail development and a similar result was obtained in vitamin A-deficient quail embryos suggesting a lack of response of this gene to RA (Kostetskii et al., 1996). RXRα is present ubiquitously in the chick embryo (Seleiro et al., 1995), whereas the RXRγ gene is expressed in the migrating neural crest and peripheral nervous system (dorsal and ventral roots, dorsal root ganglia, sympathetic ganglia, enteric ganglia) but, interestingly, not in the neural tube itself (Rowe et al., 1991b; Seleiro et al., 1995). The expression pattern of this gene showed a striking correlation with that of CRABP I (see above), at least in the neural crest and its derivatives.

VI. THE CHICK LIMB BUD

A. Excess RA

When a chromatography bead that had been soaked in tRA was implanted under the apical ectodermal ridge (AER) on the anterior margin of the limb bud then the normal 3-digit limb consisting of digits 2,3,4 (Figure 6A) was transformed into a 6-digit, mirror-imaged, double posterior limb consisting of digits 4,3,2,2,3,4 (Figure 6B) (Tickle et al., 1982). Anterior cells were thus **posteriorized**. Reducing the concentration of RA produced fewer additional digits—the effect is thus dose-dependent. Unlike the regenerating amphibian limb (where retinoids are administered systemically) in the chick limb bud the effect is position-dependent because applications of RA to the posterior side of the limb bud have no effect and applications to the apex of the bud have lesser effect (Tickle et al., 1985). Like amphibians, the effect is stage-dependent (Summerbell, 1983) and RA acts via the mesenchyme rather than the epidermis (Tickle et al., 1989). There is a minimum period of about 10-12 hours' treatment time needed to induce posteriorization (Eichele et al., 1985), but conversely embryos can be treated as early as stage 10 and the duplications will still appear several days later (Wilde et al., 1987). Not only does tRA act in this fashion, but so does ddRA (Thaller and Eichele, 1990), 9-*cis*-RA, which is considerably more potent than tRA (Thaller et al., 1993), and a whole range of synthetic retinoid analogues (Tamura et al., 1990; Maden et al., 1991a), many of which are also more potent than tRA.

These experiments on the chick limb bud gave rise to the idea that RA was the endogenous morphogen that generated pattern formation in the AP axis since it was demonstrated that exogenously applied RA rapidly diffused across the limb bud and within a few hours after administration an exponential gradient had been generated (Tickle et al., 1985). RA could also completely reverse the AP orientation of the digits that grew from an anterior half limb bud. Normally an anterior half limb bud will give rise to no digits, but when supplied with a RA bead on the anterior margin, such a limb bud will give digits with their AP orientation reversed, that is, a digit sequence of 4,3,2 instead of 2,3,4 (Eichele, 1989). These effects of RA precisely mimicked the behavior of an organizing region in the chick limb bud that had been discovered more than a decade previously (Saunders and Gasseling,

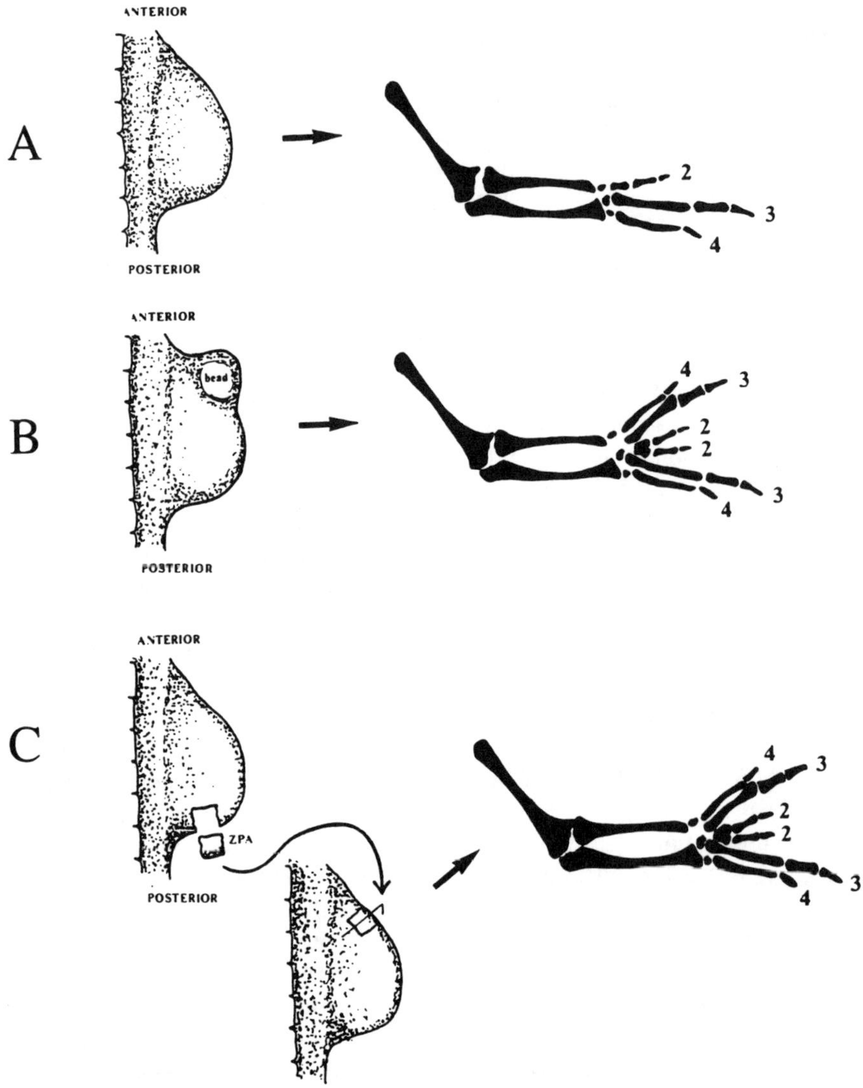

***Figure 6*.** The effects of RA on the chick limb bud. (A) A stage 21 limb bud is drawn on the left with its anterior and posterior margins marked. This gives rise to a normal 3-digit limb with digits numbered 2, 3, & 4. (B) After implantation of a RA soaked bead into the anterior margin a 6-digit, mirror-imaged limb is produced whose digit sequence is 4,3,2,2,3,4. (C) An identical result to (B) is produced by grafting a group of cells (the ZPA) from the posterior margin of a donor limb bud to the anterior margin of a host limb bud. The resulting limb is a 6-digit, mirror-imaged limb.

1968). On the posterior side of the limb bud is a group of cells that, when removed and grafted to the anterior side of another limb bud, produces a 6-digit, mirror-imaged duplicated limb consisting of digits 4,3,2,2,3,4 (Figure 6C). This region became known as the zone of polarizing activity (ZPA) and it displays an almost identical pattern of behavior after grafting to the administration of RA: its effects are stage-dependent, position-dependent, concentration-dependent; the grafted ZPA instructs anterior cells to change their specification and there is a minimum time period of 12-18 hours needed to induce duplications (review, Summerbell and Honig, 1982). It was suggested, therefore, that the ZPA released RA, which established a gradient and specified AP pattern (although diffusion or autocatalysis cannot, apparently, explain the RA phenomenology [Papageorgiou and Almirantis, 1992]). Alternatively, it may be that RA is simply inducing a new ZPA and that we know nothing about the "true" morphogen.

In fact, there are two instances in which it has subsequently been demonstrated that RA administration is *not* identical to a ZPA graft. In one instance a RA implant induces the appearance of the retinoic acid receptor β in surrounding anterior tissues whereas a ZPA graft does not (Noji et al., 1991). In the second instance, tissue adjacent to a RA implant has the ability to itself induce duplications when grafted to the anterior margin of another limb bud (Summerbell and Harvey, 1983; Wanek et al., 1991), whereas tissue adjacent to a grafted ZPA does not have this ability (Smith, 1979). The alternative interpretation, that RA simply induces a new ZPA on the anterior side, which then organizes a new AP axis in mirror-image to the existing one, seems more likely.

This debate has been brought sharply into focus by the cloning of the *sonic hedgehog* gene (Riddle et al., 1993). The expression pattern of this gene exactly coincides with the ZPA. Ectopic expression of *shh* on the anterior side of the limb bud induces a mirror-image duplication (e.g. Figure 6B) and RA induces an ectopic domain of *shh*. It was proposed, therefore, that the protein product of the *shh* gene is the morphogen released by the ZPA, rather than RA, but it has been shown that SHH protein colocalizes with the message to a small, discrete area in the posterior margin (Lopez-Martinez et al., 1995; Marti et al., 1995) so this is an untenable idea.

Most likely RA acts to induce *shh* expression, which is one component of the ZPA, and then there is a cascade of genes that establishes pattern domains across the AP axis. Certainly there are a large number of genes that have been shown to be active during limb development (review Maden, 1994b). These genes include the *Fgfs* in the apical ectodermal ridge, the 5' genes of the *Hoxd* cluster, the genes of the *Hoxa* cluster, the *Msx* genes in the ectoderm and mesoderm, the BMPs, *Wnt*-7a in the dorsal ectoderm, *Lmx*-1 and the interactions between them are just beginning to be unravelled; RA induces or suppresses many of them.

B. Endogenous RA and the Retinoid Pathway

Thaller and Eichele (1987), using HPLC, identified endogenous all-*trans*-RA in the chick limb bud, and, furthermore, it was detected at 2.5-fold higher levels on the posterior side than the anterior side. Subsequently, it was found that the predominant retinoic acid in the chick limb bud was not all-*trans*-RA but didehydroretinoic acid as it was present at concentrations four to six times higher (Thaller and Eichele, 1990). When the synthesis of RA was determined by culturing limb buds in [^{3}H] all-*trans*-retinol it was found that the major retinoic acid synthesized under these circumstances was ddRA and not tRA (Stratford et al., 1996). Since ddRA is equipotent at inducing duplications in the chick limb bud (Thaller and Eichele, 1990), it seems that ddRA is the major form of retinoic acid present in the chick limb bud. Furthermore, in the metabolism studies of Stratford and colleagues, it was observed that the posterior side of the limb bud synthesizes ddRA at a higher rate than the anterior side.

The identification of endogenous retinoids has recently been confirmed in the chick (Scott et al., 1994) and extended to the mouse limb bud (Scott et al., 1994; Horton and Maden, 1995). Interestingly, in the mouse limb bud no didehydroretinoids were present at all, only the all-*trans* forms. Using the embryonal carcinoma cell reporter system, Wagner and colleagues (Wagner et al., 1992) showed that chick limb bud mesoderm produces RA and in our studies using this system (Maden and Gale, unpublished data) we observe that posterior half limb buds produce more RA than anterior half limb buds.

Confirmation that the synthesis of endogenous RA and the components of the retinoid pathway are crucial to chick limb development has been obtained in two recent experiments. In the first (Stratford et al., 1996) the synthesis of RA was inhibited with disulphiram, a compound that inhibits the action of aldehyde dehydrogenase enzymes (Vallari and Pietruszko, 1982: McCaffery et al., 1992; 1993). When this inhibitor is administered to the flank of the chick embryo before the limb bud appears, no limb develops on that side. The domain of expression of *shh* fails to appear, although *Fgf*-8 in the ectoderm of the limb bud is expressed according to the normal program. The outgrowth of the limb bud and the initiation of expression of *shh* can be rescued with a bead soaked in RA, confirming the role that RA synthesis plays in the regulation of *shh*. In the second experiment (Helms et al., 1996) RAR and RXR pan-specific antagonists were used. Beads soaked in a mixture of these two compounds were implanted into the anterior and posterior boundaries of the limb field and this resulted in a range of abnormalities from missing digits to missing limbs. RA bead implants could restore the limb pattern and when the RA was placed on the anterior of the limb bud then the restored pattern had the reversed anteroposterior asymmetry. Like the inhibition of RA synthesis these antagonists caused the down-regulation of *shh* expression.

C. Retinoid Binding Proteins and Receptors

High levels of CRABP were detected in the chick limb bud by radiolabelled binding studies at all stages examined (Maden and Summerbell, 1986; Scott et al., 1994), whereas only very low levels of CRBP were measured (Maden et al., 1989b;

Scott et al., 1994), and in immunocytochemical studies, CRBP protein could not be detected in the limb bud. CRABP I, however, showed a fascinating distribution being expressed more intensely on the anterior side of the limb bud than the posterior side (Maden et al., 1988; Maden, 1994a). This gradient is the opposite to the endogenous gradient of all-*trans*-RA (Thaller and Eichele, 1987) and leads to the suggestion that the function of CRABP I was to increase the gradient of free RA available for binding to the nuclear receptors. Calculations of the binding affinities and levels of RA and CRABP have suggested the gradient of free RA would steepen from 2.5:1 to 10:1 on the posterior versus the anterior side when the CRABP distribution is taken into account (Smith et al., 1989). Recent repeat measurements of these variables taking into account ddRA as well a tRA have come to a similar conclusion but with lower estimates of the excess of free RA on the posterior side compared to the anterior side—2:1 (Scott et al., 1994). However, none of these calculations takes into account the presence of CRABP II, which is in a 2-fold excess over CRABP I and whose distribution in the limb bud is more uniform than that of CRABP I (Maden, 1994a; Scott et al., 1994), although some data suggest that there is an excess of CRABP II in the posterior part of the limb bud (Miyagawa-Tomita et al., 1992).

The distribution of the retinoid receptors in the limb bud has not been studied to the same degree as these cytoplasmic binding proteins, perhaps because they are more uniformly distributed. RARβ, however, is the exception. The induction of RARβ on the anterior margin of the limb bud by a RA bead, but the lack of induction by a ZPA graft, was evidence that a RA bead and a ZPA graft are not identical in their behavior in inducing duplications in the chick limb bud (Noji et al., 1991). The lack of an endogenous anteroposterior gradient of RARβ in the limb bud was also cited as evidence against the RA gradient theory. The expression of RARβ, by *in situ* hybridization, does actually show a gradient but it is a proximodistal one with high levels proximally diminishing to low levels distally (Noji et al., 1991; Smith and Eichele, 1991). The gradient seems to be established by repression distally, as proximal tissue grafted into the distal tip has its RARβ expression level repressed (Schofield et al., 1992). However, using the more sensitive method of RNase protection it seems that there is, indeed, an anteroposterior differential distribution of RARβ (Smith et al., 1995). RARα and RXRα, on the other hand, are uniformly expressed in the limb bud (Noji et al., 1991; Seleiro et al., 1995), the latter being on both in the mesenchyme and the ectoderm and RXRγ is only expressed in the nerves that enter the growing limb bud.

REFERENCES

Agarwal, V.R., & Sato, S.M. (1993). Retinoic acid affects central nervous system development of *Xenopus* by changing cell fate. Mech. Dev. 44, 167-173.

Akimenko, M-A., & Ekker, M. (1995). Anterior duplication of the *Sonic hedgehog* expression pattern in the pectoral fin buds of zebrafish treated with retinoic acid. Dev. Biol. 170, 243-247.

Alexandre, D., Clarke, J.D.W., Oxtoby, E., Yan, Y-L., Jowett, T., & Holder, N. (1996). Ectopic expression of Hoxa-1 in the zebrafish alters the fate of the mandibular arch neural crest and phenocopies a retinoic acid-induced phenotype. Development 122, 735-746.

Azuma, M., Seki, T., & Fujishita, S. (1990). Changes of egg retinoids during the development of *Xenopus laevis*. Vision Res. 30, 1395-1400.

Balkan, W., Colbert, M., Bock, C., & Linney, E. (1992). Transgenic indicator mice for studying activated retinoic acid receptors during development. Proc. Natl. Acad. Sci. USA 89, 3347-3351.

Bally-Cuif, L., Gulisano, M., Broccoli, V., & Boncinelli, E. (1995). *c-otx2* is expressed in two different phases of gastrulation and is sensitive to retinoic acid treatment in chick embryo. Mech. of Dev. 49, 49-63.

Bertwistle, D., Walmsley, M.E., Read, E.M., Pizzey, J.A., & Patient, R.K.P. (1996). GATA factors and the origins of adult and embryonic blood in *Xenopus*: responses to retinoic acid. Mech. Dev. 57, 199-214.

Blumberg, B., Mangelsdorf, D.J., Dyck, J.A., Bittner, D.A., Evans, R.M., & De Robertis, E.M. (1992). Multiple retinoid-responsive receptors in a single cell: families of retinoid "X" receptors and retinoic acid receptors in the *Xenopus* egg. Proc. Natl. Acad. Sci. USA 89, 2321-2325.

Boylan, J.F., & Gudas, L.J. (1991). Overexpression of the cellular retinoic acid binding protein-I (CRABP-I) results in a reduction of differentiation-specific gene expression in F9 teratocarcinoma cells. J. Cell Biol. 112, 965-979.

Brennan, S.M. (1992). Retinoic acid prevents accumulation of a mesoderm-specific mRNA in the amphibian embryo. Mech. Dev. 38, 17-24.

Brivanlou, A.H., & Harland, R.M. (1989). Expression of an engrailed-related protein is induced in the anterior neural ectoderm of early *Xenopus* embryos. Development 106, 611-617.

Brockes, J. P. (1992) Introduction of a retinoid reporter gene into the urodele limb blastema. Proc. Natl. Acad. Sci. USA 89, 11386-11390.

Cadi, R., Dhouailly, D., & Sengel, P. (1983). Use of retinoic acid for the analysis of dermal-epidermal interactions in the tarsometatarsal skin of the chick embryo. Dev. Biol. 100, 489-495.

Cadi, R., Pautou, M-P., & Dhouailly, D. (1984). Structure-activity realtionships of retinoids in the morphogenesis of cutaneous appendages in the chick embryo. J. Invest. Dermatol. 83, 105-109.

Chen, Y., & Solursh, M. (1992). Comparison of Hensen's node and retinoic acid in secondary axis induction in the early chick embryo. Dev. Dynamics 195, 142-151.

Chen, Y., Huang, L., Russo, A.F., & Solursh, M. (1992). Retinoic acid is enriched in Hensen's node and is developmentally regulated in the early chicken embryo. Proc. Natl. Acad. Sci. USA. 89, 10056-10059.

Chen, Y., Huang, L., & Solursh, M. (1994). A concentration gradient of retinoids in the early *Xenopus laevis* embryo. Dev. Biol. 161, 70-76.

Chen, Y., Dong, D., Kostetskii, I., & Zile, M.H. (1996). Hensen's node from vitamin A-deficient quail embryo induces chick limb bud duplication and retains its normal asymmetric expression of *Sonic hedgehog (Shh)*. Dev. Biol. 173, 256-264.

Cho, K.W.Y., & De Robertis, E. (1990). Differential activation of *Xenopus* homeobox genes by mesoderm-inducing growth factors and retinoic acid. Genes Dev. 4, 1910-1916.

Cho, K.W.Y., Blumberg, B., Steinbeisser, H., & De Robertis, E.M. (1991). Molecular nature of Spemann's organizer: the role of the *Xenopus* homeobox gene *goosecoid*. Cell 67, 1111-1120.

Chuong, C-M. (1993). The making of a feather: homeoproteins, retinoids and adhesion molecules. BioEssays 15, 513-521.

Chuong, C-M., Ting, S.A., Widelitz, R.B., & Lee, Y-S. (1992). Mechanism of skin morphogenesis. II Retinoic acid modulates axis oritentation and phenotypes of skin appendages. Development 115, 839-952.

Costaridis, P., Horton, C., Zeitlinger, J., Holder, N., & Maden, M. (1996). Endogenous retinoids in the zebrafish embryo and adult. Dev. Dynam. 205, 41-51.

Crawford, K., & Stocum, D.L. (1988). Retinoic acid coordinately proximalizes regenerate pattern and blastema differential affinity in axolotl limbs. Development 102, 687-698.

Crawford, M.J., Liversage, R.A., & Varmuza, S.L. (1995). Two isoforms of *Xenopus* retinoic acid receptor $\gamma 2$ (B) exhibit differential expression and sensitivity during embryogenesis. Dev. Genet. 17, 291-302.

Creech Kraft, J., Schuh, T., Juchau, M.R., & Kimelman, D. (1994a). Temporal distribution, localization and metabolism of all-*trans*-retinol, didehydroretinol and all-*trans*-retinal during *Xenopus* development. Biochem. J. 301, 111-119.

Creech Kraft, J., Schuh, T., Juchau, M., & Kimelman, D. (1994b). The retinoid X receptor ligand, 9-*cis*-retinoic acid, is a potential regulator of early *Xenopus* development. Proc. Natl. Acad. Sci. USA 91, 3067-3071.

Creech Kraft, J., & Juchau, M.R. (1995). *Xenopus laevis*: A model system for the study of embryonic retinoid metabolism: III. Isomerization and metabolism of all-*trans*-retinoic acid and 9-*cis*-retinoic acid and their dysmorphogenic effects in embryos during neurulation. Drug Metab. Dispos. 23, 1058-1071.

Creech Kraft, J.C., Kimelman, D., & Juchau, M.R. (1995a). *Xenopus laevis*: A model system for the study of embryonic retinoid metabolism: I. Embryonic metabolism of 9-*cis*- and all-*trans*-retinals and retinols to their corresponding acid forms. Drug Metab. Dispos. 23, 72-82.

Creech Kraft, J.C., Kimelman, D., & Juchau, M.R. (1995b). *Xenopus laevis:* A model system for the study of embryonic retinoid metabolism: II. Embryonic metabolism of all-*trans*-3,4-didehydroretinol to all-*trans*-3,4-didehydroretinoic acid. Drug Metab. Dispos. 23, 83-89.

Currie, P.D., Qiao, T., & Ingham, P. (1995). Analysis of *Hedgehog* gene family members in zebrafish. Proceedings of the European Developmental Biology Organisation, Toulouse. p. 77.

Dekker, E.-J., Pannese, M., Houtzager, E., Boncinelli, E., & Durston, A. (1992). Colinearity in the *Xenopus laevis* Hox-2 complex. Mech Dev. 40, 3-12.

Dekker, E.-J., Vaessen, M.-J., van den Berg, C., Timmermans, A, Godsave, S., Holling, T., Nieuwkoop, P., van Kessel, A.G., & Durston, A. (1994). Overexpression of a cellular retinoic acid binding protein (xCRABP) causes anteroposterior defects in developing *Xenopus* embryos. Development 120, 973-985.

Dersch, H., & Zile, M.H. (1993). Induction of normal cardiovascular development in the vitamin A-deprived quail embryo by natural retinoids. Dev. Biol. 160, 424-433.

Dhouailly, D. (1983). Early events in reinoic acid-induced ptilopody in the chick embryo. Roux's Arch. Dev. Biol. 192, 21.

Dhouailly, D., & Hardy, M.H. (1978). Retinoic acid causes the development of feathers in the scale-forming integument of the chick embryo. Wilhelm Roux's Archives 185, 195.

Dong, D., & Zile, M.H. (1995). Endogenous retinoids in the early avian embryo. Biochem. Biophys. Res. Comm. 217, 1026-1031.

Drysdale, T.A., & Crawford, M.J. (1994). Effects of localized application of retinoic acid on *Xenopus laevis* development. Dev. Biol. 162, 394-401.

Durston, A.J., Timmermans, J.P.M., Hage, W.J., Hendriks, H.F.J., de Vries, N.J., Heideveld, M., & Nieuwkoop, P.D. (1989). Retinoic acid causes an anteroposterior transformation in the developing central nervous system. Nature (London) 340, 140-144.

Eichele, G. (1989). Retinoic acid induces a pattern of digits in anterior half wing buds that lack the zone of polarizing activity. Development 107, 863-867.

Eichele. G, Tickle, C., & Alberts, B.M. (1985). Studies on the mechanism of retinoid-induced pattern duplications in the early chick limb bud: temporal and spatial aspects. J. Cell Biol. 101, 1913-1920.

Ellinger-Ziegelbauer, H., & Dreyer, C. (1991). A retinoic acid receptor expressed in the early development of *Xenopus laevis*. Genes Dev. 5, 94-104.

Ellinger-Ziegelbauer, H., & Dreyer, M. (1993). The pattern of retinoic acid receptor γ(RARγ) expression in normal development of *Xenopus laevis* and after manipulation of the main body axis. Mech. Dev. 41, 33-46.

Ferretti, P., & Geraudie, J. (1995). Retinoic acid-induced cell death in the wound epidermis of regenerting zebrafish fins. Dev. Dynam. 202, 271-283.

Ferretti, P., Brockes, J.P., & Brown, R. (1991). A newt type II keratin restricted to normal and regenerating limbs and tails is responsive to retinoic acid. Development 111, 497-507.

Fjose, A., Weber, U., & Mlodzik, M. (1995). A novel vertebrate *svp*-related nuclear receptor is expressed as a step gradient in developing rhombomeres and is affected by retinoic acid. (1995). Mech. of Dev. 52, 233-246.

Gale, E., Prince, V., Lumsden, A., Clarke, J., Holder, N., & Maden, M. (1996). Late effects of retinoic acid on neural crest and aspects of rhombomere identity. Development 122, 783-793.

Gardiner, D.M., Blumberg, B., Komine, Y., & Bryant, S.V. (1995). Regulation of HoxA expression in developing and regenerating axolotl limbs. Development 121, 1731-1741.

Geraudie, J., Brulfert, A., Monnot, M.J., & Ferretti, P. (1994). Teratogenic and morphogenetic effects of retinoic acid on the regenerating pectoral fin in zebrafish. J. Exp. Zool. 269, 12-22.

Geraudie, J., Monnot, M.J., Brulfert, A., & Ferretti, P. (1995). Caudal fin regeneration in wild type and *long-fin* mutant zebrafish is affected by retinoic acid. Int. J. Dev. Biol. 39, 373-381.

Giguere, V., Ong, E.S., Evans, R.M., & Tabin, C.J. (1989). Spatial and temporal expression of the retinoic acid receptor in the regenerating amphibian limb. Nature 337, 566-569.

Giguere, V., Lyn, S., Yip, P., Siu, C.-H., & Amin, S. (1990). Molecular cloning of cDNA encoding a second cellular retinoic acid-binding protein. Proc. Natl. Acad. Sci. USA 87, 6233-6237.

Gont, L.K., Steinbeisser, H., Blumberg, B., & De Robertis, E.M. (1993). Tail formation as a continuation of gastrulation: the multiple cell populations of the *Xenopus* tailbud derive from the late blastopore lip. Development 119, 991-1004.

Griffith, C.M., & Wiley, M.J. (1989). Direct effect of retinoic acid on the development of the tail bud in chick embryos. Teratology 39, 261-275.

Griffith, C.M., & Wiley, M.J. (1991). Effects of retinioc acid on chick tail bud development. Teratology 43, 217-224.

Hart, R.C., McCue, P.A., Ragland, W.L., Winn, K.J., & Unger, E.R. (1990). Avian model for 13-*cis*-retinoic acid emnryopathy: demonstration of neural crest related defects. Teratology 41, 463-472.

Hart, R.C., Winn, K.J., & Unger, E.R. (1992). Avian model for 13-*cis*-retinoic acid embryopathy: morphological characteristics of ventricula septal defects. Teratology 46, 533-539.

Hassell, J.R., Greenberg, J.H., & Johnston, M.C. (1977). Inhibition of cranial neural crest cell development by vitamin A in the cultured chick embryo. J. Embryol. exp. Morph. 39, 267-271.

Heine, U.I., Roberts, A.B., Munoz, E.F., Roche, N.S., & Sporn, M.B. (1985). Effects of retinoid deficiency on the development of the heart and vascular system of the quail embryo. Virchow's Arch. B [Cell Pathol.] 50, 135-152.

Helms, J.A., Kim, C.H., Eichele, G., & Thaller, C. (1996). Retinoic acid signalling is required during early chick limb development. Development 122, 1385-1394.

Henion, P.D., & Weston, J.A. (1994). Retinoic acid selectively promotes the survival and proliferation of neurogenic precursors in cultured neural crest populations. Dev. Biol. 161, 243-250.

Heyman, R.A., Mangelsdorf, D.J., Dyck, J.A., Stein, R., Eichele, G., Evans, R.M., & Thaller, C. (1992). 9-*cis*-Retinoic acid is a high-affinity ligand for the retinoid X receptor. Cell 68, 397-406.

Hill, D.S., Ragsdale, C.W., & Brockes, J.P. (1993). Isoform-specific immunological detection of newt retinoic acid receptor δ1 in normal and regenerating limbs. Development 117, 937-945.

Hill, J., Clarke, J.D.W., Vargesson, N., Jowett, T., & Holder, N. (1995). Exogenous retinoic acid causes specific alterations in the midbrain and hindbrain of the zebrafish embryo including positional respecification of the Mauthner neuron. Mech. of Dev. 50, 3-16.

Ho, L., Mercola, M., & Gudas, L.J. (1994). *Xenopus laevis* cellular retinoic acid-binding protein: temporal and spatial expression during early embryogenesis. Mech. of Dev. 47, 53-64.

Hogan, B.L.M., Thaller, C., & Eichele, G. (1992). Evidence that Hensen's node is a site of retinoic acid synthesis. Nature (London) 359, 237-241.

Holder, N., & Hill, J. (1991). Retinoic acid modifies development of the midbrain-hindbrain border and affects cranial ganglion formation in zebrafish embryos. Development 113, 1159-1170.

Horton, C., & Maden, M. (1995). The endogenous distribution of retinoids during normal development and teratogenesis in the mouse embryo. Dev. Dynam. 202, 312-323.

Hyatt, G.A., Schmitt, E.A., Marsh-Armstrong, N., & Dowling, J.E. (1992). Retinoic acid-induced duplication of the zebrafish retina. Proc. Natl. Acad. Sci. USA 89, 8293-8297.

Johnson, K.J., & Scadding, S.R. (1992). Effects of tunicamycin on retinoic acid induced respecification of positional values in regenerating limbs of the larval axolotl, *Ambystoma mexicanum*. Dev. Dynam. 193, 185-192.

Jones, B.B., Ohno, C.K., Allenby, G., Boffa, M.B., Levin, A.A., Grippo, J.F., & Petkovich, M. (1995). New retinoid X receptor subtypes in zebra fish (*Danio rerio*) differentially modulate transcription and do not bind 9-*cis* retinoic acid. Mol. Cell Biol. 15, 5226-5234.

Joore, J., van der Lans, G.B.L.J., Lanser, P.H., Vervaart, J.M.A., Zivkovic, D., Speksnijder, J.E., & Kruijer, W. (1994). Effects of retinoic acid on the experssion of retinoic acid receptors during zebrafish embryogenesis. Mech. of Dev. 46, 137-150.

Ju, B-G., & Kim, W-S. (1994). Pattern duplication by retinoic acid treatment in the regenerating limbs of Korean salamander larvae, *Hynobius leechii,* correlates well with the extent of dedifferentiation. Dev. Dynam. 199, 253-267.

Kawamura, K., Hara, K., & Fujiwara, S. (1993). Developmental role of endogenous retinoids in the determination of morphallactic field in budding tunicates. Development 117, 835-845.

Keeble, S., & Maden, M. (1986). Retinoic acid-binding protein in the axolotl: distribution in mature tissues and time of appearance during limb regeneration. Dev. Biol. 117, 435-441.

Keeble, S., & Maden, M. (1989). The relationship among retinoid structure, affinity for retinoic acid-binding protein and ability to respecify pattern in the regenerating axolotl limb. Dev. Biol. 132, 26-34.

Kim, W-S., & Stocum, D.L. (1986a). Effects of retinoids on regenerating limbs: comparison of retinoic acid and arotinoid at different amputation levels. Roux's Arch. Dev. Biol. 195, 455-463.

Kim, W-S., & Stocum, D.L. (1986b). Retinoic acid modifies positional memory in the anteroposterior axis of regenerating axolotl limbs. Dev. Biol. 114, 170-179.

Kostetskii, I., Linask, K.K., & Zile, M.H. (1996). Vitamin A deficiency and the expression of retinoic acid receptors during early cardiogenesis in quail embryo. Roux's Arch. Dev. Biol. 205, 260-271.

Lampron, C.C., Rochette-Egly, Gorry, P., Dolle, P., Mark, M., Lufkin, T., LeMeur, M., & Chambon, P. (1995). Mice deficient in cellular retinoic acid binding protein II (CRABPII) or in both CRABPI and CRABPII are essentially normal. Development, 121, 539-548.

Leonard, L., Horton C., Maden, M., & Pizzey, J.A. (1995). Anteriorization of CRABP-I expression by retinoic acid in the developing mouse central nervous system and its relationship to teratogenesis. Dev. Biol. 168, 514-528.

Leroy, P., & De Robertis, E.M. (1992). Effects of lithium chloride and retinoic acid on the expression of genes from the *Xenopus laevis* Hox 2 complex. Dev. Dyn. 194, 21-32.

Lheureux, E., Thoms, S.D., & Carey, F. (1986). The effects of two retinoids in limb regeneration in *Pleurodeles waltl* and *Triturus vulgaris*: J. Embryol. Exp. Morph. 92, 165-182.

Lopez, S.L., Dono, R., Zeller, R., & Carrasco, A.E. (1995). Differential effects of retinoic acid and a retinoid antagonist on the spatial distribution of the homeoprotein HoxB-7 in vertebrate embryos. Dev. Dynam. 204, 457-471.

Lopez-Martinez, A., Chang, D.T., Chiang, C. Porter, J.A., Ros, M.A., Simandl, B.K., Beachy, P.A., & Fallon, J.F. (1995). Limb-patterning activity and restricted posterior localization of the amino-terminal product of Sonic hedgehog cleavage. Current Biol. 5, 791-796.

Ludolph, D.C., Cameron, J.A., & Stocum, D.L. (1990). The effects of retinoic acid on positional memory in the dorsoventral axis of regenerating axolotl limbs. Dev. Biol. 140, 41-52.

Ludolph, D.C., Cameron, J., Neff, A.W., & Stocum, D.L. (1993). Cloning and tissue specific expression of the axolotl cellular retinoic acid binding protein. Develop. Growth & Differ. 35, 341-347.

Maden, M. (1982). Vitamin A and pattern formation in the regenerating limb. Nature 295, 672-675.

Maden, M. (1983a). The effect of vitamin A on the regenerating axolotl limb. J. Embryol. Exp. Morph. 77, 273-295.

Maden, M. (1983b). The effect of vitamin A on limb regeneration in *Rana temporaria*. Dev. Biol. 98, 409-416.

Maden, M. (1984). Does vitamin A act on pattern formation via the epidermis or mesenchyme? J. Exp. Zool. 230, 387-392.

Maden, M. (1993). The homeotic transformation of tails into limbs by retinoids in *Rana temporaria*. Dev. Biol. 159, 379-391.

Maden, M. (1994a). Distribution of cellular retinoic acid-binding proteins I and II in the chick embryo and their relationship to teratogenesis. Teratology 50, 294-301.

Maden, M. (1994b). The limb bud—part two. Nature 371, 560-561.

Maden, M., & Summerbell, D. (1986). Retinoic acid-binding protein in the chick limb bud: identification at developmental stages and binding affinities of various retinoids. J. Embryol. Exp. Morph. 97, 239-250.

Maden, M., & Keeble, S. (1987). The role of cartilage and fibronectin during retinoic acid-induced specification of pattern in the regenerating limb. Differentiation 36, 175-184.

Maden, M., & Holder, N. (1991). The involvement of retinoic acid in the develoment of the vertebrate central nervous system. Development (suppl) 87-94.

Maden, M., & Holder, N. (1992). Retinoic acid and development of the central nervous system. BioEssays 14, 431-438.

Maden, M., & Corcoran, J. (1996). The role of thyroid hormone and retinoid receptors in the homeotic transformation of tails into limbs in frogs. Dev. Genet.19, 85-93.

Maden, M. Ong, D.E., Summerbell, D., & Chytil, F. (1988). Spatial distribution of cellular protein binding to retinoic acid in the chick limb bud. Nature 335, 733-735.

Maden, M., Ong, D.E., Summerbell, D., & Chytil, F. (1989a). The role of retinoid-binding proteins in the generation of pattern in the developing limb, the regenerating limb and the nervous system. Development (suppl) 109-119.

Maden, M., Gale, E., Kostetskii, I., & Zile, M.H. (1996). Vitamin A-deficient quail embryos have half a hindbrain and other neural defects. Current Biology 6, 417-426.

Maden, M., Ong, D.E., Summerbell, D., Chytil, F., & Hirst, E.A. (1989b). Cellular retinoic acid-binding protein and the role of retinoic acid in the development of the chick embryo. Dev. Biol. 135, 124-132.

Maden, M., Summerbell, D., Maignan, J., Darmon, M., & Shroot, B. (1991a). The respecification of limb pattern by new synthetic retinoids and their interaction with cellular retinoic acid-binding protein. Differentiation 47, 49-55.

Maden, M., Hunt, P., Eriksson, U., Kuriowa, A., Krumlauf, R., & Summerbell, D. (1991b). Retinoic acid-binding protein, rhombomeres and the neural crest. Development 111, 35-44.

Mahapatra , P.K., & Mohanty-Hejmadi, P. (1994). Vitamin A-mediated homeotic transformation of tail to limbs, limb suppression and abnormal tail regeneration in the Indian jumping from *Polypedates maculatus*. Develop. Growth & Differ. 36, 307-317.

Marklew, S., Smith, D.P., Mason, C.S., & Old, R.W. (1994). Isolation of a novel RXR from *Xenopus* that most closely resembles mammalian RXRβ and is expressed throughout early development. Biochim. Biophys. Acta (Gene Struct. Express.) 1218, 267-272.

Marshall, H., Nonchev, S., Sham, M.H., Muchamore, I., Lumsden, A., & Krumlauf, R. (1992). Retinoic acid alters hindbrain Hox code and induces transformation of rhombomeres 2/3 into a 4/5 identity. Nature 360, 737-741.

Marsh-Armstrong, N., McCaffery, P., Gilbert, W., Dowling, J.E., & Drager, U.C. (1994). Retinoic acid is necessary for development of the ventral retina in zebrafish. Proc. Natl. Acad. Sci. USA 91, 7286-7290.

Marsh-Armstrong, N., McCaffery, Hyatt, G., Alonso, L., Dowling, J.E., Gilbert, W., & Drager, U.C. (1995). Retinoic acid in the anteroposterior patterning of the zebrafish trunk. Roux's Arch. Dev. Biol. 205, 103-113.

Marti, E., Takuda, R., Bumcrot, D.A., Sasaki, H., & McMahon, A.P. (1995). Distribution of Sonic hedgehog peptides in the developing chick and mouse embryo. Development 121, 2537-2547.

McCaffery, P., Lee, M., Wagner, M.A., Sladek, N.E., & Drager. U.C. (1992). Asymmetrical retinoic acid synthesis in the dorsoventral axis of the retina. Development 115, 371-382.

McCaffery, P., Posch, K.C., Napoli, J.L., Gudas, L., & Drager, U.C. (1993). Changing patterns of the retinoic acid system in the developing retina. Dev. Biol. 158, 390-399.

McCormick, A.M., Shubeita, H.E., & Stocum, D.L. (1988). Cellular retinoic acid binding protein: detection and quantitation in regenerating axolotl limbs. J. exp. Zool. 245, 270-276.

Minucci, S., Zand, D.J., Dey, A., Marks, M.S., Nagata, T., Grippo, J.F., & Ozato, K. (1994). Dominant negative retinoid X receptor β inhibits retinoic acid-responsive gene regulation in embryonal carcinoma cells. Mol. Cell. Biol. 14, 360-372.

Minucci, S., Saint-Jeannet, J.-P., Toyama, R., Scita, G., DeLuca, L.M., Taira, M., Levin, A.A., Ozato, K., & Dawid, I.B. (1996). Retinoid X receptor-selective ligands produce malformations in *Xenopus* embryos. Proc. Natl. Acad. Sci. USA 93, 1803-1807.

Miyagawa-Tomita, S., Kitamoto, T., Momma, K., Takao, A., & Momoi, T. (1992). Cellular retinoic acid binding protein type II was preferentially localised in medium and posterior parts of the progress zone of the chick limb bud. Biochem. Biophys. Res. Comm. 185, 217-223.

Mohanty-Hejmadi, P., Dutta, S.K., & Mahapatra, P. (1991). Limbs generated at the site of amputation in marbled balloon frog after vitamin A treatment. Nature 355, 352-353.

Monkemeyer, J., Ludolph, D.C., Cameron, J.A., & Stocum, D.L. (1992). Retinoic acid-induced change in anteroposterior positional identity in regenerating axolotl limbs is dose-dependent. Dev. Dynam. 193, 286-294.

Moriya, N., Uchiyama, H., & Asashima, M. (1993). Induction of pronephric tubules by activin and retinoic acid in presumptive ectoderm of *Xenopus laevis.* Dev. Growth. Diff. 35, 123-128.

Moro Balbas, J.A., Gato, A., Alonso Revuelta, M.I., Pastor, J.F., Represa, J.J., & Barbosa, E. (1993). Retinoic acid induces changes in the rhombencephalic neural crest cells migration and extracellular matrix composition in chick embryos. Teratology 48, 197-206.

Morriss-Kay, G.M., Murphy, P., Hill, R.E., & Davidson, D.R. (1991). Effects of retinoic acid excess on expression of Hox-2.9 and Krox-20 and on morphological segmentation in the hindbrain of mouse embryos. EMBO J. 10, 2985-2995.

Niazi I, A., & Saxena, S. (1978) Abnormal hind limb regeneration in tadpoles of the toad, *Bufo andersoni,* exposed to excess vitamin A. Folia Biol. (Krakow) 26, 3-8.

Niazi, I.A., & Saxena, S. (1979). Relationship beween inhibiting influence of vitamin A and developmental stage of regenerating tail in toad tadpoles (*Bufo andersonii*). Ind. J. Exp. Biol. 17, 866-868.

Niazi, I.A., Pescitelli, M.J., & Stocum, D.L. (1985). Stage dependent effects of retinoic acid on regenerating urodele limbs. Roux's Arch. Dev. Biol. 194, 355-363.

Nohno, T., Muto, K., Noji, S., Saito, T., & Taniguchi, S. (1991). Isoforms of retinoic acid receptor β expressed in the chicken embryo. Biochem. Biophys. Acta. 1089, 273-275.

Noji, S., Nohno, T., Koyama, E., Muto, K., Ohyama, K., Aoki, Y., Tamura, K., Ohsugi, K., Ide, H., Taniguchi, S., & Saito, T. (1991). Retinoic acid induces polarising activity but is unlikely to be a morphogen in the chick limb bud. Nature 350, 83-87.

Osmond, M.K., Butler, A.J., Voon, F.C.T., & Bellairs, R. (1991). The effects of retinoic acid on heart formation in the early chick embryo. Development 113, 1405-1417.

Papageorgiou, S., & Almirantis, Y. (1992). Diffusion or autocatalysis of retinoic acid cannot explain pattern formation in the chick wing bud. Dev. Dynam. 194, 282-288.

Papalopulu, N., Clarke, J.D.W., Bradley, L., Wilkinson, D., Krumlauf, R., & Holder, N. (1991a). Retinoic acid causes abnormal development and segmental patterning of the anterior hindbrain in *Xenopus* embryos. Development 113, 1145-1158.

Papalopulu, N., Lovell-Badge, R., & Krumlauf, R. (1991b). The expression of murine Hox-2 genes is dependent on the differentiation pathway and displays a colinear sensitivity to retinoic acid in F9 cells and *Xenopus* embryos. Nucl. Acids Res. 19, 5497-5506.

Pecorino, L.T., Lo, D.C., & Brockes, J.P. (1994). Isoform-specific induction of a retinoid-responsive antigen after biolistic transfection of chimaeric retinoic acid/thyroid hormone receptors into a regenerating limb. Development 120, 325-333.

Pecorino, L.T., Entwistle, A., & Brockes, J.P. (1996). Activation of a single retinoic acid receptor isoform mediates proximodistal respecification. Current Biol 6, 563-569.

Pfeffer, P.L., & De Robertis, E.M. (1994). Regional specificity of RARg isoforms in *Xenopus* development. Mech. Dev. 45, 147-153.

Pietsch, P. (1987). The effects of retinoic acid on mitosis during tail and limb regeneration in the axolotl larva, *Ambystoma mexicanum*. J. exp. Zool. 196, 169-175.

Pijnappel, W.W.M., Hendriks, H.F.J., Folkers, G.E., van der Brink, C.E., Dekker, E.J., Edelenbosch, C., van der Saag, P.T., & Durston, A.J. (1993). The retinoid ligand 4-oxo-retinoic acid is a highly active modulator of positional specification. Nature (London) 366, 340-344.

Ragsdale, C.W., Petkovich, M., Gates, P.B., Chambon, P., & Brockes, J.P. (1989). Identification of a novel retinoic acid receptor in regenerating tissues of the newt. Nature 341, 654-657.

Ragsdale, C.W., Gates, P.B., & Brockes, J.P. (1992a). Identification and expression pattern of a second isoform of the newt alpha retinoic acid receptor. Nucleic Acids Res. 20, 5851.

Ragsdale, C.W., Gates, P.B., Hill, D.S., & Brockes, J.P. (1992b). Delta retinoic acid receptor isoform d1 is distinguished by its exceptional N-terminal sequence and abundance in the limb regeneration blastema. Mech. of Dev. 40, 99-112.

Riddle, R.D., Johnson, R.L, Laufer, E., & Tabin, C. (1993). Sonic hedgehog mediates the polarizing activity of the ZPA. Cell 75, 1401-1416.

Rossant, J., Zirngibl, R., Cado, D., Shago, M., & Giguere, V. (1991). Expression of a retinoic acid response element-hsplacZ transgene defines specific domains of transcriptional activity during mouse embryogenesis. Genes & Dev. 5, 1333-1344.

Rowe, E., Eager, N.S.C., & Brickell, P.M. (1991a). A member of the RXR nuclear receptor family is expressed in neural crest-derived cells of the developing chick peripheral nervous system. Development 111, 771-778.

Rowe, A., Richman, J.M., & Brickell, P.M. (1991b). Retinoic acid treatment alters the distribution of retinoic acid receptor-β transcripts in the embryonic chick face. Development 111, 1007-1016.

Rowe, A., Richman, J.M., & Brickell, P.M. (1992). Development of the spatial pattern of retinoic acid receptor-β transcripts in embryonic chick facial primordia. Development 114, 805-813.

Ruberte, E., Friederich, V., Morriss-Kay, G., & Chambon, P. (1992). Differential distribution patterns of CRABP I and CRABP II transcripts during mouse embryogenesis. Development 115, 973-978.

Ruiz i Altaba, A., & Jessell, T. (1991a). Retinoic acid modifies mesodermal patterning in early *Xenopus* embryos. Genes Dev. 5, 175-187.

Ruiz i Altaba, A., & Jessell, T. (1991b). Retinoic acid modifies the pattern of cell differentiation in the central nervous system of neurula stage *Xenopus* embryos. Development, 112, 945-958.

Ruiz i Altaba, A., & Jessell, T.M. (1992). *Pintallavis*, a gene expressed in the organizer and midline cells of frog embryos: involvement in the development of the neural axis. Development, 116, 81-93.

Saunders, J.W., & Gasseling, M.T. (1968). Ectodermal-mesenchymal interactions in the origin of limb symmetry. In *Epithelial-Mesenchymal Interactions* (R.Fleischmajer & R.E.Billingham, Eds.), pp. 78-97, Williams & Wilkins Co., Baltimore.

Seleiro, E.A.P., Rowe, A., & Brickell, P.M. (1995). The chicken retinoic-X-receptor-α (RXR-α) gene and its expression in the developing limb. Roux's Arch. Dev. Biol. 204, 244-249.

Scadding, S.R. (1987). Vitamin A inhibits amphibian tail regenerates. Canad. J. Zool. 65, 457-459.

Scadding, S.R. (1990). Histological effects of vitamin A on limb regeneration in the larval axolotl, *Ambystoma mexicanum*. Canad. J. Zool. 68, 159-167.

Scadding, S.R., & Maden, M. (1986). Comparison of the effects of vitamin A on limb development and regeneration in *Xenopus laevis* tadpoles. J. Embryol. Exp. Morph. 91, 35-53.

Scadding, S.R., & Maden, M. (1994). Retinoic acid gradients in limb regeneration. Dev.Biol. 162, 608-617.

Schilthius, J.G., Gann, A.A.F., & Brockes, J.P. (1993). Chimeric retinoic acid/thyroid hormone receptors implicate RAR-a1 as mediating growth inhibition by retinoic acid. EMBO J. 12, 3459-3466.

Schofield, J.N., Rowe, A., & Brickell, P.M. (1992). Position-dependence of retinoic acid receptor-β gene expression in the chick limb bud. Dev. Biol. 152, 344-353.

Scott, W.J., Walter, R., Tzimas, G., Sass, J.O., Nau, H., & Collins, M.D. (1994). Endogenous status of retinoids and their cytosolic binding proteins in limb buds of chick vs. mouse embryos. Dev. Biol. 165, 379-409.

Sengel, P. (1976). Morphogenesis of skin. In *Developmetal and Cell Biology* series (Abercrombie, M., Newth, D.R., Torrey, J.G., Eds.), p. 1. Cambridge Univ. Press, New York.

Sharpe, C.R. (1991). Retinoic acid can mimic endogenous signals involved in transformation of the *Xenopus* nervous system. Neuron 7, 239-247.

Simeone, A., Avantaggiato, V., Moroni, M.C., Mavilio, F., Arra, C., Cotelli, F., Nigro, V., & Acampora, D. (1995). Retinoic acid induces stage-specific antero-posterior transformation of rostral central nervous system. Mech. of Dev. 51, 83-98.

Sive, H.L., Draper, B.W., Harland, R.M., & Weintraub, H. (1990). Identification of a retinoic acid-sensitive period during primary axis formation in *Xenopus laevis*. Genes Dev. 4, 932-942.

Sive, H.L., & Cheng, P.F. (1991). Retinoic acid perturbs the expression of Xhox.lab genes and alters mesodermal determination in *Xenopus laevis*. Genes Dev. 5, 1321-1332.

Smith, D.P., Mason, C.S., Jones, E., & Old, R. (1994). Expression of a dominant negative retinoic acid receptor γ in *Xenopus* embryos leads to partial resistance to retinoic acid. Roux's Arch. Dev. Biol. 203, 254-265.

Smith, J.C. (1979). Evidence for a positional memory in the development of the chick wing bud. J. Embryol. Exp. Morph. 52, 105-113.

Smith, S.M., & Eichele, G. (1991). Temporal and regional differences in the expression pattern of distinct retinoic acid receptor-β transcripts in the chick embryo. Development 111, 245-252.

Smith, S.M., Pang, K., Sundin, O., Wedden, S.E., Thaller, C., & Eichele, G. (1989). Molecular approaches to vertebrate limb morphogenesis. Development (suppl) 121-131.

Smith, S.M., Kirsten, I.J., Wang, Z-S., Fallon, J.F., Kelley, J., & Bradshaw-Rouse, J. (1995). Differential expression of retinoic acid receptor-β isoforms during chick limb ontogeny. Dev. Dynam. 202, 54-66.

Smith-Thomas, L. Lott, I., & Bronner-Fraser, M. (1987). Effects of isotretinoin on the behavior of neural crest cells *in vitro*. Dev. Biol. 123, 276-281.

Stainier, D.Y.R., & Fishman, M.C. (1992). Patterning the zebrafish heart tube: acquisition of anteroposterior polarity. Dev. Biol. 153, 91-101.

Stocum, D.L., & Thoms, S.D. (1984). Retinoic acid induced pattern completion in regenerating double anterior limbs of urodeles. J. Exp. Zool. 232, 207-215.

Stratford, T., Horton, C., & Maden, M. (1996). Retinoic acid regulates sonic hedgehog expression and initiates outgrowth in the chick limb bud. Current Biol. 6, 1124-1133.

Summerbell, D. (1983) The effect of local application of retinoic acid to the anterior margin of the developing chick limb. J. Embryol. Exp. Morph. 78, 269-298.

Summerbell, D., & Honig, L. S. (1982). The control of pattern across the antero-posterior axis of the chick limb bud by a unique signalling region. Am. Zool.,22, 105-116.

Summerbell, D., & Harvey, F. (1983). Vitamin A and the control of pattern in developing limbs. In *Limb Development and Regeneration. Part A*, pp. 109-118, Alan R. Liss, New York.

Sundin, O., & Eichele, G. (1992). An early marker of axial pattern in the chick embryo and its respecification by retinoic acid. Development 114, 841-852.

Taira, M., Otani, H., Jamrich, M., & Dawid, I.B. (1994). Expression of the LIM class homeobox gene Xlim-1 in pronephros and CNS cell lineages of *Xenopus* embryos is affected by retinoic acid and exogastrulation. Development, 120, 1525-1536.

Tamarin, A., Crawley, A., Lee J., & Tickle C. (1984). Analysis of upper beak defects in chicken embryos following with retinoic acid. J. Embryol. exp. Morph. 84, 105-123.

Tamura, K., Kagechika, H., Hashimoto, Y., Shudo, K., Oshugi, K., & Ide, H. (1990). Synthetic retinoids, retinobenzoic acids, Am80, Am580 and Ch55 regulate morphogenesis in chick limb bud. Cell Diff., & Dev. 32, 17-26.

Tassava, R.A. (1992). Retinoic acid enhances monoclonal antibody WE3 reactivity in the regenerate epithelium of the adult newt. J. Morphol. 213, 159-169.

Thaller, C., & Eichele, G. (1987). Identification and spatial distribution of retinoids in the developing chick limb bud. Nature 327, 625-628.

Thaller, C., & Eichele, G. (1990). Isolation of 3,4-didehydroretinoic acid, a novel morphogenetic signal in the chick wing bud. Nature (London) 345, 815-819.

Thaller, C., Hofmann, C., & Eichele, G. (1993). 9-*cis*-retinoic acid, a potent inducer of digit pattern duplications in the chick wing bud. Development 118, 957-965.

Thompson, J.N., Howell, J.M., Pitt, G.A.J., & McLaughlin, C.I. (1969). The biological activity of retinoic acid in the domestic fowl and the effects of vitamin A deficiency on the chick embryo. Br. J. Nutr. 23, 471-490.

Thoms, S.D., & Stocum, D.L. (1984). Retinoic acid-induced pattern duplication in regenerating urodele limbs. Dev. Biol. 103, 319-328.

Thorogood, P., Smith, L., Nicol, A., McGinty, R., & Garrod, D. (1982). Effects of vitamin A on the behaviour of migratory neural crest cells in vitro. J. Cell Sci. 57, 331-350.

Tickle, C., Crawley, A., & Farrar, J. (1989). Retinoic acid application to chick wing buds leads to a dose-dependent reorganization of the apical ectodermal ridge that is mediated by the mesenchyme. Development 106, 691-705.

Tickle, C., Lee, J., & Eichele, G. (1985). A quantitative analysis of the effects of all-*trans*-retinoic acid on the pattern of chick wing development. Dev. Biol. 109, 82-95.

Tickle, C., Alberts, B., Wolpert, L., & Lee, J. (1982). Local application of retinoic acid to the limb bud mimics the action of the polarizing region. Nature 296, 564-565.

Tsonis, P.A., Washabaugh, C.H., & Rio-Tsonis, K.D. (1994). Morphogenetic effects of 9-*cis*-retinoic acid on the regenerating limbs of the axolotl. Roux's Arch. Dev. Biol. 203, 230-234.

Vaessen, M-J., Meijers, J.H.C., Bootsma,. D., & van Kessel, A.G. (1990). The cellular retinoic acid-binding protein is expressed in tissues associated with retinoic acid-induced malformations. Development 110, 371-378.

Vallari, R.C., & Pietruszko, R. (1982). Human aldehyde dehydrogenase: mechanism of inhibition by disulphiram. Science 216, 637-639.

Viviano, C.M., Horton, C., Maden, M., & Brockes, J.P. (1995). Synthesis and release of 9-*cis* retinoic acid by the urodele wound epidermis. Development 121, 3753-3762.

Wagner, M., Thaller, C., Jessell, T., & Eichele, G. (1990). Polarizing activity and retinoid synthesis in the floor plate of the neural tube. Nature 345, 819-822.

Wagner, M., Han, B., & Jessell, T.M. (1992). Regional differences in retinoid release from embryonic neural tissue detected by an *in vitro* reporter assay. Development 116, 55-66.

Wanek, J., Gardiner, D.M., Muneoka, K., & Bryant, S.V. (1991). Conversion by retinoic acid of anterior cells into ZPA cells in the chick wing bud. Nature 350, 81-83.

Wedden, S.E. (1987). Epithelial-mesenchymal interactions in the development of chick facial primordia and the target of retinoid action. Development 99, 341-351.

Wedden, S.E., Lewin-Smith, M.R., & Tickle, C. (1987). The effects of retinoids on cartilage differentiation in micromass cultures of chick facial primordia and the relationship to a specific facial defect. Dev. Biol. 122, 78-89.

White, J.A., Boffa, M.B., Jones, B., & Petkovich, M. (1994). A zebrafish retinoic acid receptor expressed in the regenerating caudal fin. Development 120, 1861-1872.

Wigmore, P. (1990). Serially duplicated regenerates from the anterior half of the axolotl limb after retinoic acid treatment. Roux's Arch. Dev. Biol. 198, 252-256.

Wilde, S.M., Wedden, S.E., & Tickle, C. (1987). Retinoids reprogramme pre-bid mesenchyme to give changes in limb pattern. Development 100, 723-733.

Yaoita, Y., & Brown, D.D. (1990). A correlation of thyroid hormone receptor gene expression with amphibian metamorphosis. Genes & Dev. 4, 1917-1924.

Yutzey, K.E., Rhee, J.T., & Bader, D. (1994). Expression of the atrial-specific myosin heavy chain AMHC1 and the establishment of anteroposterior polarity in the developing heart. Development 120, 871-883.

Yutzey, K.E., Gannon, M., & Bader, D. (1995). Diversification of cardiomyogebnic cell lineages *in vitro*. Dev. Biol. 170, 531-541.

Zhang, Z., Balmer, J.E., Lovlie, A., Fromm, S.H., & Blomhoff, R. (1996). Specific teratogenic effects of different retinoic acid isomers and analogs in the developing anterior central nervous system of zebrafish. Dev. Dynam. 206, 73-86.

RETINOID AND GROWTH FACTOR SIGNAL TRANSDUCTION

G.V. Sherbet and M.S. Lakshmi

	Abstract	142
I.	Introduction	142
II.	The Role of Integrins in Signal Transduction	143
	A. Retinoids and Expression of Integrins	143
	B. Mechanics of Integrin-Mediated Signal Transduction	144
	C. Phosphorylation of Cellular Substrates/Focal Adhesion Kinases	146
III.	Cytoskeletal Involvement in Signal Transduction	146
IV.	Growth Factor and Retinoid Signal Transduction	148
	A. *src* Homology (SH) Proteins	148
	B. Modulation of Growth Factor Effects by Retinoids	149
	C. Retinoic Acid-Mediated Regulation of Transcription Factors	149
	D. Ca^{2+} Signaling Pathway	150
	E. Heat Shock Proteins in Retinoid Signal Transduction	151
	Acknowledgment	152
	References	153

Advances in Organ Biology
Volume 3, pages 141-159.

ISBN: 0-7623-0285-2

ABSTRACT

Retinoids play a significant role in cell growth and differentiation and suppress cancer invasion and metastasis. Retinoids are also able to modulate cellular responses to growth factors. In this review we discuss the pathways of transduction of signals imparted to the cell by retinoids and growth factors. Three pathways of flow of information are identified, defined, and discussed. The receptors mediated pathway of information flow is briefly discussed with reference to the regulation of transcription factors by retinoic acids.

Information transfer also occurs via the agency of integrin receptors. The intracellular down-stream events involve phosphorylation of substrate proteins such as the focal adhesion kinases. The cytoskeleton is involved in the flow of information that results from retinoic acid or growth factor exposure of cells. There are clear indications of disruption of the tubulin and actin cytoskeletal elements, which appears to lead to the progression of the cell cycle into the S-phase.

The transduction of growth factor binding signals involves a number of intracellular nonreceptor kinases that act as signaling proteins. These contain src homology domains. These src homology domain proteins appear to be involved also in interactions with the cytoskeletal assembly.

Growth factors and hormones alter levels of intracellular calcium, which controls many cellular functions. The Ca^{2+} signaling pathway may also be involved in genetic responses to retinoic acid, either directly or by modulating the expression of genes that code for calcium binding proteins such as the 18A2/mts1 protein. Other genetic targets can also be identified. There is much circumstantial evidence that the expression of heat shock proteins may be regulated by retinoic acid. Heat shock proteins might conceivably play an important part in recruiting essential components to the cytoskeleton for intracellular translocation or simply in altering the cytoskeletal dynamics, which has a significant bearing on cell shape, motility and invasion, and signal transduction.

I. INTRODUCTION

Retnioids play a significant role in epithelial cell growth and cell differentiation (Roberts and Sporn, 1984). Retinoic acid (RA) has been shown to possess the ability to suppress tumor invasion and growth (Edward et al., 1992, and this volume; Hendrix et al., 1990; Situ et al., 1993; Helige et al., 1993), and plays a major part in embryonic development, differentiation, and pattern formation (this volume). Several physical attributes of the cell, such as shape and adhesiveness, are also modified by RA. Therefore, it would be reasonable to assume that these phenotypic changes are a consequence of a wide spectrum of alterations in gene expression. The pathways by which the signal imparted to the cell by retinoids is transduced into the cell have been a major area of research in the past few years. Three pathways of flow of information can be identified: (1) via specific receptors that recognize specific DNA sequences and regulate the transcription of target genes. Retinoic acid receptors belong to a family of transcription regulators that includes among others steroid hormone receptors and vitamin D3; (2) a second pathway may be mediated by integrin receptors on the cell surface together with cytoskeletal involvement, phosphorylation of cellular substrate proteins, and intracellular Ca^{2+} modulation; and (3) via interaction and modulation of peptide growth factor receptors.

II. THE ROLE OF INTEGRINS IN SIGNAL TRANSDUCTION

A. Retinoids and Expression of Integrins

A major effect of retinoids is the induction of differentiation and the modulation of the effects of growth factors. RA induces the differentiation of mouse teratocarcinoma cells, melanoma cells, and human promyelocytic leukemia and neuroblastoma cells (Strickland et al., 1980; Breitman et al., 1980; Haussler et al., 1983). This is generally attributed to an enhanced transcription of specific genes.

Several studies have shown, in the context of RA-mediated suppression of cancer invasion and metastasis, that RA increases the expression of several genes that code for the components of the extracellular matrix (ECM), such as fibronectin, laminin, and collagens III and IV (Wang et al., 1985; Horton et al., 1987; Chiocca et al., 1988; LaRosa and Gudas, 1988a; 1988b). RA inhibits fibronectin release induced by the tumor promoter 12-*O*-tetradecanolyphorbol-13-acetate (TPA) (Scita and Wolf, 1994) and diacylglycerols (DAG), which are its intracellular ligands (Zerlauth and Wolf, 1984; Scheidl et al., 1991). In contrast, a reduced adhesiveness after retinoic acid treatment has been demonstrated in several systems (e g., Loeser, 1994) and this may be achieved by alterations in the expression of the integrin receptors of these extracellular ligands. During RA-induced differentiation of F9 embryonal carcinoma cells, changes are noticed in the pattern of expression of integrin receptors that mediate cellular adhesion to fibronectin and laminin (Burdsal et al., 1994; Elias et al., 1994). F9 cells, which adhere avidly to laminin, prominently express $\alpha_6\beta_1$ integrin. But when induced to differentiate into parietal endoderm by RA or dibutyryl cyclic adenosine monophosphate (N-6, 2′-O-dibutyryl adenosine 3′:5′-cyclic monophosphate (dBcAMP), these cells show a reduced expression of this integrin together with reduced adhesion to laminin (Elias et al., 1994). Hamster tracheal epithelial cells exposed to RA have been reported to show reduced adhesion to laminin rather than corresponding control cells (Lancillotti et al., 1992). But according to Ross and colleagues (Ross et al., 1994), F9 cells respond to both RA and dBcAMP with enhanced laminin synthesis and enhanced cell adhesiveness. Ross and coworkers have also stated that RA decreased the amount of laminin binding protein-37, but it increased the amount of β_1 integrin. Burdsal and colleagues found a 10-25-fold increase in $\alpha_3\beta_1$ integrin but without a corresponding increase in the levels of mRNA transcript, whereas the $\alpha_5\beta_1$ protein decreased several fold together with a decrease in $\alpha_5\beta_1$ gene transcripts. This suggests that

RA can alter the expression of these ECM proteins either at the post-transcriptional level or by down regulation of transcription. Alterations in integrin expression may also occur in neuroblastoma cells during neurite outgrowth induced by dBcAMP. There is a decrease in the binding of collagen type IV, but not type I, to $\alpha_1\beta_1$ integrin receptors and there is a general increase in the turnover of β_1 integrins (Carmeliet et al., 1994). The modulation of T-cell maturation by RA is known to be accompanied by reduced fibronectin expression (Meco et al., 1994). These biological effects are mediated by RA specific receptors and indeed appear to follow the induction of specific receptors by RA treatment (Meco et al., 1994; Ross et al., 1994).

Among other extracellular matrix components, thrombospondins (TSP) have also been reported to be regulated by retinoids. TSPs are multidomain ECM proteins containing calcium binding domains and EGF receptor-like repeats. Their expression is developmentally regulated (Qabar et al., 1994). TSPs induce angiogenesis (Nicosia and Tuszynski, 1994; Benezra et al., 1993). Liska and colleagues (Liska et al., 1994) showed that synthesis of TSPs is increased together with increases in the respective mRNA species, during the differentiation of P19EC cells into neurons, glial cells, and fibroblasts. RA differentially regulates the expression of a variety of adhesion molecules on the surface of acute myeloid leukaemia blast cells (Dinoto et al., 1994). The induction by RA of differentiation of F19 embryonal carcinoma cells along the neural pathway has been found to be associated with the activation of the gene coding for vascular cell adhesion molecule-1 (VCAM-1) (Sheppard et al., 1995).

RA induces maturation of U937 cells into monocytic phenotypes. The induction of differentiation is associated with a total suppression of interferon-gamma- (IFN-gamma)-stimulated expression of intercellular adhesion molecule 1 (ICAM-1) and RA also can inhibit ICAM-1 stimulation by IFN-gamma in isolated human blood monocytes (Weber et al., 1995). But in 4 cell lines from human melanoma, glioma, teratocarcinoma, and a neuroblastaoma, Bouillon and colleagues (Bouillon et al., 1994) earlier reported an enhancement of ICAM-1 expression in response to RA treatment. In neuroblastomas, exposure of cells to differentiation-inducing agents such as RA, 5′-bromodeoxyuridine, phorbol ester and cytokines (tau-interferon, α-tumour necrosis factor) caused a variable up-regulation of expresssion of a range of cell adhesion molecules (with the exception of N-CAM) (Gross et al., 1992).

Retinoids also regulate the transcription of genes coding for enzymes, such as stromelysin, which affect the turnover of ECM proteins (Nicholson et al., 1990). Indeed, they not only modulate the production of metalloproteinases but also the production of their inhibitors (Braunhut and Moses, 1994).

B. Mechanics of Integrin-Mediated Signal Transduction

Integrins are heterodimeric transmembrane proteins, composed of two noncovalently linked subunits α and β. At least 14 distinct α and 8 or more β subunits have been characterized. A combination of these occurs in the large number of integrin receptors that have been characterized to date. The transmembrane portion provides the link between the extracellular domain and the intracellular domain. The extracellular domains of the two subunits associate to function as the receptor and it would appear that specificity of ligand binding is determined by the type of α/β combination. The rather shorter intracellular portion of the macromolecule is linked to cytoskeletal elements (Ruoslahti, 1991).

Interaction of the cytoplasmic portion of the integrin molecule with cytoskeletal proteins is an essential ingredient of signal transduction along this pathway. Interaction occurs by means of cytoskeletal elements such as actin, filamin, tubulin, and so forth in the processes of cell adhesion and migration. Leukocyte extravasation, for example, is mediated by the binding of β_2 integrins to filamin (ABP-280) (Sharma et al., 1995). The protein links between microfilaments and the membrane may be mediated by actin-binding α-actinin and vinculin (Wachsstock et al., 1987). Vinculin binds with talin (Burridge and Mangeat, 1984) and talin with the integrin (Horwitz et al., 1986). A high affinity binding between α-actinin and the cytoplasmic domain of integrin has also been demonstrated (Otey et al., 1990). Interaction of integrins with cytoskeletal elements occurs in migrating fibroblasts and also neuronal growth and migration (Schmidt et al., 1993, 1995). Integrins containing mutations at the sites of binding with the cytoskeletal elements show significant changes in directed migration of fibroblasts, which indicates the importance of this interaction (Schmidt et al., 1993; 1994). An impairment of binding between the cytoplasmic portion of integrins and the cytoskeleton may lead to the loss of tissue integrity in cancers (Glukhova et al., 1995). Peter and Otoole (1995) have elegantly demonstrated the interaction between the cytoplasmic tail of $\alpha_L\beta_2$ integrin in the activation-dependent adhesion of leukocytes. They constructed a cDNA coding for a chimeric protein composed of the intracellular portion of $\alpha_L\beta_2$ and the extracellular (ligand-binding) domain of $\alpha_{IIb}\beta_3$. Upon transfection of this cDNA stably into Chinese hamster ovary cells, the chimeric receptors were able to mediate adhesion to fibrinogen. This could be enhanced by phorbol myristate acetate but abolished by cytochalasin D. It is obvious therefore that in this model the cytoskeleton is actively involved in the adhesion process mediated by this integrin. Deletion and/or mutations in the cytoplasmic tail also abolished adhesion, together with impaired spreading of the transfected cells and decrease in the ability of these cells to organize the cytoskeleton into stress fibers (Peter and Otoole, 1995). The cytoskeletal dynamics might be altered by affecting phosphorylation of its components. For instance, when neural adhesion molecules are triggered by ligand binding, the phosphorylation of tyrosyl residues of α and β tubulins associated with the cell membrane is inhibited (Atashi et al., 1992). Microtubule polymerization may be regulated by cell adhesion to the ECM by a process of autoregulation of tubulin monomers (Mooney et al., 1994). These observations are consistent with the association of the tubulin cytoskeleton with the regulation of cell shape and adhesion. But the tubulin cytoskeleton may mediate transport of components required for the maintenance of cell shape rather than function as rigid structural elements (Rodionov et al., 1993).

C. Phosphorylation of Cellular Substrates/Focal Adhesion Kinases

The initial consequence of membrane receptor stimulation is the generation of second messengers or biochemical signals. Phosphorylation of certain intracellular proteins is an important event in the signal transduction cascade. This can occur via three different pathways. The receptors for neurotransmitter and certain hormones are coupled via guanine nucleotide proteins (G proteins) to adenylate cyclase, which causes cyclic AMP production and consequent activation of (cAMP)-dependent protein kinases. Peptide growth factors such as the epidermal growth factor (EGF) and the platelet derived growth factor (PDGF) induce tyrosine phosphorylation of phospholipase C, which generates two second messengers: diacylglycerol (DAG) and inositol trisphosphate (IP3). DAG activates protein kinase C whereas IP3 stimulates the release of intracellular Ca^{2+}; this in turn results in the activation of several different Ca^{2+}/calmodulin-dependent kinases.

The binding of integrin receptors are now known to phosphorylate at least two molecular components; one of them is an 80kDa protein known as paxillin (Burridge et al., 1992). Paxillin interacts with an adaptor molecule containing *src* homology (SH) domains SH2-SH3, a route of transduction of some growth factor signals (Hempstead et al., 1994). There were early indications that signal transduction by integrins may involve substrate protein phosphorylation. Antibody-mediated clustering of β_1 integrins produced increased tyrosine phosphorylation of a cellular 110-130kDa protein (Kornberg et al., 1991). A second protein phosphorylated by cellular adhesion process is the 125kDa protein (Guan et al., 1991; Kornberg et al., 1991). This is itself a protein kinase (Shaller et al., 1992; Hanks et al., 1992) called focal adhesion kinase (FAK). Tyrosine phosphorylation of FAK is induced when cells are allowed to adhere to fibronectin, type IV collagen, or laminin (Kornberg et al., 1992). Platelet aggregation brings in its wake a series of phosphorylations. Fibrinogen binding to platelet-specific integrin $\alpha_{IIb}\beta_3$ results in the phosphorylation of FAK (Farrell and Martin, 1989; Golden et al., 1991; Lipfert et al., 1992), but FAK is not phosphorylated in activated platelets from patients with Glanzmann's thrombosthenia (Golden et al., 1991; Lipfert et al., 1992), which lack $\alpha_{IIb}\beta_3$ integrins. This further demonstrates a direct link between transduction of extracellular signals via integrin receptors and phosphorylation of FAK.

III. CYTOSKELETAL INVOLVEMENT IN SIGNAL TRANSDUCTION

Extraneous agents such as retinoids and growth factors have long been known to modulate several cellular attributes. Retinoids suppress the invasive ability of cancer cells (Fazely and Ledinko, 1990). Some peptide growth factors and cell surface glycosides have been shown to modulate cell adhesion and its migratory behavior. Effects of retinoids on the polymerization of cytoskeletal elements have been demonstrated by Lakshmi and colleagues (Lakshmi et al., 1993). There is much circumstantial evidence that a cytoskeletal mediation is essential for growth factor signal transduction. For instance, the *18A2/mts1* gene whose product promotes depolymerization of microtubules is constitutively highly expressed in cells that have high constitutive levels of growth factor receptor expression (Sherbet et al., 1995). Growth factors activate a serine/threonine-specific protein kinase called the mitogen-activated protein kinase (MAPK). This kinase phosphorylates the microtubule associated protein 2 (MAP2). It has been shown that this phosphorylation reduces the ability of MAP2 to induce tubulin polymerization (Hoshi et al., 1992). In cell systems, such as gliomas in culture, the modulation of invasive properties by transforming growth factor α and surface gangliosides is also accompanied by modulation in the expression of laminin (Merzak et al., 1994a; Koochekpour et al., 1995).

Ligand-induced activation of high affinity EGF receptors are known to associate with cytoskeletal elements (Roy et al., 1989; 1991; Van Bergen en Henegouwen et al., 1989) suggesting the mediation of cytoskeletal elements in EGF signal transduction. Since activation of the receptors is followed by their dimerization and subsequent autophosphorylation, it follows that genetic activity that promotes depolymerization of the cytoskeletal elements would promote the process of transduction of the signal. Furthermore, there is evidence that cytoskeletal depolymerization itself could function as a signal for transition of cells from the G_0/G_1 to the S-phase of the cell cycle (Crossin and Carney, 1981a; 1981b; Tsuji et al., 1992).

There is an obvious overlap in the pathways of signal transduction by EGF receptors and integrin receptors. It would appear that RA may influence the flow of information from ligand binding to integrins or growth factor receptors. Several hormones and peptide growth factors are known to influence focal adhesions. Hormones that raise intracellular levels of cAMP often cause disassembly of stress fibers and focal adhesions, together with changes in morphology and adhesion to substratum. Similar effects are observed when cells are stimulated with mitogenic growth factors such as EGF and PDGF. Tumor promoters such as TPA also produce disassembly of stress fibres. Retinoic acid has the opposite effect and appears to prevent cytoskeletal depolymerization (Lakshmi et al., 1993). Furthermore, HL-60 cells induced into terminal differentiation by RA have been shown to contain larger amounts of microtubules (Leung et al., 1992). Leung and Sartorelli (1992) have further reported that colchicine and vincristine, which disrupt microtubles, delay the differentiation of HL-60 in response to dimethylsulphoxide (DMSO). Although colchine somewhat inexplicably enhanced RA-induced differentiation if the cells were exposed to colchicine before RA treatment, colchicine had no effect on cell differentiation if given after the cells were exposed to RA. These studies suggest that microtubules are involved in the transduction of the differentiation signal imparted by RA.

IV. GROWTH FACTOR AND RETINOID SIGNAL TRANSDUCTION

A. *src* Homology (SH) Proteins

The binding of growth factors such as EGF and PDGF to their specific membrane receptors results in the inactivation, dimerization, and autophosphorylation of the receptors. The activated receptor kinases phosphorylate intracelluar proteins and induction of DNA synthesis.

A group of nonreceptor intracellular kinases occur that have *src, fps,* and *abl* oncogenes as prototypes (Pawson , 1988). The *src* and related kinases bind to the activated peptide growth factor receptors (Kypta et al., 1990). These signaling proteins contain noncatalytic domains called *src* homology (SH) domains. The *src* family genes, c-*src*, c-*yes*, *fyn,* and *lyn* are expressed in a variety of cell types, whereas other members such as *lck, hck,* c-*fgr,* and *blk* are selectively expressed. Genetic analyses of *src* kinases have revealed a similarity in gene organization of the members of this family, with a myristylation and membrane binding sequence at the N-terminal end. Adjacent to this a unique sequence occurs. Next to this occur noncatalytic domains called the *src* homology domains SH2/SH3. These domains share homology with other nonreceptor protein kinases and certain other cellular proteins (Pawson, 1988; Stahl et al., 1988; McCormick, 1989; Mayer et al., 1988; Drubin et al., 1990).

The activation of EGF receptor and its subsequent autophosphorylation allows the binding of proteins containing the SH2 domain (Margolis et al., 1990). Other proteins such as Grb2 (growth factor receptor bound), which contain an SH2 and two SH3 domains, may function as adaptor molecules between the activated receptor and the ras nucleotide releasing factor, son of sevenless (Sos), with the SH2 domain interacting with the activated receptor and SH3 to Sos. The Sos protein was first identified in the signaling pathway of *Drosophila* eye development (Simon et al., 1991). Its mammalian homologue possesses the ability to activate ras by exchanging GTP for GDP (Gale et al., 1993; Chardin et al., 1993; Pawson and Schlessinger, 1993). The EGF receptor signal is then routed from ras through the raf/MAP kinase kinase/MAP kinase cascade (Blenis, 1993; Davis, 1993). Raf is associated with the cytoskeleton, thus ras activation leads to the recruitment of raf to the membrane and to an association with the cytoskeleton and its subsequent activation (Stokoe et al., 1994).

Another SH domain protein is Grb7 and it has been found to be coamplified with erbB2 tyrosine kinase receptor protein in human breast cancer cell lines as well as in tumor tissue (Stein et al., 1994). The proto-oncogene c-*erbB2 (neu)* codes for a growth factor receptor kinase. Elevated expression of c-*erbB2* has been noted in a significant proportion of primary breast cancer (Slamon et al., 1987; Gullick et al., 1991; Paterson et al., 1991). The pathway of c-*erbB2* signal transduction appears to involve an interaction of SH2 domains of src with tyrosine-phosphorylated erbB2 (Muthuswamy et al., 1994).

The activation of EGF and PDGF receptors leads to the phosphorylation of phospholipase C (PLC) and PLC-gamma is an SH2/SH3 containing protein. PLC-gamma might interact with cytoskeletal elements and SH3 domains may be responsible for targeting PLC-gamma to the cytoskeleton (Bar-Sagi et al., 1993). There are other SH3 containing proteins, for example, myosin Ib and an actin binding yeast protein (ABP1p), which have been implicated in the regulation of cytoskeletal assembly (Drubin et al., 1990). SH2 containing proteins such as tensin also form links between adhesion foci and actin filaments (Davis et al., 1991).

B. Modulation of Growth Factor Effects by Retinoids

Retinoids have been found to be powerful modulators of growth and differentiation. Retinoic acid induces the differentiation of a variety of cell lines such as human promyelocytic leukaemia cell line HL60 and F9 teratocarcinoma cells. In the B16 melanoma, progression of the cells in early G_1 phase of the cell cycle is blocked by retinoic acid and the cells lose the ability to grow in soft agar (Niles, 1987). Rosenbaum and Niles (1992) found increases in protein kinase C levels, related mainly to protein kinase Cα. There have been several reports that retinoic acid enhances EGF binding and it has been suggested that this is due to an enhancement of the EGF receptor (Jetten, 1980). Kawaguchi and colleagues (Kawaguchi et al., 1994) have recently reported that retinoic acid is able to potentiate the effects of EGF on the proliferation of rat glomerular epithelial cells and enhance EGF binding to the cells. In the rat phaeochromocytoma cell line PC12 also, retinoic acid is able to enhance the expression of the nerve growth factor receptor and the sensitivity of the cells to the nerve growth factor (Scheibe and Wagner, 1992). These authors have argued that the regulation of NGF receptor expression occurs very rapidly and without any concurrent protein synthesis and therefore it may be a direct effect mediated by retinoic acid receptors.

C. Retinoic Acid-Mediated Regulation of Transcription Factors

The biological effects of retinoids are achieved by the binding and activation of receptors belonging to the steroid/thyroid receptor superfamily. As discussed in another chapter of this volume, there are two classes of RA receptors (RARs), one class is composed of RAR α, β, and Γ (gamma) and the other of retinoid-X receptors (RXRs). These ligand-induced regulators initiate transcription from promoters of RA target genes by interacting with *cis*-acting DNA response elements. The RARs regulate gene transcription by activating the expression of genes containing retinoic acid response elements or by modulating the expression of transcription factors.

The transcription factor AP1 is able to control expression of genes required for the growth of cells. It is a homo- and heterodimer of *fos* and *jun* proteins. AP1 is involved in the activation of several genes such as those coding for collagenases, stromelysin, interlukin-2, and transforming growth factor β1, among others. C-*fos* and c-*jun* genes are highly responsive to extracellular stimuli and can be regulated either positively or negatively. Both *fos* and *jun* proteins are subject to control by phosphorylation and the kinases that achieve this are activated by growth factors; furthermore, glucocorticoid and RA receptors interact with c-*jun* and c-*fos* and thus affect AP-1 mediated transcription of target genes (Karin et al., 1993). Retinoic acid induces the expression c-*fos* and c-*jun* genes during induction of differentiation (Scheibe et al., 1991; De Groot et al., 1991). WEHI-3B D+ myelomonocytic leukemia cells transfected with c-*jun* show characteristics of differentiation when the extraneous c-*jun* is switched on. In these transfected cells RA enhances the induction of differentiation by c-*jun* (Li et al., 1994). RA treatment produces a strong enhancement of c-*jun* promoter activity that is presumably mediated by RARβ. And consistent with bi-directional regulation of AP1, RA may enhance expression of genes coding for ECM proteins such as fibronectin, laminin, and collagens III and IV, whereas expression of the stromelysin gene may be repressed (Nicholson et al., 1990).

The EGFr promoter is negatively regulated by retinoic acid and thyroid hormone (T3) (Hudson et al., 1990). The T3 receptor forms a heterodimer with RXR and this binds to a DNA domain containing a half-site T3 response element and SP1 binding site (Thompson et al., 1992). Functional antagonism exists between RA and AP1-directed gene transcription in serum stimuated cell proliferation. RA represses the induction of c-*fos* and c-*jun,* but not other immediate early transcription factors such as *jun*B, c-*myc*, and egr-1 upon serum stimulation (Simonson, 1994). Thus retinoic acid signal transduction may follow two different paths—one involving DNA-receptor interaction, and another involving growth factor-induced changes in gene expression mediated by the activation of the transcription factor AP1.

D. Ca^{2+} Signaling Pathway

Growth factors and hormones are known to increase the levels of intracellular Ca^{2+} (Ca^{2+}_i) of their target cells, either by promoting an influx of Ca^{2+} or by release from intracellular Ca^{2+} stores. Ca^{2+}_i controls many different cellular functions, mainly by mediating the phosphorylation of cellular effector proteins (Berridge and Irvine, 1984, 1989; Petersen and Wakui, 1990). The calcium signals are transduced to specific subcellular compartments by calcium binding proteins such as calmodulin and S-100 family proteins.

A major mechanism of activation of the Ca^{2+} signaling pathway is the generation of secondary messengers, for example, inositol 1,4,5-trisphosphate (IP3), in response to the extracellular stimuli. The transduction of retinoic acid signal leading to the differentiation of HL-60 cells appears to involve the regulation of IP3 gene expression. These cells show a progressive increase in IP3 receptors and increased Ca^{2+} mobilization upon retinoic acid treatment (Bradford et al., 1992). Similarly, retinoic acid treatment produced an increase in the basal level of Ca^{2+}_i in U937 cells during monocytic differentiation (Tanaka et al., 1992).

The Ca^2 signals may be coupled to Ca^{2+}-binding proteins to regulate gene expression. Bouillon and Audette (1994) have shown that expression of intercellular adhesion molecule-1 (ICAM-1) is stimulated in the human neuroblastoma cell line SK-N-SH and this process is mediated by calmodulin. They showed that calmodulin inhibitors, W7 and calmidazolium, inhibit RA-stimulated expression of ICAM-1. Retinoic acid is known to inhibit invasion and metastasis in both human and murine cell systems (Fazely and Ledinko, 1990; Edward and Mackie, 1990; Edward et al., 1989; Wood et al., 1990). Cell motility is Ca^{2+}-dependent. For instance, the motility of neutrophils on vitronectin substratum can be inhibited by inhibitors of calcineurin, a Calcium/calmodulin-dependent phosphatase (Hendey and Maxfield, 1993). It has been recently demonstrated that a target gene modulated by RA is the *18A2/mts1*, an S-100 family gene, coding for a Ca^{2+}-binding protein. RA down-regulates the expression of this gene in metastatic variants of the B16 murine melanoma. It would appear that RA regulation of this gene is Ca^{2+}/calmodulin mediated, since melanocyte stimulating hormone, which causes a rapid influx of Ca^{2+} into the cell, enhances expression of *18A2/mts1* (Parker et al., 1991). Furthermore, the calmodulin inhibitor W7 down-regulates it (Parker and Sherbet, 1992).

E. Heat Shock Proteins in Retinoid Signal Transduction

Heat shock proteins (hsp) are transiently heat-inducible or stress-inducible proteins and constitute a large family of proteins that subserve a variety of cellular functions. Some members of this family appear to play an important role in cell proliferation. The expression of hsp 70, for instance, is regulated at the G_1-S boundary of the cell cycle (Milarsky and Morimoto, 1986). Hsp 70 is also known to bind to the cell cycle regulatory phosphoprotein p53 (Sturzbecher et al., 1987; Finlay et al., 1988; Clarke et al., 1988). Furthermore, hsp 70 has been implicated in the process of tubulin polymerization (Gupta, 1990). They cross-link cellular proteins to the cytoskeletal network (Tsang, 1993). Some hsps are regulated during embryonic development, for example, hsp 70 and hsp 30 (Bienz, 1984a; 1984b; Krone and Heikkila, 1988).

There are indications that hsps might be involved in the transduction of retinoid signals. The differentiation of HL-60 leukaemia cells by retinoic acid is mediated by the down-regulation of expression of the serine proteinase myeloblastin (mbn). The induction of differentiation of NB4 promyelocytic leukaemia cells by retinoic acid has also been found to be accompanied by an initial suppression followed by an up-regulation of hsp 28 protein and these changes occur parallel to alterations in the expression of mbn and this has suggested the possibility that hsp 28 might be a target of mbn proteinase. Hsp 28 is indeed hydrolyzed by mbn. Furthermore, the latter does not hydrolyze other proteins and it would therefore seem that hsp 28 is the target protein of mbn (Spector et al., 1995). These observations also raise the possibility therefore that hsp 28 might be an important component of the transduction of differentiation signal imparted by retinoic acid.

There is some circumstantial evidence that regulation of other gene targets by RA might also involve hsps. The S-100 family gene *18A2/mts1* is another example. This gene is associated with cancer invasion, metastasis, and cell cycle proliferation (Merzak et al., 1994b; Parker et al. 1994a; 1994b). Expression of this gene is also conducive to the depolymerization of cytoskeletal elements (Lakshmi et al.,1993). In transfectant B16 melanoma cells carrying the exogenous *18A2/mts1*, switching on the exogenous gene enhances lung colonization of the cells, promotes cytoskeletal depolymerization and drives the cells into the S-phase of the cell cycle (Parker et al. 1994a; 1994b). It has been found that modulation of expression of *18A2/mts1* produces changes in hsp 72/73 proteins. For instance, its up-regulation in B16 melanoma cells by exposing them to melanocyte stimulating hormone produces an increase in hsp 72/73 expression, but treatment of variants, where *18A2/mts1* is constitutively highly expressed, with retinoic acid not only down-regulates gene expression but in parallel reduces hsp 72/73 expression. Furthermore, the B16 variant line BL6 exposed to heat shock and the heat-resistant variant of BL6 both show a down-regulation of *18A2/mts1* together with a variety of changes in the pattern of hsp expression (e.g., hsp28, see below) but without concomitant changes in levels of hsp transcripts (Cajone et al., 1994). This suggests that the hsp expression is regulated at the post-transcriptional level and this observation is compatible with the findings of Spector and colleagues (Spector et al., 1995). Among other changes in hsp expression occurring concomitantly with *18A2/mts1*, most prominent have been the alterations in hsp28. We have found that hsp28 expression increases markedly (2-3 fold) upon exposure of B16 melanoma cells to hyperthermia at 40° for 1 hour and also in the heat resistant B16 variant. The hsp28 expression is closely related to inhibition of cell proliferation and G_1-arrest (Spector et al., 1992; 1993). Therefore, the up-regulation of hsp28 and the down-regulation of *18A2/mts1* occurring in these cell lines in response to heat shock may be related.

Tsang (1993) has reported that hsp 70 is constitutively bound to actin microfilaments through hydrophobic bonds. Hsp 70 appears to be involved also with tubulin folding and dimerization (Paciucci, 1994). We postulate therefore that a possible mechanism by which the *18A2/mts1* protein affects cytoskeletal depolymerization may be mediated by their being recruited to the cytoskeletal assembly by hsps. Hsp 70 has a calmodulin binding site and is known to complex with calmodulin (Stevenson and Calderwood, 1990), and, in analogy with this, it is conceivable that hsps form complexes with 18A2/mts1, which is also a Ca^{2+}-binding protein.

ACKNOWLEDGMENT

The authors thank the North of England Cancer Research Campaign for financial support.

REFERENCES

Atashi, J.R., Klinz, S.G., Ingraham, C.A., Matten, W.T., Schachner, M., & Maness, P.F. (1992). Neural cell adhesion molecules modulate tyrosine phosphorylation of tubulin in nerve growth cone membranes. Neuron 8, 831-842.

Bar-Sagi, D., Rotin, D., Batzer, A., Mandiyan, V., & Schlessinger, J. (1993). SH3 domains direct cellular localization of signaling molecules. Cell 74, 83-91.

Benezra, D., Griffin, B.W., Maftzir, G., & Aharonov, O. (1993). Thrombospondin and in vivo angiogenesis induced by basic fibroblast growth factor or lipopolysaccharide. Invest. Opthalmol. Visual Sci. 34, 3601-3608.

Berridge, M.J., & Irvine, R.F. (1984). Inositol trisphosphate, a novel second messenger in cellular signal transduction. Nature 312, 315-321.

Berridge, M.J., & Irvine, R.F. (1989). Inositol phosphates and cell signalling. Nature, 341, 197-205.

Bienz, M. (1984a). *Xenopus hsp* 70 genes are constitutively expressed in injected oocytes. EMBO J. 3, 2477-2483.

Bienz, M. (1984b). Deveopemental control of the heat shock response in *Xenopus*. Proc. Natl. Acad. Sci. USA 81, 3138-3142.

Blenis, L. (1993). Signal transduction via the MAP kinase. Proceed at your own RSK. Proc. Natl. Acad. Sci. USA 90, 5889-5892.

Bouillon, M., & Audette, M. (1994). Retinoic acid-stimulated intercellular adhesion molecule-1 expression on SK-N-SH cells—Calcium/calmodulin-dependent pathway. Cancer Res. 54, 4144-4149.

Bouillon, M., Tessier, P., Boulianne, R., Destrempe R., & Audette, M. (1994). Regulation by retinoic acid of ICAM-1 expression on human tumour cell lines. Biochim. Biophys. Acta 1097, 95-102.

Bradford, P.G., Wang, X.H., Jin, Y.Y., & Hui, P. (1992). Transcriptional regulation and increased functional expression of the inositol trisphosphate receptor in retinoic acid-treated HL-60 cells. J. Biol. Chem. 267, 20959-20964.

Braunhut, S.J., & Moses, M.A. (1994). Retinoids modulate endothelial cell production of matrix-degrading proteiases and tissue inhibitors of metalloproteinases (TIMP). J. Biol. Chem. 269, 13472-13479.

Breitman, T.R., Selonick, S., & Collins, S.J. (1980). Induction of differentiation of the human promyelocytic leukaemia cell line HL-60 by retinoic acid. Proc. Natl. Acad. Sci. USA 77, 2936-2940.

Burdsal, C.A., Lotz, M.M., Miller, J., & McClay, D.R. (1994). Quantitative switch in integrin expression accompanies differentiation of F9 cells treated with retinoic acid. Develop. Dynamics 201, 344-353.

Burridge, K., & Mangeat, P. (1984). An interaction between vinculin and talin. Nature 308, 744-746.

Burridge, K., Turner C.E., & Romer, L.H. (1992). Tyrosine phosphorylation of paxillin land $pp125^{FAK}$ accompanies cell adhesion to extracellular matrix: A role in cytoskeletal assembly. J. Cell Biol. 119, 893-903.

Cajone, F., Debiasi, S., Parker, C., Lakshmi, M.S., & Sherbet, G.V. (1994). Metastasis-associated *mts1* gene expression is down-regulated by heat shock in variant cell lines of the B16 murine melanoma. Melanoma Res. 4, 143-150.

Carmeliet, G., Himpens, B., & Cassiman, J.J. (1994). Selective increase in the binding of alpha(1) beta(1)-integrin for collagen type-IV during neurite outgrowth of human neuroblastoma TR-14 cells. J. Cell Sci. 107, 3379-3392.

Chardin, P., Camonis, J.H., Gale, N.W. Van Aeist, L., Schlessinger J., Wigler, M.H., & Bar-Sagi, D. (1993). Human Sos1: a guanine nucleotide exchange factor for Ras that binds to GRB2. Science 260, 1338-1343.

Chiocca, E.A., Davies, P.J.A., & Stein, J.P. (1988). The molecular basis of retinoic acid action. Transcriptional regulation of tissue transglutaminase gene expression in macrophages. J. Biol. Chem. 263, 11584-11589.

Clarke, C.F., Cheng, K., Frey, A.B., Stein, R., Hinds, P.W., & Levine, A.J. (1988). Purification of complexes of nuclear oncogene p53 with rat and *Escherichia coli* heat shock proteins: *in vitro* dissociation of hsc 70 and dnaK from murine p53. Mol. Cell. Biol. 8, 1206-1215.

Crossin, K.L., & Carney, D.H. (1981a). Evidence that microtubule depolymerisation early in the cell cycle is sufficient to initiate DNA synthesis. Cell 23, 61-71.

Crossin, K.L., & Carney, D.H. (1981b). Microbutule stabilisation by taxol inhibits initiation of DNA synthesis by thrombin and by epidermal growth factor. Cell 27, 341-350.

Davis, R.J. (1993). The mitogen-activated protein kinase signal transduction pathway. J. Biol. Chem. 268, 14553-14556.

Davis, S., Lu, M.L., Lo, S.H., Lin, S., Butler, J.A., Druker, B.J., Roberts, T.M., An, Q., & Chen, L.B. (1991). Presence of an SH2 domain in the actin-binding protein tensin. Science 252, 712-715.

De Groot, R.P., Pals, C., & Kruijer, W. (1991). Transcriptional control of c-*jun* by retinoic acid. Nucleic Acids Res. 19, 1585-1591.

Dinoto, R., Schivvone, E.M., Ferrara, F., Manzo, C., Lopardo, C., & Delvecchio, L. (1994). All-trans-retinoic acid promotes a differential regulation of adhesion molecules on acute myeloid leukemia blast cells. Br. J. Haematol. 88, 247-255.

Drubin, D.G., Mulholland, J., Zhu, Z., & Botstein, D. (1990). Homology of a yeast actin-binding protein to signal transduction and proteins nad myosin-l. Nature 343, 288-290.

Edward, M., & Mackie, R.M. (1990). Retinoic acid-induced inhibition of lung colonisation and changes in the synthesis properties of glycosaminoglycans of metastatic B16 melanoma cells. J. Cell Sci. 94, 537-543.

Edward, M., Gold, J.A., & Mackie, R.M. (1989). Modulation of melanoma cell adhesion to basement membrane components by retinoic acid. J. Cell Sci. 93, 155-161.

Edward, M., Gold, J.A., & Mackie, R.M. (1992). Retinoic acid-induced inhibition of metastatic melanoma cell lung colonisation and adhesion to endothelium and subendothelial extracellular matrix. Clin. Expt. Metastasis 10, 61-67.

Elias, M.C.Q.B., Chammas, R., Giorgi, R.R., Frochtengarten, F., Santos, C.L.S., & Brentani, R.R. (1994). Adhesion to laminin is down-regulated upon retinoic acid-induced F9 cell differentiation. A role for alpha6/beta1 integrin. Brazilian J. Med. Biol. Res. 27, 21181-21184.

Farrell, J.E., & Martin, G.S. (1989). Tyrosine specific protein phosphorylation is regulated by glycoprotein IIB-IIIa in platelets. Proc. Natl. Acad. Sci. USA 86, 2234-2238.

Fazely, F., & Ledinko, N. (1990). Retinoic acid and epidermal growth factor binding in retinoid-mediated invasion suppressed lung carcinoma cells. Anticancer Res. 10, 667-670.

Finlay, C.A., Hinds, P.W., Tan, T., Eliyahu, D., Oren, M., & Levine, A.J. (1988). Activating mutations for transformation by p53 produce a gene product that forms an hsc70-p53 complex with altered half-life. Mol. Cell Biol 8, 531-539.

Gale, N.W., Kaplan, S., Lowenstein, E.J., Schlessinger, J., & Bar-Sagi, D. (1993). Grb2 mediates the EGF-dependent activation of guanine nucleotide exchange of Ras. Nature 363, 88-92.

Glukhova, M., Koteliansky, V., Sastre, X., & Thiery, J.P. (1995). Adhesion systems in normal breast and in invasive breast carcinoma. Amer. J. Pathol. 146, 706-716.

Golden A., Brugge, J.S., & Shattil, S. (1991). Role of platelet membrane glycoprotein IIb-IIIa in agonist-induced tyrosine phosphorylation of platelet proteins. J. Cell Biol. 111, 3117-3127.

Gross, N., Favre, S., Beck, D., & Meyer, M. (1992). Differentiation-related expression of adhesion molecules and receptors on human neuroblastoma tissues, cell lines and variants. Int. J. Cancer 52, 85-91.

Guan, J., Trevithick, J.E., & Hynes, R.O. (1991). Fibronectin integrin interaction induces tyrosine phosphorylation of 120kDa protein. Cell Regulation, 2, 951-964.

Gullick, W.J., Love, S.B., Wright, C., Barnes, D.M., Gusterson, B., Harris, A.L., & Altman, D.G. (1991). C-erbB2 protein overexpression in breast cancer is a risk factor in patients with involved and uninvolved lymph nodes. Br. J. Cancer 63, 434-438.

Gupta, R.S. (1990). Microtubules, mitochondria, and molecular chaperones: a new hypothesis for *in vitro* assembly of microtubules. Biochem. Cell. Biol. 68, 1352-1363.

Hanks, S.K., Callab, M.B., Harper, M.C., & Patel, S.K. (1992). Focal adhesion protein tyrosine kinase phosphorylated in response to cell attachment to fibronectin. Proc. Natl. Acad. Sci. USA 89, 8487-8491.

Haussler, M., Sidell, N., Kelly, M., Donaldson, C., Altman, A., & Mangelsdorf, D. (1983). Specific high affinity binding and biologic action of retinoic acid in human neuroblastoma cell lines. Proc. Natl. Acad. Sci. USA 80, 5525-5529.

Helige, C., Smolle, J., Zellnig, G., Hartmann, E., Finkpuches, R., Kerl, H., & Tritthart, H.A. (1993). Inhibition of K1735-M2 melanoma cell invasion in vitro by retinoic acid. Clin. Expt. Metastasis 11, 409-418.

Hempstead, B.L., Birge, R.B., Fajardo, J.E., Mahadeo D., Glassman R., Kraemer R., & Hanafusa, H. (1994). Expression of v-crk oncogene product in PC12 cells results in rapid differentiation by both nerve growth factor and epidermal growth factor dependent pathways. Mol. Biol. Cell 14, 1964-1971.

Hendey B., & Maxfield, F.R. (1993). Regulation of neutrophil motility and adhesion by intracellular calcium transients. Blood Cells 19, 143-164.

Hendrix, M.J.C., Wood, R., Seftor, E.A., Lotan, D., Nakajima, M., Misiorowski, R.I., Seftor, R.E.B., Stetler-Stevenson, W.G., Bevilacqua, S.J., Liotta, L.A., Sobel, M.R., Raz, A., & Lotan, R. (1990). Retinoic acid inhibition of human melanoma cell invasion through a reconstituted basement membrane and its relation to decreases in the expression of proteolytic enzymes and motility factor receptor. Cancer Res. 50, 4121-4130.

Horton, W.E., Yamada, Y., & Hassell, J.R. (1987). Retinoic acid rapidly reduces cartilage matrix synthesis by altering gene transcription in chrondrocytes. Develop. Biol. 123, 508-516.

Horwitz, A., Duggan, K., Buck, C., Beckerle, M.C., & Burridge, K. (1986). Interaction of plasma membrane fibronectin receptor with talin—a transmembrane linkage. Nature 320, 531-533.

Hoshi, M., Ohta, K., Gotoh, Y., Mori, A., Murofushi, M., Sakai, H., & Nishida, E. (1992). Mitogen-activated protein kinase-catalyzed phosphorylation of microtubule-associated proteins, microtubule-associated protein-2 and microtubule-associated protein-4, induces an alteration in their function. Eur. J. Biochem. 203, 43-52.

Hudson, L.G., Santon, J.B., Glass, C.K., & Gill, G.N. (1990). Ligand-activated thyroid hormone and retinoic acid receptors inhibit growth factor receptor promoter expression. Cell 62,1165-1175.

Jetten, A.M. (1980). Retinoids specifically enhance the number of epidermal growth factor receptors. Nature 284, 626-629.

Karin, M., Yang-Yen, H.F., Chambard, J.C., Deng, T., & Saatcioglu, F. (1993). Various modes of gene regulation by nuclear receptors for steroid and thyroid hormones. Eur. J. Clin. Pharmacol. 45, suppl.1, S9-S15.

Kawaguchi, M., Kawashima, F., Ohshima K., Kawaguchi, S., & Wada H. (1994). Retinoic acid enhances the number of epidermal growth factor receptors in rat glomerular epithelial cells *in vitro*. Nephron 68, 97-103.

Koochekpour S., Merzak, A., & Pilkington, G.J. (1995). Growth factors and gangliosides stimulate laminin production by human glioma cells *in vitro*. Neurosci. Lett. 186, 53-56.

Kornberg, L.J., Earp, H.S., Turner, C.E., Prockop, C., & Juliano, R.L. (1991). Signal transduction by integrins: Increased protein tyrosine phosphorylation caused by clustering of β_1 integrins. Proc. Natl. Acad. Sci. USA 88, 8392-8396.

Kornberg, L., Earp, H.S., Parsons, J.T., Schaller, M., & Juliano, R.L. (1992). Cell adhesion or integrin clustering increases phosphorylation of a focal adhesion-associated tyrosine kinase. J. Biol. Chem. 267, 23439-23442.

Krone, P.H., & Heikkila, J.J. (1988). Analysis of hsp 30, hsp 70 and ubiquitin gene expression in *Xenopus laevis* tadpoles. Development 103, 59-67.

Kypta, R.M., Goldberg, Y., Ulug, E.T., & Courtneidge, S.A. (1990). Association between the PDGF receptor and members of the src family of tyrosine kinases. Cell 62, 481-492.

Lakshmi, M.S., Parker, C., & Sherbet, G.V. (1993). Metastasis associated MTS1 and NM23 genes affect tubulin polymerisation in B16 melanomas. A possible mechanism of their regulation of metastatic behaviour of tumours. Anticancer Res. 13, 299-304.

Lancillotti, F., Darwiche, N., Celli, G., & Deluca, L.M. (1992). Retinoid status and the control of keratin expression and adhesion during the histogenesis of squamous metaplasia of tracheal epithelium. Cancer Res. 52, 6144-6152.

LaRosa, G.J., & Gudas, L.J. (1988a). An early effect of retinoic acid. Cloning of a messenger RNA (ERA-1) exhibiting rapid and protein synthesis-independent induction during teratocarcinoma stem-cell differentiation. Proc. Natl. Acad. Sci. USA 85, 329-333.

LaRosa, G.J., & Gudas, L.J. (1988b). Early retinoic acid induced F9 teratocarcinoma stem cell gene *ERA-1*. Alternate splicing creates transcripts for a homeobox-containing protein and one lacking the homeobox. Mol. Cell. Biol. 8, 3906-3917.

Leung, M.F., & Sartorelli, A.C. (1992). The effects of microtubule disrupting drugs on the differentiation of HL-60 leukemia cells. Leukemia Res. 16, 929-935.

Leung, M.F., Sokoloski, J.A., & Sartorelli, A.C. (1992). Changes in microtubules, microtubule associated proteins, and intermediate filaments during the differentiation of HL-60 leukemia cells. Cancer Res. 52, 949-954.

Li, J.M., King, I., & Sartorelli, A.C. (1994). Differentiation of WEHI-3B D+ myeomonocytic leukemia cells induced by ectopic expression of the protooncogene c-JUN. Cell Growth Differen. 5, 743-751.

Lipfert, L., Haimovich, B., Shaller M.D., Cobb, B.S., Parsons, J.T., & Brugge, J.S. (1992). Integrin-dependent phosphorylation and activation of protein tyrosine kinase pp125^{FK} in plateltes. J. Cell Biol. 119, 905-912.

Liska, D.J., Hawkins, R., Wikstrom, K., & Bornstein, P. (1994). Modulation of thrombospondin expression during differentiation of embryonal carcinoma cells. J. Cell. Physiol. 158, 494-505.

Loeser, R.F. (1994). Modulation of integrin-mediated attachment of chondrocytes to extracellular matrix proteins by cations, retinoic acid, and transforming growth factor beta. Exp. Cell Res. 211, 17-23.

Margolis, B., Li, N., Koch, A., Mohammadi, M., Hurwitz, D., Zilberstein, A., Ullrich, A., Pawson, T., & Schlessinger, J. (1990). The tyrosine phosphorylated carboxyterminus of the EGF receptor is a binding site for GAP and PLC-gamma. EMBO J. 9, 4375-4380.

Mayer, B.J., Hamaguchi, M., & Hanafusa, H. (1988). A novel viral oncogene with structural similarity to phospholipase C. Nature , 332, 272-275.

McCormick, F. (1989) Ras GTPase activating protein: signal transmitter and signal terminator. Cell 56, 5-8.

Meco, D., Scarpa, S., Napolitano, M., Maroder, M., Bellavia, D., Demaria, R., Raganocaracciolo, M., Frati, L., Modesti, A., Gulino, A., & Screpanti, I. (1994). Modulation of fibronectin and thymic stromal cell-dependent thymocyte maturation by retinoic acid. J. Immunol. 153, 73-83.

Merzak, A., McCrea, S., Koochekpour S., & Pilkington, G.J. (1994a). Control of human glioma cell growth, migration and invasion *in vitro* by transforming growth factor β_1. Br. J. Cancer 70, 199-203.

Merzak, A., Parker, C., Koochekpour, S., Sherbet, G.V., & Pilkington, G.J. (1994b). Overexpression of the *18A2/mts1* gene and down-regulation of the *TIMP-2* gene in invasive human glioma cell lines *in vitro*. Neuropath. Appl. Neurobiol. 20, 614-619.

Milarsky, K., & Morimoto, R.I. (1986). Expression of human hsp 70 during the synthetic phase of the cell cycle. Proc. Natl. Acad. Sci. USA 82, 9517-9521.

Mooney, D.J., Hansen, L.K., Langer, R., Vacanti, J.P., & Ingber, D.E. (1994). Extracellular matrix controls tubulin monomer levels in hepatocytes by regulating protein turnover. Mol. Biol. Cell 5, 1281-1288.

Muthuswamy, S.K., Siegel, P.M., Dankort, D.L., Webster, M.A., & Muller, W.J. (1994). Mammary tumours expressing the *neu* proto-oncogene possess elevated c-src tyrosine kinase activity. Mol. Cell. Biol. 14, 735-743.

Nicholson, R.C., Mader, S., Nagpal, S., Leid, M., Rochette-Egly, C., & Chambon, P. (1990). Negative regulation of the rat stromelysin gene promoter by retinoic acid is mediated by an AP1 binding site. EMBO J. 13, 4443-4454.

Nicosia, R.F., & Tuszynski, G.P. (1994). Matrix bound thrombospondin promotes angiogenesis *in vitro*. J. Cell Biol. 124, 183-193.

Niles, R. (1987). Retinoic acid-induced arrest of mouse melanoma cells in G1 without inhibition of protein synthesis. In Vitro 23, 803-804.

Otey C.A., Pavalko, F.M., & Burridge, K. (1990). An interaction between α-actinin and the β_1 integrin subunit *in vitro*. J. Cell Biol. 111, 721-729.

Paciucci, R. (1994). Role of 300-kDa complexes as intermediates in tubulin folding and dimerization. Characterisation of a 25-kDa cytosolic protein involved in the GTP-dependent release of monomeric tubulin. Biochem. J. 301, 105-110.

Parker, C., & Sherbet, G.V. (1992). Modulators of intracellular Ca^{2+} and the calmodulin inhibitor W-7 alter the expression of metastasis-associated genes *mts1* and *nm23* in metastatic variants of the B16 murine melanoma. Melanoma Res. 2, 337-343.

Parker, C., Whittaker, P.A., Weeks, R.J., Thody, A.J., & Sherbet, G.V. (1991). Modulators of metastatic behaviour alter the expression of metastasis-associated genes MTS1 and NM23 in metastatic variants of the B16 murine melanoma. Clin. Biotechnol. 3, 217-222.

Parker, C., Lakshmi, M.S., Piura, B., & Sherbet, G.V. (1994a). Metastasis-associated *mts1* gene expression correlates with increased p53 detection in the B16 murine melanoma. DNA & Cell Biol. 13, 343-351.

Parker, C., Whittaker, P.A., Usmani, B.A., Lakshmi, M.S., & Sherbet, G.V. (1994b). Induction of *18A2/mts1* gene expression and its effects on metastasis and cell cycle control. DNA & Cell Biol. 13, 1021-1028.

Paterson, M.C., Dietrich, K.D., Danyluk, J., Paterson, A.H., Lees, S.W., Jamil, N., Hanson, J., Jenkins, H., Krause B.E., McBlain, W.A., Slamon, D.J., & Fourney, R.M. (1991). Correlation between c-erbB2 amplification land risk of recurrent disease in node-negative breast cancer. Cancer Res. 51, 556-567.

Pawson, T. (1988). Non-catalytic domains of cytoplasmic protein tyrosine kinases: regulatory elements in signal transduction. Oncogene 3, 245-262.

Pawson, T., & Schlessinger, J. (1993). SH2 and SH3 domains. Curr. Biol. 3, 434-442.

Peter, K., & Otoole, T.E. (1995). Modulation of cell adhesion by changes in alpha(L)beta (2) (LFA-1, CD11A/CD18) cytoplasmic domain/cytoskeleton interaction. J. Exp. Med. 181, 315-326.

Petersen, O.H., & Wakui, M. (1990). Oscillating intracellular Ca^{2+} signals evoked by activation of receptors linked to insositol lipid hydrolysis. Mechanism of generation. J. Membr. Biol. 118, 93-105.

Qabar, A.N., Lin, Z.W., Wolf, F.W., Oshea, K.S., Lawler, J., & Dixit, V.M. (1994). Thrombospondin-3 is a developmentally regulated heparan-binding protein. J. Biol. Chem. 269, 1262-1269.

Roberts, A.B., & Sporn, M.B. (1984). Cellular biology and biochemistry of the retinoids. In: Sporn, M.B., Roberts, A.B., & Goodman, D.S. (eds) *The Retinoids*, Vol. 2, Academic Press, Orlando, pp. 209-286.

Rodionov, V.I., Gyoeva, F.K., Tanaka, E., Bershadsky, A.D., Vasiliev, J.M., & Gelfand, V.I. (1993). Microtubule-dependent control of cell shape and pseudopodial activity is inhibited by the antibody to kinesin motor domain. J. Cell Biol. 123, 1811-1820.

Rosenbaum, S.E., & Niles, R.M. (1992). Regulation of protein kinase C gene expression by retnioic acid in B16 mouse melanoma cells. Arch. Biochem. Biphys. 294, 123-129.

Ross, S.A., Ahrens, R.A., & Deluca, L.M. (1994). Retinoic acid enhances adhesiveness, laminin and integrin beta1 synthesis, and retinoic acid receptor expression in F9 teratocarcinoma cells. J. Cell. Physiol. 159, 263-273.

Roy, L.M., Gittinger, C.K., & Landreth, G.E. (1989). Characterisation of epidermal growth factor receptor associated with cytoskeletons of A431 cells. J. Cell Physiol. 140, 295-304.

Roy, L.M., Gittinger, C.K., & Landreth, G.E. (1991). Epidermal growth factor treatment of A431 cells alters the binding capacity and electrophoretic mobility of the cytoskeletally associated epidermal growth factor receptor. J. Cell Physiol. 146, 63-72.

Ruoslahti, E. (1991). Integrins. J. Clin. Invest. 87, 1-5.

Scheibe, R.J., & Wagner, J.A. (1992). Retinoic acid regulates both expression of nerve growth factor receptor and sensivitity to nerve gfowth factor. J. Biol. Chem. 267, 17611-17616.

Scheibe, R.J., Ginty, D.D., & Wagner, J.A. (1991). Retinoic acid stimulates the differentiation of PC12 cells that are deficient in cAMP-dependent protein kinase. J. Cell Biol. 113, 1173-1182.

Scheidl, H., Chandra, T., Gmeiner, G., Zerlauth, G., Scita, G., & Wolf, G. (1991). The effect of diacylglycerols on fibronectin release and its reversal by retinoic acid in cell culture. Carcinogenesis 12, 1791-1794.

Schmidt, C.E., Horwitz, A.F., Lauffenburger, D.A., & Sheetz, M.P. (1993). Integrin cytoskeletal interactions in migrating fibroblasts are dynamic, asymmetric and regulated. J. Cell Biol. 123, 977-991.

Schmidt, C.E., Chen, T., & Lauffenburger, D.A. (1994). Stimulation of integrin-cytoskeletal interactions in migrating fibroblasts. Biophy. J. 67, 461-474.

Schmidt, C.E., Dai, J.W., Lauffenburger, D.A., Sheetz, M.P., & Horwitz, A.F. (1995). Integrin-cytoskeletal interactions in neuronal growth cones. J. Neurosci. 15, 3400-3407.

Scita, G., & Wolf, G. (1994). Retinoic acid and beta carotene inhibit fibronectin synthesis and release by fibroblasts antagonism to phorbol ester. Carcinogenesis 15, 1043-1048.

Shaller M.D., Borgman, C.A., Cobb, B.S., Vines, R.R., Reynods, A.B., & Parsons, J.T. (1992). pp^{125}, a structurally distinctive protein tysosine kinase associated with focal adhesion. Proc. Natl. Acad. Sci. USA 89, 5192-5196.

Sharma, C.P., Ezzell, R.M., & Arnaout, M.A. (1995). Direct interaction of filamin (ABP-280) with the beta-2 integrin subunit of CD18. J. Immunol. 154, 3461-3470.

Sheppard, A.M., Maquillan J.J., Iademarco, M.F., & Dean, D.C. (1995). Control of vascular cell adhesion molecule-1 gene promoter activity during neural differentiation. J. Biol. Chem. 270, 3710-3719.

Sherbet, G.V., Parker, C., Usmani, B.A., & Lakshmi, M.S. (1995). Epidermal growth factor receptor status correlates with cell proliferation related *18A2/mts1* gene expression in human carcinoma cell lines. Ann. N.Y. Acad. Sci. 768, 272-275.

Simon, M.A., Bowtell, D.D.L., Dodson, G.S., Laverty, T.R., & Rubin, G.M. (1991). Ras1 and a putative guanine nucleotide exchange factor perform crucial steps in signalling by the sevenless protein tyrosine kinase. Cell 67, 701-716.

Simonson, M.S. (1994). Anti-AP-1 activity of all-*trans* retinoic acid in glomerular mesangial cells. Amer. J. Physiol. 36, F805-F815.

Situ, R., Inman, D.R., Fligiel, S.E.G., & Varani, J. (1993). Effects of all-*trans*-retinoic acid on melanocyte adhesion and motility. Dermatology 186, 38-44.

Slamon, D.J., Clark, G.M., Wong, S.G., Levin, W.J., Ullrich, A., & McGuire, W.L. (1987). Human breast cancer: correlation of relapse and survival with amplification of Her-2/*neu* oncogene. Science 235, 177-182.

Spector, N.L., Samson, W., Ryan, C, Gribben, J., Urba, W., Welch, W.J., & Nadler, L.M. (1992). Growth arrest of human lymphocytes-B is accompanied by induction of the low molecular weight mammalian heat shock protein (hsp28). J. Immunol. 148, 1668-1673.

Spector, N.L., Ryan, C., Samson, W., Levine, H., Nadler, L.M., & Arrigo, A.P. (1993). Heat shock protein is a unique marker of growth arrest during macrophage differentiation of HL-60 cells. J. Cell. Physiol. 156, 619-625.

Spector, N.L., Hardy, L., Ryan, C., Miller, W.H., Humes, J.L., Nadler, L.M., & Luedke, E. (1995). 28kDa mammalian heat shock protein, a novel substrate of a growth-regulatory protease involved in differentiation of human leukemia cells. J. Biol. Chem. 270, 1003-1006.

Stahl, M.L., Ferenz, C.R., Kelleher, K.L., Kriz, R.W., & Knopf, J.L. (1988). Sequence similarity of phospholipase C with the non-catalytic region of *src*. Nature 332, 269-272.

Stein, D., Wu, J., Fuqua, S.A.W., Roonprapunt, C., Yajnik, V., D'Eustachio, P., Moskow, J.J., Buchberg, A.M., Osborne, C.K., & Margolis, B. (1994). The SH2 domain protein GRB7 is co-amplified, overexpressed in a tight complex with HER2 in human breast cancer. EMBO J. 13, 1331-1340.

Stevenson, M.A., & Calderwood, S.K. (1990). Members of the 70-kilodalton heat shock protein family contain a highly conserved calmodulin binding domain. Mol. Cell. Biol. 10, 1234-1238.

Stokoe, D., Macdonald, S.G., Cadwallader, K., Symons, M., & Hancock, J.F. (1994). Activation of Raf as a result of recruitment to the plasma membrane. Science 264, 1463-1467.

Strickland, S., Smith, K.K., & Marotti, K.R. (1980). Hormonal induction of differentiation in teratocarcinoma stem cells: Generation of parietal endoderm by retinoic acid. Cell 21, 347-355.

Sturzbecher, H.W., Chumakov, P., Welch, W.J., & Jenkins, J.R. (1987). Mutant p53 protein binds hsp 72/73 cellular heat shock related protein in SV40 transformed monkey cells. Oncogene 1, 201-211.

Tanaka Y., Shima, M., Yamaoka, K., Okada, S., & Seino, Y. (1992). Synergistic effect of 1, 25-dihydroxyvitamin-D3 and retinoic acid in inducing U937 cell differentiation. J. Nutri. Sci. Viteminol. 38, 415-426.

Thompson, K.L., Santon J.B., Shephard, L.B., Walton, G.M., & Gill, G.N. (1992). Nuclear protein is required for thyroid hormone receptor binding to an inhibitory half-site in the epithelial growth factor receptor protomer. Mol. Endocrinol. 6, 627-635.

Tsang, T.C. (1993). New model for 7-kDa heat shock proteins potential mechanism of function. FEBS Lett. 323, 1-3.

Tsuji, K., Ueno, A., & Ide, T. (1992). Inhibitory effect of taxol, a microtubule stabilizing agent, induction of DNA synthesis is dependent upon cell lines and growth factors. Cell Str. Function 17, 139-144.

Van Bergen en Henegouwen, P.M.P., Defize, L.H.K., Dekroon, J., Van Damme H., Verkleij, A.J., & Boonstra, J. (1989). Ligand-induced association of epidermal growth factor receptor to cytoskeleton of A431 cells. J. Cell. Biochem. 39, 455-465.

Wachsstock, D.H., Wilins, J.A., & Lin, S. (1987). Specific interaction of vinculin with α-actinin. Bichem. Biophys. Res. Commun. 146, 554-560.

Wang, S.Y., LaRosa, G.J., & Gudas, L.J. (1985). Molecular cloning of gene sequences transcriptionally regulated by retinoic acid and dibutyryl cyclic AMP in cultured mouse teratocarcinoma cells. Develop. Biol. 197, 75-86.

Weber, C., Calzadawack, J.C., Goretzki, M., Pietsch, A., Johnson, J.P., & Zieglerheitbrock, H.W.L. (1995). Retinoic acid inhibits basal and interferon gamma induced expression of intercellular adhesion molecule-1 in monocytic cells. J. Leukocyte Biol. 57, 401-406.

Wood, W.R., Seftor, E.A., Lotan, D., Nakajima, M., Misiorowski, R.L., Seftor, R.E.B., Lotan, R., & Hendrix, M.J.C. (1990). Retinoic acid inhibits human melanoma tumour cell invasion. Anticancer Res. 10, 423-432.

Zerlauth G., & Wolf, G. (1984). Kinetics of fibronectin release from fibroblasts in response to 12-*O*-tetradecanoylphorbol-13-acetate and retinoic acid. Carcinogenesis 5, 863-868.

RETINOIDS AND APOPTOSIS

Li-Xin Zhang and Anton M. Jetten

	Abstract	161
I.	Introduction	162
II.	Apoptosis: A Programmed Cell Death	163
	A. Characteristics of Apoptosis	163
	B. Genes Involved in Apoptosis	164
	C. Cell Cycle-Related genes	168
III.	Regulation of Apoptosis by Retinoids	170
IV.	Mechanisms of the Regulation of Apoptosis by Retinoids	171
	A. Retinoid Receptor-Dependent Mechanisms	171
	B. Retinoid Receptor-Independent Mechanisms	180
V.	Concluding Remarks	184
	References	186

ABSTRACT

Apoptosis, a genetically controlled process of cell death, plays an important role in embryogenesis, development, homeostasis, and in many diseases, such as cancer, certain neurodegenerative, and immune disorders. An increasing number of studies have demonstrated that retinoids, natural and synthetic analogs of vitamin A, have

Advances in Organ Biology
Volume 3, pages 161-196.

ISBN: 0-7623-0285-2

profound effects on the apoptotic process. The effect of retinoids on apoptosis is dependent on the chemical nature of the retinoid and the cell type. In most instances, retinoids induce apoptosis, while in a few cell types, retinoids exert an inhibitory effect. The mechanisms by which retinoids regulate apoptosis are just emerging during the last year. In most cases, activation of nuclear retinoid receptors, either retinoic acid (RA) receptors (RARs) or retinoid X receptors (RXRs) or both, is involved in retinoid-regulated apoptosis. The induction of apoptosis by N-(4-hydroxyphenyl)retinamide (4-HPR) is not mediated by retinoid receptors and may involve the generation of reactive oxygen intermediates. Induction of apoptosis by the retinoid 6-[3-(1-adamantyl)-4-hydroxyphenyl]-2-naphthalene carboxylic acid (AHPN) also occurs through a retinoid receptor-independent pathway. Induction of apoptosis may be one mechanism by which retinoids control morphogenesis during normal embryonic development. It may also be responsible for some of the teratogenic effects induced by retinoids at high, nonphysiological concentrations. In addition, the induction of apoptosis may be part of the mechanism by which retinoids inhibit cell proliferation in various carcinoma cells *in vitro* and suppress the development of tumors. Defects in retinoid signaling pathways have been shown to cause resistance of certain carcinoma cells to retinoid-induced growth-inhibition and apoptosis, and contribute to the malignant phenotype of cancer cells. Insight into the mechanisms by which retinoids regulate apoptosis may lead to more effective strategies in the prevention and therapy of disease as well as a better understanding of the regulatory role retinoids play in development.

I. INTRODUCTION

Apoptosis or programmed cell death is an active, genetically regulated process of cell suicide and represents a mechanism of cell clearance in both physiological and pathological situations where the elimination of cells is required (reviewed by Wyllie et al., 1980; Williams and Smith, 1993; Sen, 1992; Steller, 1995). Apoptosis plays a fundamental role in the morphological and functional maturation in a variety of tissues during normal development. It is a defense mechanism against genomic instability and cancer, and modulates viral pathogenesis and latency. Apoptosis also plays an important role in homeostasis. Programmed cell death is important in maintaining a balance between cell loss and cell gain and therefore the normal architecture of the tissue. Apoptosis is relevant to several disease processes, including the emergence of cancer. Genetic defects in genes that promote apoptosis result in an increased ability of cancer cells to survive and propagate (Martin and Green, 1995). In contrast, too much cell death can lead to impaired development and degenerative diseases.

Retinoids, natural and synthetic derivatives of vitamin A, effect many biological processes, including development, cellular proliferation, and differentiation (reviewed by De Luca, 1991; Gudas et al., 1994). An increasing number of laboratories have demonstrated that retinoids can either induce or inhibit apoptosis (Piacentini et al., 1991; Delia et al., 1993; Fesus et al., 1995; Lotan, 1995; Zhang and Jetten,

1995; Kalemkerian et al., 1995; Shao et al., 1995; Seewaldt et al., 1995; Yang et al., 1995a; 1995b). In this chapter, we will review the regulation of apoptosis by retinoids in a variety of cell types and discuss the different possible mechanisms by which retinoids exert their effects on programmed cell death. In addition, we analyze the relationship between retinoid-regulated cell death and cell proliferation and differentiation, and discuss the implications for chemoprevention and therapy of certain human diseases, particularly cancer.

II. APOPTOSIS: A PROGRAMMED CELL DEATH

A. Characteristics of Apoptosis

There are three morphologically and biochemically distinct forms of cell death, namely apoptosis, necrosis, and differentiation-associated cell death (reviewed by Wyllie et al., 1980; Sen, 1992; Fesus, 1993; Polakowska and Haake, 1994). The early process of apoptosis is often associated with characteristic morphological changes, including cell shrinkage, loss of cell-cell contacts, and the condensation and segregation of chromatin at the margin of the nuclear membrane, while the plasma membrane and organelles retain their integrity (Figure 1). For example, mitochondria stay functional until late in the apoptotic process. At a later stage of apoptosis, the entire cell is fragmented into small membrane-bound vesicles containing condensed chromatin fragments and other cellular constituents, which are termed apoptotic bodies. Without causing inflammation, the apoptotic bodies are then rapidly phagocytosed by neighboring cells or macrophages and subsequently digested by the host cell. In contrast to apoptosis, necrosis is evoked by a general failure of cellular homeostatic regulation following injury and charac-

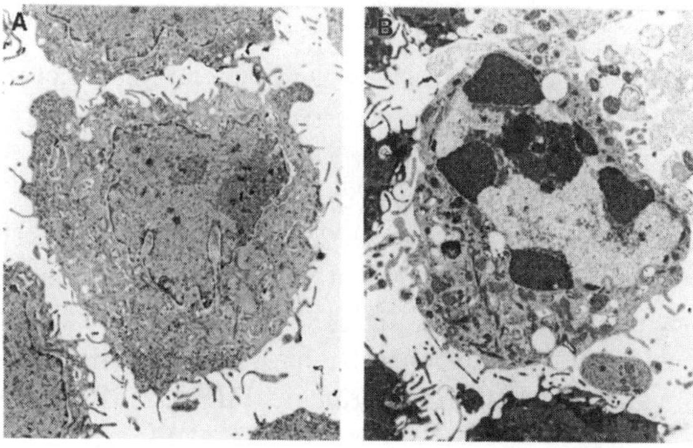

Figure 1. Electron micrograph of rat tracheal epithelial cells. (A) a normal cell; (B) a typical apoptotic cell at an early stage of apoptosis in cultures treated with 10^{-8} M RAR-selective retinoid for 4 days (from Zhang et al., 1995, with permission).

terized by the loss of integrity of plasma membrane and organelles, especially rupture of mitochondria and lysosomes leading to gross tissue inflammation (reviewed by Wyllie et al., 1980; 1984). Differentiation in the skin is an example of differentiation-associated cell death when keratinocytes undergo cornification in the last stages of squamous cell differentiation (Holbrook and Hennings, 1983; Jetten et al., 1992; Polakowska and Haake, 1994).

Apoptosis can be determined morphologically by electron microscopy (Figure 1). In addition, multistage fragmentation of chromatin into pieces of different oligonucleosomal length in apoptotic cells will appear as a "DNA ladder" after standard agarose gel electrophoresis (Wyllie et al., 1984; Eastman, 1995). This technique has been widely used as an assay for apoptosis, although in a number of instances cells can undergo apoptosis without showing a "DNA ladder" (Schwartz et al., 1993). Many methods to assess DNA strand breaks are based on staining of the cellular DNA with fluorochromes or enzymatic labeling of DNA with tagged nucleotides either by DNA polymerase (nick translation) or terminal deoxynucleotidyl transferase (TdT) (Meyaard et al., 1992; Ben-Sasson et al., 1995). The end-labeling method has also been termed TUNEL (**T**dT-mediated **dUTP** **n**ick **e**nd **l**abeling)(Ben-Sasson et al., 1995). The labeled DNA is then analyzed by flow cytometry, or by fluorescence or light microscopy (Darzynkiewicz et al., 1992). In order to conclude whether cells are undergoing apoptosis, a combination of these methods is often being used. Single apoptotic cells can be distinguished from necrotic cells by the *in situ* tailing translation techniques (Gold, 1994).

Apoptosis is an extremely varied and complex biological process. In principle, almost all mammalian cells could die by apoptosis (Fesus, 1993; Steller, 1995). Many external and internal signals have been identified that can lead to apoptosis. These include treatment with various retinoids, withdrawal of growth factors, ionizing irradiation, various hormones, and many genotoxic drugs (Sen, 1992; Thompson, 1994).

B. Genes Involved in Apoptosis

The mechanism by which apoptosis is triggered has not yet been precisely defined but it is clear that it occurs as a multi-step process. Recent molecular and genetic studies, particularly in the nematode *Caenorhabditis elegans* and in the fruit fly *Drosophila melanogaster*, have led to the discovery of specific sets of genes that are regulating the initiation, execution, and completion of cell death. Mutations that affect each of these stages have been identified, and 14 genes that function in programmed cell death of *C. elegans* have been isolated (Steller, 1995). Some of these genes have sequence and/or functional homology with genes that have been demonstrated to play a role in apoptosis in mammalian cells (Hengartner and Horvitz, 1994a; 1994b).

Bcl-2 Family

In *C. elegans*, apoptosis is suppressed by the ced-9 gene and loss of its function by mutation initiates the cell death program (Hengartner and Horvitz, 1994a).

Ced-9 is a member of the Bcl-2 family, which includes bcl-2, bcl-x, Bax, Bak, Bad, and Mcl-1 (Hengartner and Horvitz, 1994b; Boise et al., 1995; White, 1996). Proteins of this family can form homo- and heterodimers and either enhance or inhibit apoptosis (Sedlak et al., 1995). Bcl-2 can serve as a suppressor of apoptosis under a wide variety of conditions (Vaux, 1993; Hawkins and Vaux, 1994). The bcl-2 gene encodes a 26 kD integral membrane protein that is associated with the mitochondrial and perinuclear membrane, and the endoplasmic reticulum (Hockenbery et al., 1990). Loss of bcl-2 function in "knock-out" mice results early after birth in death of mature T and B cells due to apoptosis (Veis et al, 1993). In contrast, inappropriate gain of bcl-2 function may inhibit apoptosis and contribute to the malignant phenotype of cancer cells. Exogenous expression of bcl-2 in haemopoietic cells has been shown to protect cells from apoptosis induced by a variety of conditions (Vaux et al., 1988; Miyashita and Reed, 1992; Vaux, 1993). Bcl-2 can also protect cells from c-myc and p53-dependent apoptosis (Chiou et al., 1994); however, it does not inhibit p53-induced G1 growth arrest and transactivation of WAF-1 and Bax (Wang et al., 1995). The repression of apoptosis by bcl-2 may be related to a reduction in the Ca^{2+}-efflux through the endoplasmic membrane (Lam et al., 1994), thereby abrogating the Ca^{2+}-signaling of apoptosis.

It has been suggested that not bcl-2 expression persé but the ratio between bcl-2 and bax, which is an inhibitor of bcl-2, is a critical factor in determining apoptosis (Oltvai et al., 1993; Oltvai and Korsmeyer, 1994; Sedlak et al., 1995). For example, the increased level of bax and the reduced level of bcl-2 induced by p53 shift the balance between bcl-2 and bax and promote apoptosis.

Bcl-x, another member of the bcl-2 family, has been shown to affect apoptosis in haemopoietic and epithelial cells. Alternative splicing results in two bcl-x-derived mRNAs, called bcl-x_L and bcl-x_S (Boise et al., 1993). Bcl-x_L appears to have functions similar to those of bcl-2 while bcl-x_S functions as a promoter of apoptosis (Boise et al., 1993; Schott et al., 1995; Boise et al., 1995).

The information on the role of members of the bcl-2 family in retinoid-induced apoptosis is rather scant at this moment. Treatment of myeloid leukemia HL-60 cells with RA induces differentiation and apoptosis, and is accompanied by a down-regulation of bcl-2 (Park et al., 1994). Overexpression of bcl-2 in these cells has no effect on the induction of differentiation into granulocytes but blocks the induction of apoptosis, suggesting that these two processes are under different controls (Benito et al., 1995). In addition, these results suggest that bcl-2 and possibly other members of the bcl-2 family may be involved in the RA-induced apoptosis in maturing granulocytes. These results contrast those obtained by Bruel and colleagues (Bruel et al., 1995), which indicate that the down-regulation of bcl-2 in the leukemic cell line NB4 is associated with differentiation and not apoptosis. Induction of apoptosis by N-(4-hydroxyphenyl)retinamide (4-HPR) has been reported to be associated with a decrease in bcl-2 mRNA levels in several leukemic cell lines; however, the levels of bcl-2 protein remained unchanged, suggesting that in these cells apoptosis is not controlled by bcl-2 (Delia et al., 1995). 4-HPR, which also induces apoptosis in several small-cell lung cancer cell lines, has been reported not to alter significantly the expression of bcl-2 in these cells (Kalemkerian et al., 1995). The induction of apoptosis by the synthetic retinoid 6-[3-(1-adamantyl)-4-hydroxyphenyl]-2-naphthalene carboxylic acid (AHPN) in mammary carcinoma MCF-7 cells has been shown to be accompanied by a decrease in bcl-2 expression and an increase in bax mRNA levels. Although these results would be in agreement with the hypothesis that changes in the expression of bcl-2 and bax are involved in the induction of apoptosis by AHPN, this appears not to be a general mechanism by which AHPN acts since in mammary carcinoma MDA-MB-231 cells bcl-2 and bax mRNA could not be detected in either AHPN-treated or untreated cells (Shao et al., 1995).

ICE Family

Further execution of programmed cell death in *C. elegans* requires the activation of ced-3 and -4. Ced-3 has substantial homology to the mammalian interleukin-1β-converting enzyme (ICE), which cleaves the 33kD form of interleukin 1β to the active 17kD form (Miura et al., 1993; Kumar, 1995; Yuan, 1995). Ced-3 and ICE belong to a family of cysteine proteases that include also Nedd-2/ich1 and Mch-2 (Kumar, 1995; Yuan, 1995). Expression of ced 3 or ICE has been shown to cause apoptosis in several cell types (Miura et al., 1993; Kondo et al., 1995). Activation of ICE is required for Fas/APO-1-mediated apoptosis (Enari et al., 1995; Los et al., 1995). Poly(ADP-ribose) polymerase (PARP), an enzyme involved in DNA repair and genomic integrity, is cleaved by ced-3 and ICE (Lazebnik et al., 1994; Nicholson et al., 1995; reviewed by Earnshaw, 1995). This prevents it from participating in DNA repair. PARP also regulates the activation of Ca^{2+}/Mg^{2+}-dependent nucleases that catalyze internucleosomal DNA cleavage (reviewed by Earnshaw, 1995). How critical the cleavage of PARP is in the apoptotic process has still to be determined. Other substrates of ICE-like proteases with a potential role in apoptosis are nuclear lamins, which are cleaved early in programmed cell death (Oberhammer et al., 1994). Currently, little is known about the role of members of the ICE family in retinoid-induced apoptosis.

Type II (Tissue) Transglutaminase

Transglutaminases (TGases) consist of a family of Ca^{2+}-dependent enzymes that catalyze the covalent cross-linking of proteins by forming (γ-glutamyl-ε-lysyl)-isopeptide bonds (Folk, 1980). The type I (epidermal) TGase and type II (tissue) TGase are two members of this gene family that have been linked to two different pathways of cell death, squamous cell differentiation and apoptosis, respectively (Fesus et al., 1987; Thacher and Rice, 1985). TGase I is mostly associated with the membrane and plays a role in the formation of the crosslinked envelope, a layer of crosslinked protein deposited beneath the plasma membrane, during squamous cell differentiation of a variety of keratinocytes (Thacher and Rice, 1985; Jetten et al., 1994). The localization of TGase II is largely restricted to the cytosol. Although the function of TGase II has not yet been precisely defined, several studies have indicated that it may have multiple functions. TGase II has been found to be involved in the activation of TGF-β1 (Kojima et al., 1993) and interleukin α (Eitan and Schwartz, 1993). It has also been demonstrated to crosslink specific components of the extracellular matrix (Martinez et al., 1994; Aeschlimann et al., 1995; Mehul et al., 1995). In addition, TGase II exhibits GTPase activity and as such may play a role in signal transduction (Nakaoka et al., 1994).

Although it is clear that TGase II expression is not always associated with apoptosis, many laboratories have demonstrated a correlation between TGase II expression and apoptosis in a variety of cell types *in vivo* and *in vitro* (Fesus et al., 1987; Davies et al., 1992; Piacentini et al., 1991; 1992; 1993; Zhang et al., 1995). Strong support for a role of TGase II in apoptosis was provided by studies examining the effect of ectopic expression of TGase II in cells. Overexpression of TGase II in human neuroblastoma SK-N-BE(2) cells renders these neural crest-derived cells highly susceptible to apoptosis while transfection with a human TGase II cDNA in the antisense orientation decreases both spontaneous and RA-induced apoptosis (Melino et al., 1994).

Although the role for TGase II in the apoptotic process has not yet been precisely established, it has been proposed that TGase II may crosslink intracellular proteins and therefore result in the formation of a detergent-insoluble scaffold that could temporarily stabilize the integrity of the dying cell before its clearance by phagocytosis (Fesus et al., 1989; 1991; Tarcsa et al., 1992; Davies et al., 1992). This may prevent the release of harmful intracellular components and the subsequent induction of inflammatory responses. Since in most cases the induction of TGase II by different triggers is correlated to the induction of apoptosis in a variety of cell types, TGase II expression has been recently widely used as an apoptotic marker. However, it should be emphasized that the association of TGase II with apoptosis is not universal.

In a number of cell types the induction of apoptosis by RA is accompanied by increased expression of TGase II (Fesus et al., 1987; Fesus, 1993, Davies et al., 1992; Knight et al., 1993; Piacentini et al., 1994; Zhang et al., 1995). Treatment of the human cervical adenoma Hela-TV (Piacentini et al., 1991), human neuroblastoma SK-N-BE(2) (Piacentini et al., 1991, 1992), rat tracheal SPOC-1 cells (Zhang et al., 1995), and several rat hepatoma cell lines (Fukuda et al., 1994) with RA has been shown to induce both programmed cell death and TGase II. However, in other cell lines the induction of apoptosis by retinoids is not associated with an increase in transglutaminase type II. For example, treatment of human acute promyelocytic leukemia NB4 cells, which contain a t(15;17) chromosomal translocation that fuses the *PML* gene on chromosome 15, to that of RARα on chromosome 17 (de Thé et al, 1991; Kakizuka et al., 1991; Bruel et al., 1995), with RA induces differentiation as indicated by an increase in NBT staining and the expression of the leukocyte

CD18 mRNA (Benedetti et al., 1996). And although these cells do not undergo apoptosis, TGase II is highly induced, suggesting that this induction of TGase II is associated with differentiation rather than apoptosis. This interpretation is supported by experiments with NB4-306 cells, which do not express PML-RARα. These cells are resistant to RA-induced differentiation and do not express TGase II after the addition of RA (Bruel et al., 1995; Benedetti et al., 1996). In contrast to RA, 4-HPR is able to induce apoptosis in NB4 cells but does not induce TGase II. These studies demonstrate that in these cells the induction of TGase II does not correlate with apoptosis.

C. Cell Cycle-Related Genes

The control of apoptosis is considered to be tightly linked to the control of the cell cycle (Meikrantz and Schlegel, 1995; King and Cidlowski, 1995; Evan et al., 1995). This is based on observations indicating that (1) apoptosis is often associated with proliferating tissues within the organism; (2) manipulation of the cell cycle can either potentiate or prevent apoptosis, depending on the point of growth arrest; Damage to the cell cycle or to DNA integrity has been shown to be a very effective trigger of apoptosis; and (3) proteins that control progression through the cell cycle are often involved in the induction of apoptosis.

The cell cycle is a very tightly controlled process that contains several checkpoints that serve as mechanisms that sense the progress through the cell cycle before continuing to the next stage. Several genes, encoding proteins that control the progression of cells through the cell cycle, have been demonstrated to be able to promote apoptosis (Sen, 1992; White, 1996; Evan et al., 1995: Darzynkiewicz, 1995). For example, overexpression of the tumor suppressor p53, the proto-oncogene c-myc, or E2F have been well documented to play important roles in the induction of programmed cell death (Shan and Lee 1994; Evan et al., 1995). Also loss of expression of the tumor suppressor Rb in Rb-/Rb- mice results in profound cell death in the nervous system (Lee et al., 1994).

p53

p53 is a multifunctional, nuclear phosphoprotein (Elledge and Lee, 1995) that modulates expression of genes important for apoptosis, the control of cell cycle check points, DNA replication and repair, and the maintenance of genome stability (Diller, 1990; Elledge and Lee, 1995). *p53* gene is not an absolute necessity for all forms of apoptosis since immature thymocytes lacking p53 die normally by apoptosis when exposed to glucocorticoids or compounds that may mimic T-cell receptor engagement. However, $p53^-$ cells are resistant to the lethal effects of ionizing radiation (Lowe et al., 1993). p53 acts as a transcriptional factor and induces growth arrest by altering the rate of transcription of specific genes, including the cdk inhibitor p21/WAF1/Cip1 (Harper et al., 1993; Elledge and Lee,

1995). Although p21/WAF1/Cip1 is induced during p53-mediated apoptosis, its role in apoptosis is not yet clear. A recent study provided evidence suggesting that increased p21/WAF1/Cip1 expression may relate to growth arrest rather than apoptosis (Kobayashi et al., 1995).

p53 plays a pivotal role in the control of the G1/S transition by switching on apoptosis in case the cell has undergone DNA damage (Yonish-Rouach et al., 1991), thereby preventing potentially mutant cells from propagating. Tumor cells that have lost functional p53 become refractory to induction of apoptosis by a range of genotoxic agents and radiation (Lowe et al., 1993). The importance of these defects is demonstrated by the fact that a majority of malignant tumors have mutations or rearrangements of both copies of the *p53* gene on human chromosome 17.

The induction of apoptosis by the retinoid AHPN in mammary carcinoma MCF-7 cells that harbor the wild-type p53, and MDA-MB-231 cells that possess a mutant p53, is independent from p53 (Shao et al., 1995). Although AHPN induces p21/ WAF1/Cip1 in these cells, this occurs by a p53-independent mechanism. The induction of apoptosis in several leukemic cell lines by 4-HPR was found to be associated with a decrease in the level of p53 mRNA while the p53 protein levels remained unchanged suggesting that in these cells apoptosis occurs independent of p53 (Delia et al., 1995). Also the induction of apoptosis in small-cell lung carcinoma cell lines by 4-HPR has been reported to be independent of p53 (Kalemkerian et al., 1995). These results indicate that at this time there is not great support for a role of p53 in retinoid-induced apoptosis.

Expression of wild-type p53 in promyelocytic leukemia HL-60 cells has been demonstrated to cause induction of apoptosis (Noguchi et al., 1995). Both RA and granulocyte macrophage colony stimulating factor (GM-CSF) inhibit this induction of apoptosis. Inhibitors of protein kinase C abolished the inhibitory effect of GM-CSF but not that of RA, suggesting that different mechanisms are involved in the action of these two agents.

c-myc

c-myc is a DNA-binding protein that resembles transcriptional factors of the basic helix-loop-helix leucine zipper family. It appears to control two different sets of genes, one involved in controlling proliferation, the other in apoptosis (Evan et al., 1995). c-myc can induce apoptosis under conditions of growth arrest, such as serum deprivation or thymidine block (Evan et al., 1992). In addition, the expression of endogenous c-myc is required for the activation-induced apoptosis of T-cell hybridomas (Shi et al., 1992).

The effects of retinoids on growth inhibition and apoptosis and its relation to c-myc expression are rather ambivalent. A correlation between the induction of apoptosis and the upregulation of c-myc gene expression by all-*trans*-RA has been observed in human ovarian carcinoma cell lines (Krupitza et al., 1995). In addition, small-cell lung carcinoma cell lines that overexpress either c-myc, L-myc, or

N-myc are more sensitive to 4-HPR-induced apoptosis than small-cell carcinoma cells that do not express myc (Kalemkerian et al., 1995). However, in mammary carcinoma MCF-7 cells RA-induced apoptosis is not linked to an effect on c-myc (Seewaldt et al., 1995). And although the induction of apoptosis in several leukemic cell lines by 4-HPR is associated with a suppression of c-myc transcription, it appears that c-myc does not control apoptosis in these cells either (Delia et al., 1995). The inhibition of proliferation of human prostate adenocarcinoma PC3 cells by 4-HPR has been reported to be due to a block in the G1 to S transition of the cell cycle and found to be associated with a decrease in c-myc expression (Igawa et al., 1994).

III. REGULATION OF APOPTOSIS BY RETINOIDS

An increasing number of reports have demonstrated that retinoids regulate apoptosis (Piacentini et al., 1991; Delia et al., 1993; Fesus et al., 1995; Lotan, 1995; Zhang and Jetten, 1995; Kalemkerian et al., 1995; Shao et al., 1995; Seewaldt et al., 1995; Yang et al., 1995a; 1995b). The effect of retinoids on apoptosis appears to be dependent on the chemical nature of the retinoid and the cell type used, and to involve different mechanisms. Retinoids can promote apoptosis in a variety of normal cell types and nontumorigenic cell lines. RA induces programmed cell death in the wound epidermis of regenerating zebrafish fins (Ferretti and Geraudie, 1995), and inhibits chondrogenesis and promotes apoptosis in cultures of mouse interdigital mesoderm (Lee et al., 1994). RA has also been reported to induce apoptosis between digits 2, 3, 4, and 5 of the limb in Hammertoe (Hm) mutant mice (Zakeri and Ahuja, 1994). Retinoids have been reported to promote apoptosis in human early erythrocytic progenitor cells (Gandrillon et al., 1994), human neuronal cells (Howard et al., 1993; Mailhos et al., 1994), human lymphocytes (Helmberg et al., 1995), rat tracheal epithelial cells (Zhang et al., 1995), mouse 3T3-L1 preadipocytes (Chawla and Lazar, 1994), and mouse limb bud core mesenchymal cells (Jiang and Kochhar, 1992; Kochhar et al., 1993).

The effect of retinoids on apoptosis has been extensively studied in human tumor cell lines. Retinoids have been reported to induce apoptosis in neuroblastoma cells (Fesus et al., 1987; Piacentini et al., 1991; 1993; Mariotti et al., 1994; Ponzoni et al., 1995), squamous carcinoma cells (Sacks et al., 1990), lung carcinoma cells (Zhang et al., 1995), ovarian carcinoma cells (Krupitza et al., 1995), cervical carcinoma cells (Oridate et al., 1995), head and neck carcinoma cells (Lotan, 1995), several human small-cell lung cancer cells (Kalemkerian et al., 1995), breast cancer cells (James et al., 1995; Shao et al., 1995; Seewaldt et al., 1995), malignant haematopoietic cells (Delia et al., 1993), hepatoma-derived HuH-7 cells (Nakamura et al., 1995), erythroblastosis virus-transformed haematopoietic progenitor cells (Gandrillon et al., 1994), several myeloid leukemia cells (Martin et al, 1990; Park et al., 1994; Tosi et al., 1994; Delia et al., 1995; Bunce et al., 1995; Nagy et al.,

1995; Wallington et al., 1995), myelomonocytic leukemia P39 cells (Anzai et al., 1994), t(15;17) acute promyelocytic leukemia NB4 cells (Bruel et al., 1995), and B lymphoma cells (Turley et al., 1995). Retinoids can also induce programmed cell death in murine embryonic carcinoma F9 (Atencia et al., 1994) and PCC4aza1R cells (Adachi and Jetten, 1996).

Study of the process of retinoid-induced apoptosis is particularly relevant to the understanding of the mechanisms by which retinoids inhibit the proliferation of tumor cells *in vitro* and *in vivo* and prevent tumor formation. Insight into these mechanisms may lead to the discovery of more effective strategies in the retinoid-treatment of cancer.

IV. MECHANISMS OF REGULATION OF APOPTOSIS BY RETINOIDS

A. Retinoid Receptor-Dependent Mechanisms

Only recently, several studies have provided some insight into the mechanisms by which retinoids promote or inhibit apoptosis. These studies demonstrated that retinoids exert their effects on apoptosis through several different mechanisms depending on the cell type and the nature of the retinoid tested. In addition, the presence or absence of other hormonal factors or cytokines has also been found a determinant in the effect of retinoids. In several cell types the induction of apoptosis by retinoids has been reported to be dependent on the activation of different nuclear retinoid receptors (RARs and/or RXRs). Although retinoids have been demonstrated to exert anti-AP1-activity, an effect that does not involve an interaction of the receptors with retinoid response elements (RAREs or RXREs), no study has yet reported a regulation of apoptosis that is mediated through this mechanism. However, several retinoids have been shown to induce apoptosis through a receptor-independent mechanism. The induction of apoptosis by 4-HPR may involve the generation of reactive oxygen while the induction of AHPN occurs through an as yet unknown mechanism.

Nuclear Retinoid Receptors

Many of the effects of retinoids on gene expression are mediated by the activation of the RAR and RXR nuclear retinoid receptors (Giguère, 1994; Mangelsdorf and Evans, 1995; Kastner et al., 1995). The RAR and RXR gene family each comprises three subtypes named α, β, and γ. These receptors belong to a superfamily of ligand-dependent transcriptional factors that include the steroid hormone and thyroid hormone receptors and an increasing number of orphan receptors for which the ligand has not yet been identified (Mangelsdorf and Evans, 1995; Beato et al., 1995). Members of this superfamily share a common modular structure consisting of four major domains (Evans, 1988; Glass, 1994). The DNA-binding domain is the most conserved among the nuclear receptors and is composed of two "zinc-finger" motifs that play a role in DNA recognition and protein-protein interactions (Truss and Beato, 1993; Glass, 1994). The ligand binding domain at the carboxyl-terminus is involved in several functions, including ligand binding, receptor dimerization, nuclear translocation, and transcriptional activation. The function of the amino-terminal domain and the hinge domain, which separates the DNA-binding and ligand-binding domains, is still poorly understood. The amino-terminal domain has been reported to contain a transactivation function.

Nuclear receptors control the transcription of target genes by binding to DNA sequences referred to as hormone response elements. Most members of this superfamily bind as homo- or heterodimers to *cis*-acting DNA sequences that contain two core motifs of PuGGTCA configured in either direct, palindromic, or inverted repeats separated by a spacer of different length (Truss and Beato, 1993; Glass, 1994). Members of the RXR family have been shown to be able to form homodimers that bind to direct repeats spanned by one nucleotide (DR1)(Giguère, 1994). In addition, RXRs form heterodimers with a variety of other members of the superfamily including RARs, the thyroid hormone, vitamin D, and peroxisome proliferator activating receptor (Mangelsdorf et al., 1995). RAR-RXR heterodimers bind with the highest affinity to direct repeats spanned by five nucleotides (DR5)(Zechel et al., 1994).

The activation of RAR and RXR receptors is mediated through the binding of specific ligands. All-*trans*-RA binds to RARs but is unable to bind to or activate RXRs, while 9-*cis*-RA is able to bind to and activate both receptor subtypes (Giguère, 1994). A number of synthetic retinoids have been synthesized that are able to selectively activate or antagonize either the RAR or RXR receptors (Delescluse et al., 1991; Lehmann et al., 1991; 1992; Apfel et al., 1992; Martin et al., 1992; Heyman et al., 1992; Levin et al., 1992; Fanjul et al., 1994). Figure 2 shows the structures of several synthetic retinoids that have been useful tools in the study of the retinoid signaling pathways.

Recently, it has been demonstrated that the retinoid receptors interact with other nuclear proteins that function either as repressors or as coactivators. The binding of the ligand causes a conformational change of the receptor and thereby alters the interactions of the receptor with other nuclear proteins. In one scenario, binding of RA to RAR induces dissociation of a repressor and allows association of a coactivator resulting in transactivation of the target gene (Hörlein et al., 1995; Chen and Evans, 1995; Onate et al., 1995; vom Bauer et al., 1996). In another scenario, binding of RA leads to binding of a repressor to the receptor, resulting in the repression of transcription of the target gene (Chen and Evans, 1995; Kurokawa et al., 1995).

The RAR and RXR receptors are expressed in a development- and tissue-specific manner (reviewed by Gudas et al., 1994; Giguère, 1994), and each may regulate the expression of different genes. This is supported by recent studies showing that retinoids regulate apoptosis, differentiation, and expression of specific genes

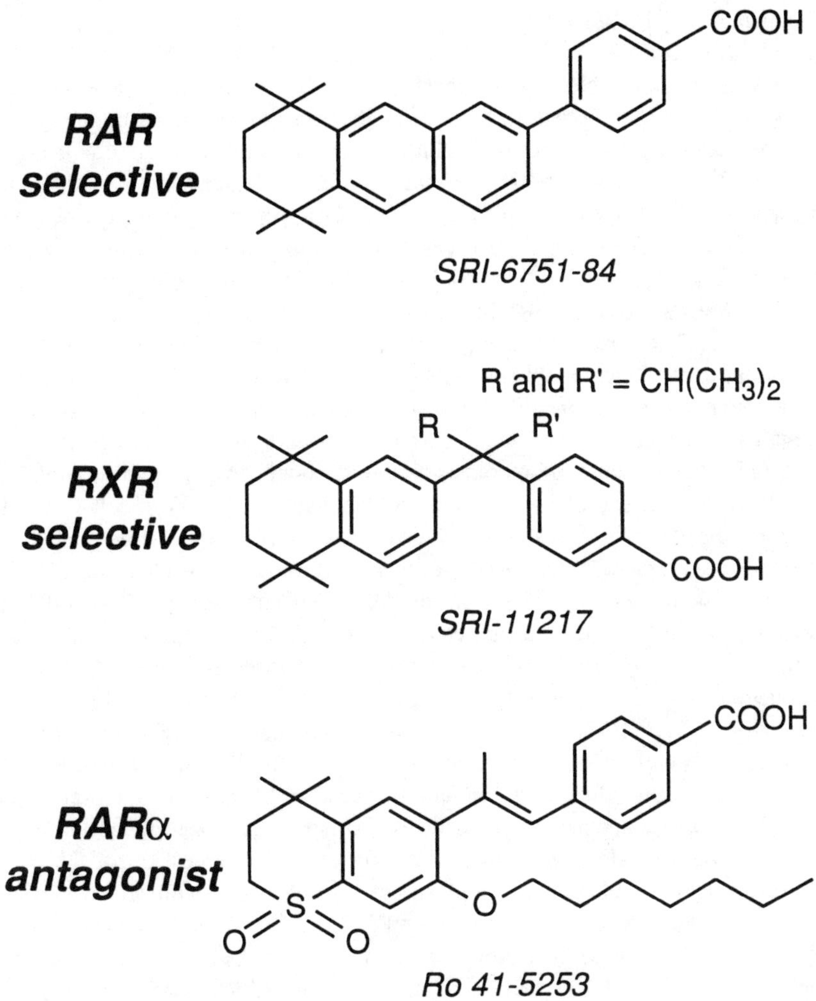

Figure 2. Chemical structure of two retinoid receptor-selective agonists and antagonist. RAR- and RXR-selective agonists, SRI-6751-84 and SRI11217, respectively. Ro40-6055, RARα-selective antagonist.

through specific retinoid receptor signaling pathways (Boylan et al., 1995; Zhang et al, 1995; Seewaldt et al., 1995; Yang et al., 1995c; Liu et al, 1996).

RARα-mediated Induction of Apoptosis

Recently, our laboratory demonstrated that RA is able to induce apoptosis in rat tracheal SPOC-1 cells and human lung carcinoma cells H-460 as indicated by the appearance of apoptotic cells by electron microsopy and the induction of DNA fragmentation ("DNA ladder")(Zhang et al., 1995). SPOC-1 cells express RARα, γ, and RXRβ mRNAs, and after RA-treatment also RXRα, while RARβ, RXRγ mRNAs are undetectable. To analyze which retinoid signaling pathway is involved in the induction of apoptosis in these cells, the effect of several retinoid receptor-selective agonists, an antagonist, and a dominant-negative truncated *RARα* gene (RARα403)(Tsai et al., 1992) was analyzed. These studies indicated that the induction of apoptosis by retinoids is mediated by a specific signaling pathway that involves RARα. This conclusion was supported by several findings. First, the RAR-selective retinoid SRI-6751-84 is very effective in inducing apoptosis while the RXR-selective retinoid SRI-11217 is unable to cause apoptosis in SPOC-1 cells (Figure 3A). These results indicate that activation of RARs rather than RXRs is important in mediating apoptosis in these cells. However, this does not imply that RXRs are not involved since RARs mediate their action as part of a heterodimeric complex with RXRs. Second, the RARα-selective antagonist Ro41-5253 is able to antagonize the induction of apoptosis by the RAR-selective retinoid (Figure 3A). Third, overexpression of the truncated *RARα* gene RARα403 in SPOC-1 cells inhibited the induction of apoptosis by the RAR-selective retinoid (Figure 3B). The RARα403 acts as a dominant-negative receptor; however the mechanism by which it exerts its dominant-negative action has still to be determined. The RARα403 has been reported to be able to form heterodimers with RXR and to bind RA allbeit with a 12-fold lower affinity (Tate et al., 1994). Overexpression of RARα403 may compete with endogenous RARs for binding to RXRs thereby inhibiting the formation of RAR-RXR heterodimers and transactivation of target genes. Fourth, in contrast to the RARγ-selective retinoid CD437, the RARα-selective retinoid Ro40-6055 was able to strongly induce apoptosis in SPOC-1 cells. Observations similar to those for apoptosis were made for the regulation of transglutaminase type II. These findings demonstrate that in these cells the expression of TGase II correlates with the induction of apoptosis and is in agreement with a role for TGase II in the apoptotic process. Since RARβ mRNA is undetectable and is not induced by RA in SPOC-1 cells, these observations indicate that the induction of apoptosis and TGase II in SPOC-1 cells occurs through an RARα-dependent signaling pathway (Zhang et al., 1995; Zhang and Jetten, 1995).

In addition to apoptosis, SPOC-1 cells can undergo another program of cell death, namely that of squamous differentiation and cornification (Zhang et al., 1995). These cells undergo squamous cell differentiation and cornification after reaching confluence and start to express several squamous cell-specific genes, including TGase I, cornifin, and keratin 13. Treatment of these cultures with RA strongly inhibits the expression of TGase I and other squamous markers (Jetten et al., 1994; Zhang et al., 1995). These results demonstrate that while inducing apoptosis in SPOC-1 cells, RA inhibits another cell death program. Thus, retinoids appear to control homeostasis in squamous differentiating tissues in a complex manner (Figure 4).

RARβ-mediated Induction of Apoptosis

RA has been reported to inhibit the growth of several mammary carcinoma cell lines (van der Burg et al., 1993; Sheikh et al., 1993; Seewaldt et al., 1995). The repression of cellular proliferation has been demonstrated to be due at least in part to an induction of apoptosis. These effects have been linked to the presence and activation of RARβ (Seewaldt et al., 1995; Liu et al., 1996).

The responsiveness of mammary carcinoma cells to growth inhibition by retinoids is dependent on the expression of the estrogen receptor (ER). Several

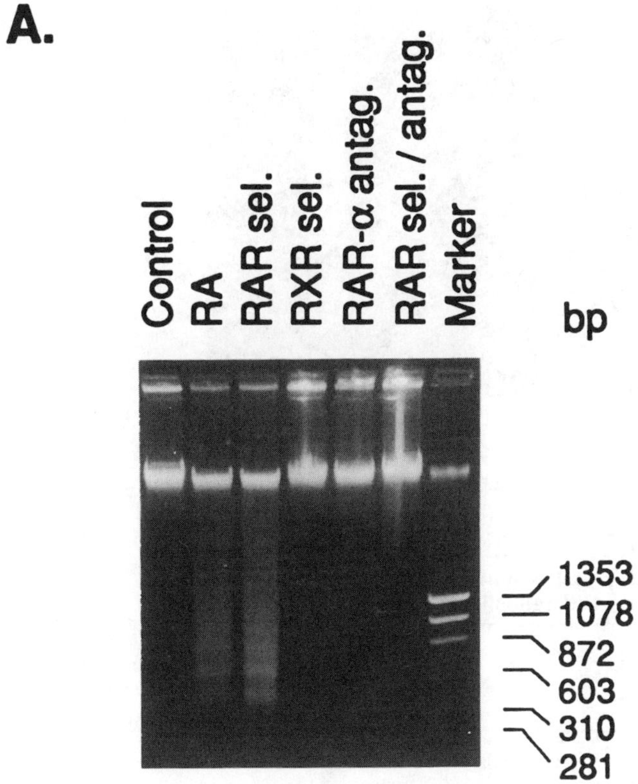

Figure 3. Induction of DNA fragmentation in SPOC-1 cells by retinoids. (A) Cells were treated with 10^{-6} M all-*trans*-RA, 10^{-8} M of an RAR- or RXR-selective retinoid (SRI-6751-84 and SRI11217, respectively) or 2X 10^{-6} M RARα antagonist (Ro40-6055) in the presence or absence of 10^{-8} M RAR-selective retinoid for 4 days. DNA was isolated and separated by 1 percent agarose gel electrophoresis. (B) SPOC-1 cells, stably transfected with the plasmid LRARα403SN encoding the dominant-negative receptor RARα403 or with the control plasmid LXSN, were treated in the presence or absence of 10^{-8} M RAR-selective retinoid for 4 days (from Zhang et al., 1995, with permission).

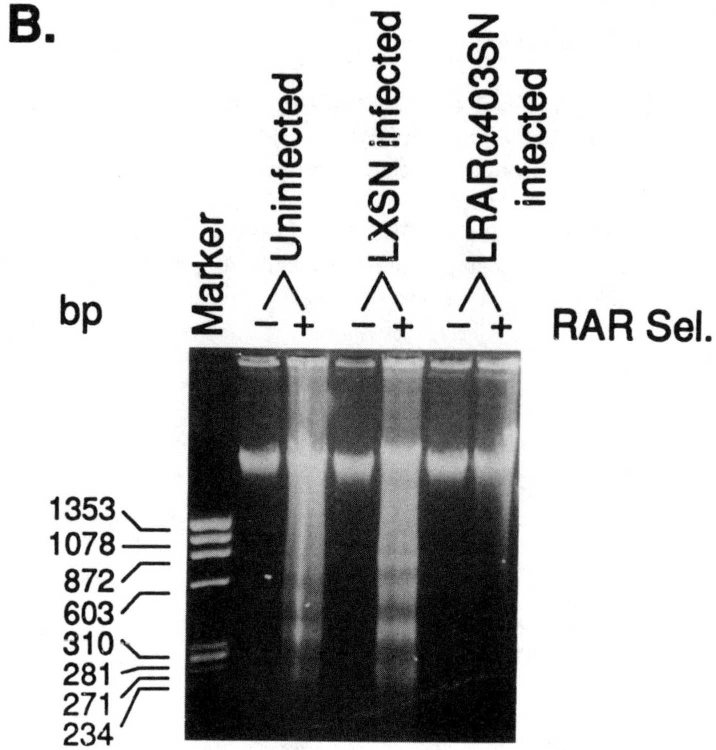

Figure 3. Continued.

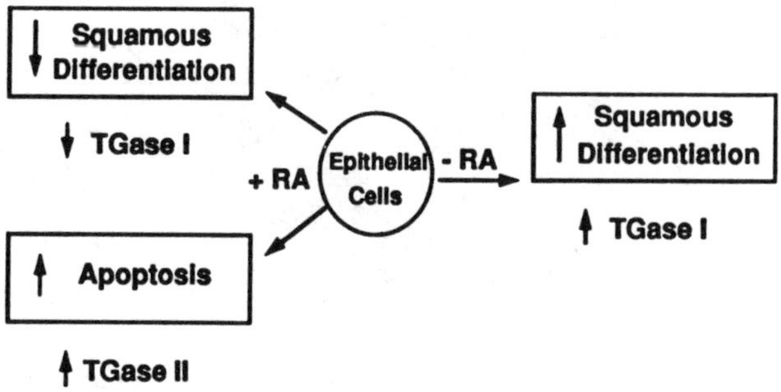

Figure 4. Regulation of two different cell-death programs by retinoids. Treatment of rat tracheal SPOC-1 cells with retinoic acid inhibits squamous differentiation as indicated by the suppression of the squamous-specific gene, TGase I, while promoting apoptosis and the expression of TGase II (from Zhang et al., 1995, with permission).

ER-positive cell lines, including MCF-7, ZR-75-1 and T-47D, are inhibited by RA, while ER-negative cell lines, including BT-20, MB-468, and MB231, are refractory to RA-induced growth inhibition. RARβ expression was greatly enhanced after RA treatment in the hormone-dependent cell lines, whereas it did not increase RARβ mRNA in the hormone-independent cell lines (van der Burg et al., 1993; Sheikh et al., 1993; Seewaldt et al., 1995). The correlation between responsiveness of mammary carcinoma cells to RA-growth inhibition and the induction of RARβ suggested that RARβ expression is important to the growth-inhibition by RA. This was supported by experiments showing that expression of RARβ in the hormone-independent cell lines restored the ability of RA to inhibit growth and apoptosis (Seewaldt et al., 1995, Liu et al., 1996). Moreover, increased expression of RARα in mammary carcinoma cells expressing low levels of RARα also resulted in growth inhibition by RA. This latter effect is probably due to the observed induction of RARβ, which has been shown to depend on the activation of RARα. A role for RARβ in RA-induced apoptosis was further indicated by the expression of RARβ antisense RNA in the ER-positive cell line ZR-75-1 (Liu et al., 1996). These cells exhibited a greatly reduced sensitivity to RA-induced growth inhibition and apoptosis. These results suggest that the induction of apoptosis and growth inhibition by RA in mammary carcinoma cells are linked to the activation of RARβ. This raises the possibility that RARβ may be an important regulator of proliferation in normal mammary epithelial cells as well. This is supported by experiments showing a correlation between increasing RARβ expression in senescing normal mammary epithelial cells (Swisshelm et al., 1994). In addition, defects in the expression of RARβ in mammary tumor cells may result in the loss of a negative growth control mechanism (van der Burg et al., 1993; Sheikh et al., 1993; Swisshelm et al., 1994) and be part of the process of tumor progression in this tissue.

Several teratogenic effects of retinoids, such as the retinoic acid-induced truncation of the embryonic limb (Kochhar et al., 1993; Jiang et al., 1994) have been reported to involve apoptosis and shown to be associated with increased expression of transglutaminase type II (Jiang and Kochhar, 1992). Evidence has been provided indicating that activation of the RARβ receptor may mediate some of these effects.

RXR-Dependent Induction of Apoptosis

The human myeloblastic leukemia cell line HL-60 can differentiate into either granulocytes or macrophages in response to a variety of stimuli (Collins et al., 1990). RA triggers differentiation of HL-60 cells into granulocytes as well as apoptosis (Martin et al., 1990; Nagy et al., 1995). It has been demonstrated that the induction of differentiation is a prerequisite for these cells to undergo apoptosis and that these two processes are mediated by two different retinoid receptor signaling pathways (Martin et al., 1990; Nagy et al., 1995). The induction of differentiation by retinoids depends on the activation of an RAR signaling pathway as indicated by experiments showing that RAR-selective retinoids, like (E)-4-[2-(5,6,7,8-tetrahydro-5,5,8,8-tetramethyl-2-naphthalenyl)-1-propenyl]benzoic acid (TTNPB), are able to induce differentiation whereas RXR-selective retinoids are inactive (Nagy et al., 1995). Since HL-60 cells have been reported to express, in addition to RXRα and RXRβ, only RARα (Nagy et al., 1995), it was concluded that activation of RARα is involved in the induction of differentiation in these cells. These studies support the conclusions reached earlier by Collins and colleagues (Collins et al., 1990) that were based on the observed correlation between the absence of an active RARα and the resistance to retinoid-induced differentiation in HL-60R cells, a subclone of HL-60 (Collins et al., 1990). In addition, these studies demonstrated that expression of RARα in HL-60R cells restored their ability to undergo differentiation after RA treatment while HL-60 cells expressing a dominant-negative RARα are unable to differentiate (Collins et al., 1990).

In contrast to differentiation, induction of apoptosis in HL-60 by retinoids involves the activation of RXR receptors. Neither treatment with RAR-selective nor RXR-selective retinoids alone is able to induce apoptosis in HL-60 cells, but only panagonists that activate both RARs and RXRs induce apoptosis (Nagy et al., 1995). HL-60 cells, allowed to differentiate after treatment with an RAR-selective retinoid, only undergo apoptosis when subsequently treated with RXR-selective retinoids or panagonists, suggesting that activation of RXRs is involved in the induction of apoptosis and that differentiation is a prerequisite for these cells to undergo apoptosis (Martin et al., 1990; Nagy et al., 1995; Boehm et al., 1995). In addition, these studies indicate that differentiation and apoptosis in HL-60 cells are under separate controls. This is supported by findings showing that overexpression of bcl-2 in HL-60 cells prevents RA-induced apoptosis but has no effect on the ability of RA to induce granulocyte differentiation (Park et al., 1994). These results also indicate that bcl-2 is involved in regulating apoptosis in maturing granulocytes. Studies by Mehta and coworkers (Mehta et al., 1996) using HL-60R cells showed that these cells are unable to undergo apoptosis after RA treatment. HL-60R-RARα cells, in which RARα is overexpressed, undergo differentiation after treatment with RA while overexpression of RXRα followed by treatment with 9-*cis*-RA induces apoptosis. Although the results of this study largely support the findings with HL-60 cells described by other investigators (Martin et al., 1990; Collins et al., 1990; Nagy et al., 1995), they differ in the respect that in HL-60R-RARa differentiation is not a requirement for apoptosis.

The results obtained with HL-60 cells contrast with those reported for another promyelocytic leukemia cell line, NB4. These cells contain a t(15;17] rearrangement involving PML and RARα resulting in the synthesis of a PML-RARα fusion protein. NB4 cells differentiate in the presence of all-*trans*-RA and 9-*cis*-RA. This differentiation is accompanied by a down-regulation of bcl-2; however, these cells do not undergo apoptosis. It is possible that the t(15;17] translocation is responsible for increased survival after maturation due to the anti-apoptotic action of PML-RARα (Bruel et al., 1995). This hypothesis is supported by evidence showing that U937 myeloid precursor cells expressing the PML-RARα protein exhibit a higher growth rate consequent to diminished apoptotic cell death (Grignani et al., 1993). In a derivative of NB4, referred to as NB4-R1, which does not undergo extensive differentiation, 9-*cis*-RA and to a lesser degree all-*trans*-RA are able to induce apoptosis. In these cells, bcl-2 is highly expressed and is not down-regulated by RA, suggesting that this induction of apoptosis is not under the control of bcl-2. In NB4-R1 cells, cAMP is able to restore apoptosis-coupled maturation in RA-treated cells (Bruel et al., 1995). The correlation between bcl-2 expression and apoptosis found in HL-60 cells (Park et al., 1994) discussed earlier are in contrast to the findings in NB4 and NB4-R1 in which down-regulation of bcl-2 appears to correlate with differentiation rather than apoptosis (Bruel et al., 1995).

RXR-Dependent Inhibition of Apoptosis

In certain cell systems retinoids exhibit a dual effect on apoptosis. In thymocytes, RA can have a stimulatory or an inhibitory effect on apoptosis (Yang et al., 1993; Iwata et al., 1992). Activation-induced apoptosis of immature thymocytes is important in the removal of potentially autoreactive T cells (Green and Martin, 1995). Activation of T-cell hybridomas via antigen-specific T-cell receptors rapidly induces a block in the G1/S phase of the cell cycle that is followed by apoptotic cell death. RA inhibits activation-induced apoptosis in human T-cell hybridomas and thymocytes (Iwata et al., 1992; Yang et al., 1993; 1995a; 1995c). Feeding mice a vitamin A-supplemented diet has been reported to cause enlargement of the thymus and an increase in the number of thymocytes (Seifter, 1981) that may be related to an inhibition of apoptosis by retinoids. RA also inhibits apoptosis in thymocytes induced by the combination of phorbol esters and ionomycin (Iwata et al., 1992) and has been shown to inhibit the *ex vivo* apoptosis of peripheral blood lymphocytes from patients infected with human immunodeficiency virus (Yang et al., 1995b). In contrast, RA has been reported to stimulate dexamethasone-induced apoptosis in thymocytes (Iwata et al., 1992).

Activation-induced apoptosis is mediated through an up-regulation of the expression of the ligand for the Fas receptor, thereby triggering the Fas signal pathway. The Fas/APO-1/CD95 is a 36 kD protein that belongs to a membrane receptor family that includes the tumor necrosis factor receptors and the nerve growth factor (Smith et al., 1994). The Fas ligand consists of a 40 kD membrane protein that occurs in a soluble form after cleavage. Both the membrane and the soluble form of the Fas ligand can bind to Fas and induce apoptosis in T cells (Itoh and Nagata, 1993). Retinoids have been shown to inhibit apoptosis in thymocytes by repressing Fas ligand expression (Yang et al., 1995). All-*trans*-RA was found to be 10 times less active in suppressing activation-induced apoptosis than 9-*cis*-RA. In addition, an RAR-selective retinoid had little effect while an RXR-selective retinoid was very effective in inhibiting apoptosis in these cells suggesting that activation of RXRs are important in mediating this action. This was further confirmed by experiments with T-cell hybridoma cells in which RXRβ was overexpressed (Yang et al., 1995). These cells were more sensitive to the suppression of apoptosis by the RXR-selective retinoid. In contrast, cells expressing a dominant-negative RXRβ could not be rescued from activation-induced cell death (Yang et al., 1995).

Inhibition of apoptosis by retinoids has been observed in several other cell systems. Retinoids suppress apoptosis induced by serum- and growth factor-deprivation in human fetal haemopoietic progenitor CD34+ cells (Zauli et al., 1995). Over-expression of wild type *p53* gene in HL-60 cells induces cell death with morphological changes typical of apoptosis. This induction of apoptosis can be inhibited by treatment with all-*trans*-RA (Noguchi et al., 1995).

B. Retinoid Receptor-Independent Mechanisms

Induction of Apoptosis by N-(4-hydroxyphenyl) Retinamide (4-HPR)

Fenretinide or N-(4-hydroxyphenyl) retinamide (4-HPR, Figure 5) is a synthetic RA derivative that has been used in cancer chemoprevention and therapy (Pollard et al., 1991; Veronesi et al., 1992; Chiesa et al., 1992; Moon et al., 1994; Oridate et al., 1995; 1996). 4-HPR reduces significantly the incidence of carcinogen-induced mammary, lung, bladder, and skin tumors in animals and protects against the development of ovarian carcinoma in humans (Welsch ct al., 1983; Moon and Mehta, 1989; De Palo et al., 1995). 4-HPR treatment has also been shown to inhibit prostate cancer in rats (Pienta et al., 1993). It is active against human ovarian carcinoma transplanted in mice and potentiates the activity of *cis*-platin (Formelli and Cleris, 1993). *In vitro*, 4-HPR at concentrations of 1 μM and higher inhibits the growth of many tumor cell lines, including mammary carcinoma (Marth et al., 1985; Bhatnagar et al., 1991), small-cell lung carcinoma (Kalemkerian et al., 1995), leukemia and lymphomas (Delia et al., 1993), prostate carcinoma (Igawa et al., 1994, Hsieh et al., 1995), and neuroblastoma (Di Vinci et al., 1994; Mariotti et al., 1994; Ponzoni et al., 1995). It is interesting to note that of the lung carcinoma cell lines examined only the small-cell carcinoma but not squamous or adenocarcinoma cell lines were sensitive to 4-HPR (Kalemkerian et al., 1995). In addition to anchorage-dependent growth, 4-HPR can also inhibit colony formation in soft agar of several carcinoma cell lines (Meyskens et al., 1983). Although the concentrations of 4-HPR are relatively high (1 μM and higher), they are in a range attainable in clinical trials without causing great toxicity (Routmensz et al., 1991). In several cell systems, the growth inhibitory action by 4-HPR has been contributed to the induction of apoptosis as shown by morphological observations, DNA fragmentation assays, flow cytometry, and TUNEL assays. The mechanism by which 4-HPR induces apoptosis has yet to be determined. Although 4-HPR appears to be a weak but highly selective activator of RARγ (Fanjul et al., 1996), its apoptosis-inducing effects appear not to be related to the activation of RARs since other retinoids that are able to activate RARs were unable to induce apoptosis. These results suggest that 4-HPR mediates its action through a retinoid receptor-independent mechanism.

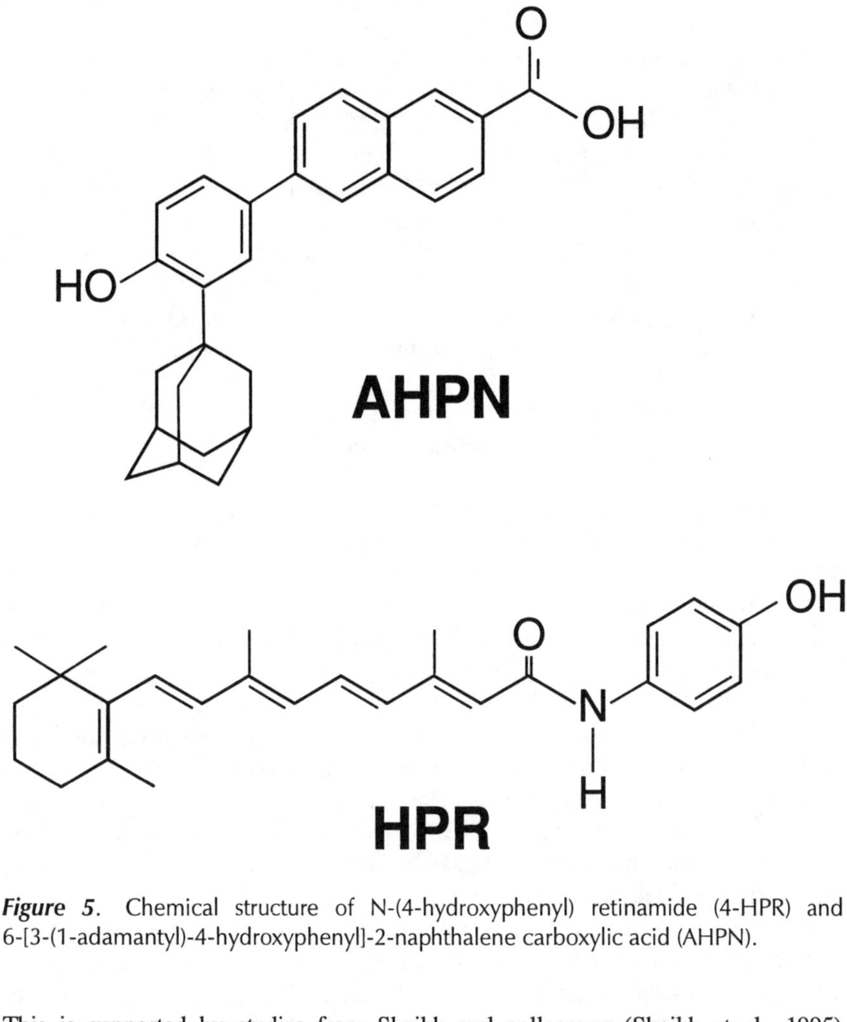

Figure 5. Chemical structure of N-(4-hydroxyphenyl) retinamide (4-HPR) and 6-[3-(1-adamantyl)-4-hydroxyphenyl]-2-naphthalene carboxylic acid (AHPN).

This is supported by studies from Sheikh and colleagues (Sheikh et al., 1995) showing that in mammary carcinoma cells 4-HPR only weakly induces RARE- and RXRE-dependent transactivation. The different mechanism of action of 4-HPR and RA is further illustrated by observations showing that the proliferation of small-cell carcinoma cells is inhibited by 4-HPR but is resistant to RA treatment (Kalemkerian et al., 1995). Moreover, in contrast to the action of RA, which induces differentiation without apoptosis in several human neuroblastoma cell lines, treatment with 4-HPR causes programmed cell death (Ponzoni et al., 1995; Di Vinci et al., 1994). Unlike RA, 4-HPR does not possess anti-AP-1 activity. Therefore, the inhibition of cell proliferation by 4-HPR seems not to involve anti-AP1 activity but appears to be mediated by a unique mechanism (Sheikh et al., 1995).

The induction of apoptosis by 4-HPR is not accompanied by any great changes in the expression of known mediators of apoptosis including p53, bcl-2, bax, or c-myc in lung carcinoma and leukemic cells (Kalemkerian et al., 1995; Delia et al., 1995). Ectopic expression of bcl-2 in HL-60 cells delays but does not prevent apoptosis after 4-HPR treatment (Delia et al., 1995). In the androgen-independent, human prostatic carcinoma cells JCA-, 4-HPR-treatment induces apoptosis and down-regulates cyclin D and E, and p34cdc2 (Hsieh et al., 1995). The precise mechanism by which 4-HPR induces apoptosis has yet to be elucidated. A recent study reported that several anti-oxidants, including N-acetyl-L-cysteine, L-ascorbic acid, and α-tocopherol, inhibit 4-HPR induced apoptosis, suggesting that activation of an oxidative pathway may be involved (Delia et al., 1995). However, further studies are needed to determine whether this is a valid mechanism.

Induction of Apoptosis by 6-[3-(1-Adamantyl)-4-hydroxyphenyl]-2-naphthalene Carboxylic Acid (AHPN)

Recently, it has been demonstrated that the synthetic retinoid AHPN (SRI11248 or CD437)(Figure 5) inhibits the proliferation of several human lung carcinoma (Adachi et al., 1997) and mammary carcinoma cell lines (Shao et al., 1995). In contrast to 4-HPR, which inhibits the growth of only small cell lung carcinoma cells, AHPN inhibits the growth of small-cell-, adeno-, as well as squamous cell carcinoma cells (Adachi et al., 1997). Most of these lung carcinoma cell lines are resistant to growth inhibition by RA. In mammary carcinoma cells, the effect of AHPN is very different from the growth-inhibitory action of RA. RA inhibits largely the proliferation of ER-positive cell lines (Roman et al., 1992; Sheikh et al., 1993), whereas AHPN inhibits the growth of both ER-positive cell lines MCF-7 and T47D as well as the ER-negative cell lines MDA-MB-231 and MDA-MB-468 (Shao et al., 1995). This growth inhibition has been related to induction of apoptosis as shown by DNA fragmentation (Shao et al., 1995). AHPN does not bind RARa or RARb and binds poorly to the RARγ receptor. It induces only very weakly RARE-dependent transactivation and is unable to induce RXRE-dependent transactivation of a CAT reporter gene. AHPN does not exhibit any anti-AP1 activity. These results suggest that the action of AHPN is not mediated through RAR or RXR receptors. The mechanism by which AHPN induces apoptosis remains to be elucidated.

AHPN increases the levels of WAF1/CIP1/p21 in both MCF-7 and MDA-MD-231 cells (Shao et al., 1995), and lung carcinoma cells (Adachi et al., 1997). Since MDA-MD-231 cells contain a mutated p53 that is unable to bind the p53 consensus sequence, it is unlikely that the increase in WAF1/CIP1/p21 occurs through a p53-dependent mechanism. WAF1/CIP1/p21 is able to form a complex with and inhibit the activity of cyclin-dependent kinase (cdk)/cyclin complexes, thereby blocking their activity and the progression of the cells through the cell cycle. The increase in WAF1/CIP1/p21 may at least in part be responsible for the G0/G1 cell cycle arrest caused by AHPN. AHPN also decreased the level of bcl-2 mRNA in

MCF-7 cells while it increased bax mRNA expression (Shao et al., 1995). These RNAs encode proteins that, respectively, inhibit and promote apoptosis. Therefore, the differential regulation of bcl-2 and bax could be involved in the induction of apoptosis in MCF-7 cells. However, bcl-2 and bax mRNA were undetectable in MDA-MD-231 cells suggesting that the regulation of bcl-2 and bax are probably not a general mechanism by which AHPN induces apoptosis in mammary carcinoma cells.

Nakamura and colleagues (Nakamura et al., 1995) have reported on another synthetic retinoid, all-*trans* 3, 7, 11, 15-tetramethyl- 2, 4, 6, 10, 14-hexadecapentaenoic acid, or acyclic retinoid. This retinoid rapidly induces apoptosis in the human hepatoma-derived cell line HuH-7. However, neither all-*trans*-RA nor 9-*cis*-RA induce apoptosis in these cells, suggesting that the action of this synthetic retinoid also occurs through a retinoid receptor-independent mechanism.

Interaction with Other Factors

In a number of cell systems, retinoids exert their effect on apoptosis through interaction with a variety of other signaling pathways, including those mediated by glucocorticoids, vitamin D derivatives, vitamin E derivatives, growth factors, cytokines, serum, lipids, and agents that activate protein kinase C and A (Mailhos et al., 1994; Krupitza et al., 1995). Many of these studies are, however, rather descriptive and provide relatively little insight into their molecular mechanisms.

Wallington and coworkers (Wallington et al., 1995) reported that 9-*cis* RA promoted significantly the differentiation of HL-60 cells into neutrophils while treatment with 1α, 25 dihydroxyvitamin D3 induced low level of monocyte differentiation. Treatment with these agents alone did not induce apoptosis; however, induction of apoptosis is dramatically enhanced when these cells are treated simultaneously with 9-*cis*-RA and 1α, 25 dihydroxyvitamin D3. A cooperative effect between RA and 1α, 25 dihydroxyvitamin D3 in HL-60 cells has also been noted by Bunce and colleagues (Bunce et al., 1990). Vitamin D derivatives in combination with 9-*cis*-RA have been shown to promote active cell death in breast cancer cells (James et al., 1995). These agents act in a cooperative manner to reduce bcl-2 and increase p53 protein in human breast cancer MCF-7 cells.

As mentioned earlier, RA alone inhibits activation-induced apoptosis in T-cell hybridomas and thymocytes (Yang et al., 1995a; 1995b). However, in the presence of glucocorticoids RA promotes apoptosis (Iwata et al., 1992). The exact mechanism of this synergism is not known.

Turley and coworkers (Turley et al., 1995) found that vitamin E succinate and RA induce apoptosis of human B lymphoma RL cells that can be blocked by antibodies to transforming growth factor β1 (TGF-β1). Since both of these inducers increase the expression of TGF-β1, these results suggest that a TGF-β autocrine loop is involved in the growth suppression and the induction of apoptosis by vitamin E succinate and RA. Similar results have been reported for Ki-1 lymphoma cells

(Su et al., 1993). In these cells, the induction of apoptosis by RA is accompanied by an increase in RARα and TGFβ expression and a decrease in c-myc expression. In a different capacity, vitamin E has been shown to inhibit the induction of apoptosis by 4-HPR (Delia et al., 1995). It has been suggested that this inhibition is due to the anti-oxidant properties of vitamin E.

Interferon α-2b and RA both inhibit the growth of human cervical carcinoma ME180 cells (Lancillotti et al., 1995). Inhibition of growth of these cells was enhanced by combined treatment with these two agents. These effects on proliferation are thought to be related to an induction of apoptosis. The expression of the 2-5A synthetase gene, which is thought to play a role in antiproliferative responses to interferon, was increased by interferon α-2b but nor by RA, suggesting different mechanisms of action.

Corbeil and colleagues (Corbeill et al., 1994) observed that RA inhibits growth and induces apoptosis in cultures of Kaposi's sarcoma cells (KSC). The inhibitory activity of RA on KSC growth is decreased in the presence of human serum (but not fetal calf serum), and partially restored by the removal of serum lipids. An effect of serum lipids on the induction of apoptosis by retinoids has also been noted in 3T3-L1 preadipocytes (Chawla and Lazar, 1994). RA induces apoptosis in these cells when cultured in delipidized serum.

Horie and coworkers (Horie and Broxmeyer, 1995) have shown that treatment of the human haematopoietic growth factor-dependent cell line MO7e with RA induces apoptotic cell death in the presence of either Steel factor or GM-CSF. This RA-induced apoptosis can be completely blocked by the combined treatment with Steel factor plus GM-CSF. This treatment has no effect on the expression of RARα gene.

V. CONCLUDING REMARKS

The action of retinoids on apoptosis is relevant to the understanding of the function that retinoids have in regulating embryonic development, the generation of teratogenic effects, and the therapeutic efficacy of retinoids (Lotan, 1995).

Apoptosis occurs normally during development and constitutes a mechanism by which morphogenesis is regulated. Retinoids have been shown to regulate embryonic development through their nuclear receptors and to cause at nonphysiological concentrations teratogenic effects. The induction of programmed cell death in the wound epidermis of regenerating zebrafish fins by retinoids (Ferretti and Geraudie, 1995), and the inhibition of chondrogenesis and promotion of apoptosis in cultures of mouse interdigital mesoderm (Kochhar et al, 1993; Lee et al., 1994), support the hypothesis that regulation of apoptosis by retinoids may play an important role in their control of morphogenesis.

Since apoptosis is involved in the maintenance of the homeostasis in many tissues, defects in apoptosis are relevant to the emergence of many disease processes. Recent progress in the understanding of apoptosis has implicated increased resistance of cells to undergo apoptosis in the development of cancer and hyperplasia (Green and Martin, 1995), while increased apoptosis may lead to degenerative disease. Genes, encoding proteins that are able to suppress apoptosis (e.g., bcl-2), function as oncogenes and have been reported to be up-regulated, while loss of expression of genes, encoding proteins promoting apoptosis (e.g, p53), is observed frequently in a variety of carcinoma cells (Fisher, 1994). Several studies have implicated defects in the control of apoptosis by retinoids in the development of cancer. The resistance of ER-negative mammary carcinoma cells to RA-growth inhibition and apoptosis appears to be related to defects in the expression of RARβ and contribute to the process of tumor progression. Another example is the t(15;17) rearrangement in acute promyelocytic leukemia cells, which results in the synthesis of the PML-RARα fusion protein. This protein has been shown to exhibit anti-apoptotic properties and may promote cell survival (Grignani et al., 1993). Defects in RA signaling pathways have also been reported in several lung carcinoma cell lines (Nervi et al., 1991; Gebert et al., 1991; Moghal and Neel, 1995) and oral lesions (Lotan et al., 1995) and may be responsible for the resistance of these cells to growth-inhibitory responses of RA.

The study of apoptosis has not only extended our understanding of carcinogenesis but has also revealed mechanisms involved in several forms of radiation and chemo-therapy (D'Amico and McKenna, 1994). As has recently been pointed out by Fisher (1994), cancer therapy operates most potently through induction of apoptosis. Likewise, the induction of apoptosis in (pre)malignant cells by retinoids may be an important mechanism in the chemoprevention and chemotherapy of cancer by these agents. Several pilot clinical trials with oral supplementation of 13-*cis*-RA have shown very promising results for the therapy of T-cell lymphomas (Su et al., 1993; Cheng et al., 1994). 13-*cis*-RA has also been shown to be effective in clinical trials in the treatment and prevention of cancer of the aerodigestive tract (Benner et al., 1995; Lippmann et al., 1995a; 1995b). Likewise, retinyl palmitate was found to suppress the development of second primary cancers in patients with lung cancer (Pastorino et al., 1993). HRP has been shown to be an effective antagonist of tumor formation in rodents (Moon et al., 1994). The 4-HPR concentrations that are effective in inhibiting cell growth and inducing apoptosis in cultured carcinoma cells have been shown to be minimally cytotoxic in humans. The efficacy of this retinoid as an anti-cancer drug has been evaluated in clinical trials examining prevention and treatment of various malignancies, including cervical, breast and ovarian cancer, oral leukoplakia, and basal cell carcinomas (Veronesi et al., 1992; Chiesa et al., 1992; Nava et al., 1992; Costa et al., 1994; De Palo et al., 1995; Oridate et al., 1995; 1996). The antagonistic effects retinoids have on tumor formation *in vivo* may be at least in part due to their property to induce apoptosis.

Vitamin A and its metabolites have also been reported to be important enhancers of the immune response (reviewed by Ross and Håmmerling, 1994; Semba, 1994). The mechanisms by which retinoids enhance the immune system have not yet been elucidated. Evidence has, however, emerged (Iwata et al., 1992; Yang et al., 1993; 1995a; 1995b) suggesting that the enhancement of the immune system by retinoids may be at least in part due to a reduction in cell death of immune cells.

The data reviewed in this article reveals that retinoids are capable of mediating apoptosis through multiple mechanisms, depending on, among other things, the chemical nature of the retinoid and cell type used. Although in several instances a role of a specific retinoid nuclear receptor has been demonstrated, the molecular events downstream of the activation of the retinoid receptors are still poorly understood. Likewise, a better understanding of the molecular mechanisms involved in the induction of apoptosis by 4-HPR and AHPN, which do not require activation of RARs or RXRs, awaits further study. The ability of retinoids to induce apoptosis in a variety of human cancer cells and to inhibit apoptosis in immune cells holds promise for the development of successful stategies in the chemoprevention and -therapy of human cancer and infectious diseases by retinoids.

REFERENCES

Adachi, H., & Jetten, A.M. (1996). Unpublished observations.

Adachi, H., Preston, G., Harvat, B., Dawson, M.I., & Jetten, A.M. (1997). Inhibition of cell proliferation and induction of apoptosis by the retinoid CD437 in human lung carcinoma cells. Am. J. Respir. Cell Mol. Biol. Submitted.

Aeschlimann, D., Kaupp, O., & Paulsson, M. (1995). Transglutaminase-catalyzed matrix cross-linking in differentiating cartilage: identification of osteonectin as a major glutaminyl substrate. J. Cell Biol. 129, 881-892.

Anzai, N., Kawabata, H., Hirama, T., Masutani, H., Ohmori, M., Yoshida, Y., & Okuma, M. (1994). Marked apoptosis of human myelomonocytic leukemia cell line P39: significance of cellular differentiation. Leukemia 8, 446-453.

Apfel, C., Crettaz, B. M., Forni, L., Kamber, M., Kaufmann, F., LeMotte, P., Pirson, W., & Klaus, M. (1992). A RA receptor α antagonist selectively counteracts RA effects. Proc. Natl. Acad. Sci. USA 89, 7129-7133.

Atencia, R., Garcia-Sanz, M., Unda, F., & Arechaga, J. (1994). Apoptosis during RA-induced differentiation of F9 embryonal carcinoma cells. Exp. Cell Res. 214, 663-667.

Beato, M., Herrlich, P., & Schütz, G. (1995). Steroid hormone receptors: many actors in search of a plot. Cell 83, 851-857.

Benedetti, L., Grignani, F., Scicchitano, B.M., Jetten, A.M., Diverio, D., Lo Coco, F., Avvisati, C., Gambacorti-Passerini, C., Adamo, S., Levin, A., Pelicci, P.G., & Nervi, C. (1996). Retinoid-induced differentiation of acute promyelocytic leukemia involves PML-RARα-mediated increase of type II transglutaminase. Blood 87, 1939-1950.

Benito, A., Grillot, D., Nunez, G., & Fernandez-Luna, J. L. (1995). Regulation and function of Bcl-2 during differentiation-induced cell death in HL-60 promyelocytic cells. Am. J. Pathol. 146, 481-490.

Benner, S.E., Lippmann, S.M. & Hong, W.K. (1995). Chemoprevention of lung cancer. Chest 107, 316S-321S.

Ben-Sasson, S. A., Sherman, Y., & Gavrieli, Y. (1995). Identification of dying cells: *In situ* staining. Methods Cell Biol. 46, 29-39.

Bhatnagar, R., Abou-Issa, H., Curley, R.W., Koolemans-Beynen, A., Moeschberger, M.L., & Webb, T.E. (1991). Growth suppression of human carcinoma cells in culture by N-(4-hydroxyphenyl)retinamide and its glucuronide and through synergism with glucarate. Biochem Pharmacol. 41, 1471-1477.

Boehm, M. F., Zhang, L., Zhi, L., McClurg, M. R., Berger, E., Wagoner, M., Mais, D. E., Suto, C. M., Davies, P. J. A., Heyman, R. A., & Nadzan, A. M. (1995). Design and synthesis of potent retinoid X receptor selective ligands that induce apoptosis in leukemia cells. J. Med. Chem. 38, 3146-3155.

Boise, L. H., Gonzalez-Garcia, M., Postema, C. E., Ding, L., Lindsten, T., Turka, L. A., Mao, X., Nunez, G., Thompson, C. B. (1993). bcl-X, a bcl-2-related gene that functions as a dominant regulator of apoptotic cell death. Cell 74, 597-608.

Boise, L.H., Gottschalk, A.R., Quintans, J., & Thompson, C.B. (1995). Bcl-2 and Bcl-2-related proteins in apoptosis regulation. Curr. Top. Microbiol. Immunol. 200, 107-121.

Boylan, J.F., Lufkin, T., Achkar, C.C., Chambon, P., & Gudas, L.J. (1995). Targeted disruption of retinoic acid receptor α (PAPα) and RARγ results in receptor-specific alterations in retinoic acid-mediated differentiation and retinoic acid metabolism. Mol. cell. Biol. 15, 853-851.

Bruel, A., Benoit, G., De-Nay, D., Brown, S., & Lanotte, M. (1995). Distinct apoptotic responses in maturation sensitive and resistant t(15;17) acute promyelocytic leukemia NB4 cells. 9-*cis* RA induces apoptosis independent of maturation and Bcl-2 expression. Leukemia. 9, 1173-1184.

Bunce, C. M., Wallington, L. A., Harrison, P., Williams, G. R., & Brown, G. (1995). Treatment of HL60 cells with various combinations of retinoids and 1 alpha,25 dihydroxyvitamin D3 results in differentiation towards neutrophils or monocytes or a failure to differentiate and apoptosis. Leukemia. 9, 410-418.

Chawla, A., & Lazar, M. A. (1994). Peroxisome proliferator and retinoid signaling pathways co-regulate preadipocyte phenotype and survival. Proc. Natl. Acad. Sci. USA 91, 1786-1790.

Chen, J. D., & Evans, R. M. (1995). A transcriptional co-repressor that interacts with nuclear hormone receptors. Nature 377, 454-457.

Cheng, A. L., Su, I. J., Chen, C. C., Tien, H. F., Lay, J. D., Chen, B. R., Pu, Y. S., Hong, R. L., Shen, M. C., & Wang, C. H. (1994). Use of RA in the treatment of peripheral T-cell lymphoma: a pilot study. J. Clin. Oncol. 12, 1185-1192.

Chiesa, F., Tradati, N., Marazza, M., Rossi, N., Boracchi, P., Mariani, L., Clerici, M., Formelli, F., Barzan, L. Carrassi, A., Pastorini, A., Camerini, T., Giardini, R., Zurrida, S., Minn, F.L., Costa, A., De Palo, G., & Veronesi, U. (1992). Prevention of relapses and new localizations of oral leukoplakias with the synthetic retinoid fenretinide (4-4-HPR). Preliminary results. Oral Oncol. Eur. J. Cancer 28B, 97-103

Chiou, S-K., Rao, L., & White, E. (1994). Bcl-2 blocks p53-dependent apoptosis. Mol. Cell. Biol. 14, 2556-2563.

Collins, S. J., Robertson, K. A., & Mueller, L. (1990). RA-induced granulocytic differentiation of HL-60 myeloid leukemia cells is mediated directly through the RA receptor (RAR-α). Mol. Cell. Biol. 10, 2154-2163.

Corbeil, J., Rapaport, E., Richman, D. D., & Looney, D. J. (1994). Antiproliferative effect of retinoid compounds on Kaposi's sarcoma cells. J. Clin. Invest. 93, 1981-1986.

Costa, A., Formelli, F., Chiesa, F., Decensi, A., De Palo, G., & Veronesi, U. (1994). Prospects of chemoprevention of human cancers with the synthetic retinoid fenretinide. Cancer Res. 54 (Suppl.), 2032s-2037s.

Darzynkiewicz, Z., Bruno, S., Del Bino, G., Gorczyca, W., Hotz, M.A., Lassota, P. & Traganos, F. (1992). Features of apoptotic cells measured by flow cytometry. Cytometry 13, 795-808.

Darzynkiewicz, Z. (1995). Apoptosis in antitumor strategies: modulation of cell cycle or differentiation. J. Cell Biochem. 58, 151-159.

Davies, P. J. A., Stein, J. P., Chiocca, E. A., Basilion, J. P., Gentile, V., Thomazy, V., & Fesus, L. (1992). Retinoid-regulated expression of transglutaminases: links to the biochemistry of programmed cell death. In: *Retinoids in Normal Development and Teratogenesis* (Morriss-Kay, G., Ed.), pp. 249-263. Oxford University Press, New York.

D'Amico, A. V., & McKenna, W. G. (1994). Apoptosis and a re-investigation of the biologic basis for cancer therapy. Radiother. Oncol. 33, 3-10.

Delescluse, C., Cavey, M. T., Martin, B., Bernard, B. A., Reichert, U., Maignan, J., Darmon, M., & Shroot, B. (1991). Selective high affinity RA receptor α or β-γ ligands. Mol. Pharmacol. 40, 556-562.

Delia, D., Aiello, A., Formelli, F., Fontanella, E., Costa, A., Miyashita, T., Reed, J. C., & Pierotti, M. A. (1995). Regulation of apoptosis induced by the retinoid N-(4-hydroxy-phenyl) retinamide and effect of deregulated bcl-2. Blood. 85, 359-367.

Delia, D., Aiello, A., Lombardi, L., Pelicci, P. G., Grignani, F., Grignani, F., Formelli, F., Menard, S. Costa, A., Veronesi, U., & Pierotti, M. A. (1993). N-(4-hydroxyphenyl) retinamide induces apoptosis of malignant hemopoietic cell lines including those unresponsive to RA. Cancer Res. 53, 6036-6041.

De Luca, L.M. (1991). Retinoids and their receptors in differentiation, embryogenesis, and neoplasia. FASEB J. 5, 2924-2933.

De Palo, G., Veronesi, U., Camerini, T., Formelli, F., Mascotti, G., & Boni, C. (1995). Can fenretinide protect women against ovarian cancer? J. Natl. Cancer Inst. 87, 146-147.

de Thé, H., Lavau, C., Marchio, A., Chomienne, C., Degos, L., & Dejean, A. (1991). The PML-RARα fusion mRNA generated by the t(15;17) translocation in acute promyelocytic leukemia encodes a functionally altered RAR. Cell 66, 675-684.

Diller, L. (1990). p53 functions as a cell cycle control protein in osteosarcomas. Mol. Cell. Biol. 10, 5772-5781.

Di Vinci, A., Geido, E., Infusini, E., & Giaretti, W. (1994). Neuroblastoma cell apoptosis induced by the synthetic retinoid N-(4-hydroxyphenyl)retinamide. Int. J. Cancer 59, 422-426.

Earnshaw, W, C, (1995). Nuclear changes in apoptosis. Curr. Opin. Cell Biol. 7, 337-343.

Eastman, A. (1995). Assays for DNA fragmentation, endonucleases, and intracellular pH and Ca^{2+} associated with apoptosis. Methods Cell Biol. 46, 41-56.

Eitan, S., & Schwartz, M. (1993). A transglutaminase that converts interleukin-2 into a factor cytotoxic to oligodendrocytes. Science 261, 106-108.

Elledge, R., & Lee, W-H. (1995). Life and death by p53. BioEssays 17, 923-930.

Enari, M., Hug, H., Nagata, S. (1995). Involvement of an ICE-like protease in Fas-mediated apoptosis. Nature 375, 78-81.

Evan, G. I., Wyllie, A. H., Gibert, C. S., Littlewood, T. D., Land, H., Brooks, M., Waters, C. M., Penn, L. Z., & Hancock, D. C. (1992). Induction of apoptosis in fibroblasts by c-myc protein. Cell 69, 119-128.

Evan, G., Brown, L., Whyte, M. & Harrington, E. (1995). Apoptosis and the cell cycle. Curr. Opin. Cell Biol. 7, 825-834.

Evans, R.M. (1988). The steroid and thyroid receptor superfamily. Science 240, 889-895.

Fanjul, A., Dawson, M. I., Hobbs, P. D., Jong, L., Cameron, J. F., Harlev, E., Graupner, G., Lu, X. P., & Pfahl, M. (1994). A new class of retinoids with selective inhibition of AP-1 inhibits proliferation. Nature 372, 107-111.

Ferretti, P., & Geraudie, J. (1995). RA-induced cell death in the wound epidermis of regenerating zebrafish fins. Dev. Dyn. 202, 271-283.

Fesus, L., Thomazy, V., & Falus, A. (1987). Induction and activation of tissue transglutaminase during programmed cell death. FEBS Lett 224, 104-108.

Fesus, L., Thomazy, V., Autuori, F., Ceru, M. P., Tarcsa, E., & Piacentini, M. (1989). Apoptotic hepatocytes become insoluble in detergents and chaotropic agents as a result of transglutaminase action. FEBS Lett. 245, 150-154.

Fesus, L., Davies, P. J. A., & Piacentini, M. (1991). Apoptosis: molecular mechanisms in programmed cell death. Eur. J. Cell Biol. 56, 170-177.

Fesus, L. (1993). Biochemical events in naturally occurring forms of cell death. FEBS Lett. 328, 1-5.

Fesus, L., Szondy, Z., & Uray, I. (1995). Probing the molecular program of apoptosis by cancer chemopreventive agents. J. Cell. Biochem., Suppl., 22, 151-161.

Fisher, D. E. (1994). Apoptosis in cancer therapy: crossing in the threshold. Cell 78, 539-542.

Folk, J.E. (1980). Transglutaminases. Annu. Rev. Biochem. 49, 517-531.

Formelli, F., & Cleris, L. (1993). Synthetic retinoid fenretinide is effective against human ovarian carcinoma xenograft and potentiates cis-platin activity. Cancer Res. 53, 5374-5376

Fukuda, K., Kojiro, M., & Chiu, J-F. (1994). Differential regulation of tissue transglutaminase in rat hepatoma cell lines McA-RH7777 and McA-RH8994: relation to growth rate and cell death. J. Cell. Biochem. 54, 67-77.

Gandrillon, O., Ferrand, N., Michaille, J. J., Roze, L., Zile, M. H., & Samarut, J. (1994). c-erbA alpha/T3R and RARs control commitment of hematopoietic self-renewing progenitor cells to apoptosis or differentiation and are antagonized by the v-erbA oncogene. Oncogene 9, 749-758.

Glass, C.K. (1994) Differential recognition of target genes by nuclear receptor monomers, dimers and heterodimers. Endocrine Rev. 15, 391-407.

Gebert, J.F., Moghal, N., Frangioni, J.V., Sugarbaker, D.J., & Neel, B.G. (1991). High frequency of retinoic acid reeptor β abnormalities in human lung cancer. Oncogene 6, 1859-1868.

Giguère, V. (1994). RA receptors and cellular retinoid binding proteins: complex interplay in retinoid signaling. Endocrine Rev. 15, 61-79.

Gold, R. (1994). Differentiation between cellular apoptosis and necrosis by the combined use of in situ tailing translation techniques. Lab. Invest. 71, 219-233.

Green, D. R., & Martin, S. J. (1995). The killer and the executioner: how apoptosis controls malignancy. Curr. Opin. Immunol. 7, 694-703.

Grignani, F., Ferrucci, P. F., Testa, U., Talamo, G., Fagioli, M., Alcalay, M., Mencarelli, A., Grignani, F., Peschle, C., Nicoletti, I., & Pelicci, P. G. (1993). The acute promyelocytic leukemia-specific PML-RARα Fusion protein inhibits differentiation and promotes survival of meyloid precusor cells. Cell 74, 423-431.

Gudas, L. J., Sporn, M. B., & Roberts, A.B. (1994). Cellular biology and biochemistry of the retinoids. In: *The Retinoids: Biology, Chemistry, and Medicine*. (Sporn, M. B., Roberts, A. B. & Goodman, D. S. Eds.), pp. 443-520, Raven Press, New York.

Harper, J.W.i, G.R., Wei, N., Keyomarsi, K., & Elledge, S.J. (1993). The p21 Cdk-interacting protein Cip 1 is a potent inhibitor of G1 cyclin-dependent kinases.Cell 75, 805-816.

Hawkins, C. J., & Vaux, D. L. (1994). Analysis of the role of bcl-2 in apoptosis. Immunol. Rev. 142, 127-139.

Helmberg, A., Auphan, N., Caelles, C., & Karin, M. (1995). Glucocorticoid-induced apoptosis of human leukemic cells is caused by the repressive function of the glucocorticoid receptor. EMBO J. 14, 452-460.

Hengartner, M.O., & Horvitz, H.R. (1994a) C. elegans cell survival gene ced-9 encodes a functional homolog of the mammalian proto-oncogene bcl-2. Cell 76, 665-676.

Hengartner, M.O., & Horvitz, H.R. (1994b) Programmed cell death in Caenorhabditis elegans. Curr. Opin. Genet. Dev. 4, 581-586.

Heyman, R.A., Mangelsdorf, D.J., Dyck, J.A., Stein, R.B., Eichele, G., Evans, R.M., & Thaller, C. (1992). 9-Cis retinoic acid is a high affinity ligand for the retinoid X receptor. Cell 68, 397-406.

Hockenbery, D., Nunez, G., Milliman, C., Schreiber, R.D., & Korsmeyer, S.J. (1990) Bcl-2 is an inner mitochondrial membrane protein that blocks programmed cell death. Nature 348, 334-336.

Holbrook, K.A., & Hennings, H. (1983). Phenotypic expression of epidermal cells in vitro: a review. J. Invest. Dermatol. 81, 11S-24S.

Horie, M., & Broxmeyer, H. E. (1995). The combination of Steel factor and GM-CSF blocks apoptosis induced by RA and upregulates AP-1 in a human growth factor-dependent cell line. Exp. Hematol. 23, 168-173.

Hörlein, A.J., Naar, A.M., Heinzel, T., Torchia, J., Gloss, B., Kurakawa, R., Ryan, A., Kamie, Y., Soderstrom, M., Glass, C.K., & Rosenfeld, M.G. (1995). Ligand-dependent repression by the thyroid hormone receptor mediated by a nuclear receptor co-repressor. Nature 377, 397-404.

Howard, M. K., Burke, L. C., & Mailhos, C., Pizzey, A., Gilbert, C. S., Lawson, W. D., Collins, M. K., Thomas, N. S., Latchman, D. S. (1993). Cell cycle arrest of proliferating neuronal cells by serum deprivation can result in either apoptosis or differentiation. J. Neurochem. 60, 1783-1791.

Hsieh, , T.C., Ng, C., & Wu, J.M. (1995). The synthetic retinoid N-(4-hydroxyphenyl) retinamide (4-HPR) exerts antiproliferative and apoptosis-inducing effects in the androgen-independent human prostatic JCA-1 cells. Biochem. Mol. Biol. Int. 37, 499-506.

Igawa, M., Tanabe, T., Chodak, G.W., & Rukstalis, D.B. (1994) N-(4-hydroxyphenyl)retinamide induces cell cycle specific growth inhibition in PC3 cells. Prostate 24, 299-305.

Itoh, N., & Nagata, S. (1993). A novel protein domain required for apoptosis. Mutational analysis of human Fas antigen. J. Biol. Chem. 268, 10932-10937.

Iwata, M., Mukai, M., Nakai, Y., & Iseki, R. (1992). RAs inhibit activation-induced apoptosis in T cell hybridomas and thymocytes. J. Immunol. 149, 3302-3308.

James, S.Y., Mackay, A. G., & Colston, K. W. (1995). Vitamin D derivatives in combination with 9-*cis* RA promote active cell death in breast cancer cells. J. Mol. Endocrinol. 14, 391-394.

Jetten, A. M., Fujimoto, W., Zhang, L, X., Marvin, K. W., Austin, S., Mills, K. J., Bernacki, S. H., & Saunders, N. A. (1994). Retinoic acid receptors and the regulation of gene expression in tracheobronchial epithelial cells. In: *Retinoids: From Basic Science to Clinical Applications.* (Livrea, M.A., and Vidali, G., Eds.) pp. 243-252, Birkhauser Verlag, Basel.

Jetten, A. M., Nervi, C., & Vollberg, T. M. (1992). Control of squamous differentiation in tracheobronchial and epidermal epithelial cells: role of retinoids. J. Natl. Cancer Inst. Monogr. 13, 93-100.

Jiang, H., Gyda, M., Harnish, D.C., Chandraratna, R.A., & Soprano, K.J. (1994). Teratogenesis by retinoic acid analogs positively correlates with elevation of retinoic acid receptor-beta2 mRNA levels in treated embryos. Teratology 50, 38-43.

Jiang, H., & Kochhar, D. M. (1992). Induction of tissue transglutaminase and apoptosis by retinoic acid in the limb bud. Teratology 46, 333-340.

Kakizuka, A., Miller, W.H., Umesono, K., Warrell, R.P., Frankel, S.R., Murty, V.V.V.S., Dmitrovsky, E., & Evans, R.M. (1991). Chromosomal translocation t(15;17) in human acute promyelocytic leukemia fuses RARα with a novel putative transcription factor, PML. Cell 66, 663-674.

Kalemkerian, G. P., Slusher, R., Sakkaraiappan, R., Gadgeel, S., & Mabry, M. (1995). Growth inhibition and induction of apoptosis by fenretinide in small-cell lung cancer cell lines. J. Natl. Cancer Inst. 87, 1674-1680.

Kastner, P., Mark, M., & Chambon, P. (1995). Nonsteroid nuclear receptors: What are the genetic studies telling us about their role in real life? Cell 83, 859-869.

King, K. L., & Cidlowski, J. A. (1995). Cell cycle and apoptosis: common pathways to life and death. J. Cell Biochem. 58, 175-180.

Knight, R.L., Hand, D., Piacentini, M., & Griffin, M. (1993). Characterization of the transglutaminase-mediated large molecular weight polymer from rat liver; its relationship to apoptosis. Eur. J. Cell Biol. 60, 210-216.

Kochhar, D. M., Jiang, H., Harnish, D. C., & Soprano, D. R. (1993). Evidence that RA-induced apoptosis in the mouse limb bud core mesenchymal cells is gene-mediated. Prog. Clin. Biol. Res. 383B, 815-825.

Kojima, S., Nara, K., & Rifkin, D. B. (1993). Requirement for transglutaminase in the activation of latent transforming growth factor-beta in bovine endothelial cells. J. Cell Biol. 121, 439-448.

Kondo, S., Barna, B. P., Morimura, T., Takeuchi, J., Yuan, J., Akbasak, A., & Barnett, G. H. (1995). Interleukin-1 beta-converting enzyme mediates cisplatin-induced apoptosis in malignant glioma cells. Cancer Res. 55, 6166-6171.

Kobayashi, T., Consoli, U., Andreeff, M., Shiku, H., Deisseroth, A. B., & Zhang, W. (1995). Activation of p21/WAF1/Cip1 expression by a temperature-sensitive mutant of human p53 does not lead to apoptosis. Oncogene 11, 2311-2316.

Krupitza, G., Hulla, W., Harant, H., Dittrich, E., Kallay, E., Huber, H., Grunt, T., & Dittrich, C. (1995). RA induced death of ovarian carcinoma cells correlates with c-myc stimulation. Int. J. Cancer 61, 649-657.

Kumar, S. (1995). ICE-like proteases in apoptosis. Trends Biochem. Sci. 20, 198-202.

Kurokawa, R., Soderstrom, M., Horlein, A., Halachmi, S., Brown, M., Rosenfeld, M. G., and Glass, C. K. (1995). Polarity-specific activities of RA receptors determined by a co-repressor. Nature 377, 451-454.

Lam, M., Dubryak, G., Chen, L., Nunez, G., Miesfeld, R. L., & Distelhorst, C. W. (1994). Evidence that BCL-2 represses apoptosis by regulating endoplasmic reticulum-associated Ca2+ fluxes. Proc. Natl. Acad. Sci. USA 91, 6569-6573.

Lancillotti, F., Giandomenico, V., Affabris, E., Fiorucci, G., Romeo, G., & Rossi, G. B. (1995). Interferon alpha-2b and RA combined treatment affects proliferation and gene expression of human cervical carcinoma cells. Cancer Res. 55, 3158-3164.

Lazebnik, Y.A., Kaufmann, S.H., Desnoyers, S., Poirier, G.G., & Earnshaw, W.C. (1994). Cleavage of poly(ADP-ribose) polymerase by a proteinase with properties like ICE. Nature 371, 346-347.

Lee, E. Y., Hu, N., Yuan, S. S., Cox, L. A., Bradley, A., Lee, W. H., & Herrup, K. (1994). Dual roles of the retinoblastoma protein in cell cycle regulation and neuron differentiation. Genes & Dev. 8, 2008-2021.

Lee, K.K., Li, F.C., Yung, W.T., Ng, J.L., & Cheah, K.S. (1994). Influence of digit, ectoderm, and retinoic acid and chondrogenesis by mouse interdigital mesoderm in culture. Dev. Dyn. 201, 297-309.

Lehmann, J.M., Dawson, M.I., Hobbs, P.D., Husmann, M., & Pfahl, M. (1991). Identification of retinoids with nuclear receptor subtype-selective activities. Cancer Res. 51, 4804-4809.

Lehmann, J. M., Jong, L., Fanjul, A., Cameron, J. F., Lu, X. P., Haefner, P., Dawson, M. I., & Pfahl, M. (1992). Retinoids selective for retinoid X receptor response pathways. Science 258, 1944-1946.

Levin, A.A., Sturzenbecker, L.J., Kazmer, S., Bosakowski, T., Huselton, C. Allenby, G., Speck, J., Kratzeisen, C., Rosenberger, M., Lovey, A., & Grippo, J.F. (1992). 9-*Cis* retinoic acid stereoisomer binds and activates the nuclear receptor RXRα. Nature 355, 359-361.

Lippmann, S.M., Spitz, M.R., Huber, M.H. & Hong, W.K. (1995a). Strategies for chemoprevention study of premalignant and second primary tumors in the head and neck. Curr. Opin. Oncol. 7, 234-241.

Lippmann, S.M., Heyman, R.A., Kurie, J.M., Benner, S.E. & Hong, W.K. (1995b). Retinoids and chemoprevention: clinical and basic studies. J. Cell. Biochem. 22, 1-10.

Liu, Y., Lee, M-O., Wang, H-G., Li, Y., Hashimoto, Y., Klaus, M., Reed, J.C., and Zhang, X. (1996). RARβ mediates the growth inhibitory effect of retinoic acid by inducing cell apoptosis in human breast cancer cells. Mol. Cell. Biol. 16, 1138-1149.

Los, M., Van-de-Craen, M., Penning, L. C., Schenk, H., Westendorp, M., Baeuerle, P. A., Droge, W., Krammer, P. H., Fiers, W., & Schulze-Osthoff, K. (1995). Requirement of an ICE/CED-3 protease for Fas/APO-1-mediated apoptosis. Nature 375, 81-83.

Lotan, R., Xu, X.C., Lippmann, S.M., Ro, J.Y., Lee, J.S., Lee, J.J. & Hong, W.K. (1995). Suppression of retinoic acid receptor-beta in premalignant oral lesions and its up-regulation by isotretinoin. N. Eng. J. Med. 332, 1405-1410.

Lotan, R. J. (1995). Retinoids and apoptosis: implications for cancer chemoprevention and therapy. J. Natl. Cancer Inst. 87, 1655-1657.

Lowe, S. W., Schmitt, E. M., Smith, S. W., Osborne, B. A., & Jacks, T. (1993). p53 is required for radiation-induced apoptosis in mouse thymocytes. Nature 362, 847-849.

Mailhos, C., Howard, M. K., & Latchman, D. S. (1994). A common pathway mediates RA and PMA-dependent programmed cell death (apoptosis) of neuronal cells. Brain Res. 644, 7-12.

Mangelsdorf, D. J., & Evans, R. M. (1995). The RXR heterodimers and orphan receptors. Cell 83, 841-850.

Mangelsdorf, D. J., Thummel, C., Beato, M., Herrich, P., Schutz, G., Umesono, K., Blumberg, B., Kastner, P., Mark, M., Chambon, P., & Evans, R. M. (1995). The nuclear receptor superfamily: The second decade. Cell 83, 835-839.

Mariotti, A., Marcora, E., Bunone, G., Costa, A., Veronesi, U., Pierotti, M. A., Della Valle, G. (1994). N-(4-hydroxyphenyl)retinamide: a potent inducer of apoptosis in human neuroblastoma cells. J. Natl. Cancer Inst. 86, 1245-1247.

Marth, C., Bock, G., & Daxenbichler, G. (1985). Effect of 4-hydroxyphenyl retinamide and retinoic acid on proliferation and cell cycle of cultured human breast cancer cells. J. Natl. Cancer Inst. 75, 871-875.

Martin, B., Bernardon, J. M., Cavey, M. T., Bernard, B., Carlavan, I., Charpentier, B., Pilgrim, W. R., Shroot, B., & Reichert, U. (1992). Selective synthetic ligands for human nuclear RA receptors. Skin Pharmacol. 5, 57-65.

Martin, S. J., Bradley, J. G., & Cotter, T. G. (1990). HL-60 cells induced to differentiate towards neutrophils subsequently die via apoptosis. Clin. Exp. Immunol.79, 448-453.

Martin, S.J. and Green, D.R. (1995). Apoptosis and cancer: the failure of controls of cell death and cell survival. Crit. Rev. Oncol. Hematol. 18, 137-153.

Martinez, J., Chalupowicz, D. G., Roush, R. K., Sheth, A., & Barsigian, C. (1994). Transglutaminase-mediated processing of fibronectin by endothelial cell monolayers. Biochem. 33, 2538-2545.

Mehta, K., McQueen, T., Neamati, N., Collins, S., & Andreeff, M. (1996). Activation of retinoid receptors RARα and RXRα induces differentiation and apoptosis, respectively, in HL-60 cells. Cell Growth Differ. 7, 179-186.

Mehul, B., Bawumia, S., & Hughes, R. C. (1995). Cross-linking of galectin 3, a galactose-binding protein of mammalian cells, by tissue-type transglutaminase. FEBS Lett. 360, 160-164.

Meikrantz, W., & Schlegel, R. (1995). Apoptosis and the cell cycle. J. Cell Biochem. 58, 160-174.

Melino, G., Annicchiarico-Petruzzelli, M., Piredda, L., Candi, E., Gentile, V., Davies, P. J., & Piacentini, M. (1994). Tissue transglutaminase and apoptosis: sense and antisense transfection studies with human neuroblastoma cells. Mol. Cell. Biol. 14, 6584 6596.

Meyaard, L., Otto, S.A., Jonker, R.R., Mijnster, M.J., Keet, R.P., & Miedema, F. (1992). Programmed death of T cells in HIV-1 infection. Science 257, 217-219.

Meyskens, F.L., Alberts, D.S., & Salmon, S.E. (1983). Effect of 13-cis retinoic acid and 4-hydroxyphenyl-all-*trans* retinamide in human tumor-colony formation in soft agar. Int. J. Cancer 32, 295-299.

Miura, M., Zhu, H., Rotello, R., Hartwieg, E. A., & Yuan, J. (1993). Induction of apoptosis in fibroblasts by IL-1 beta-converting enzyme, a mammalian homolog of the C. elegans cell death gene ced-3. Cell 75, 653-660.

Miyashita, T., & Reed, J. C. (1992). bcl-2 gene transfer increases relative resistance of S49.1 and WEHI7.2 lymphoid cells to cell death and DNA fragmentation induced by glucocorticoids and multiple chemotherapeutic drugs. Cancer Res. 52, 5407-5411.

Moghal, N., & Neel, B.G. (1995). Evidence for impaired retinoic acid receptor-thyroid hormone receptor AF-2 cofactor activity in human lung cancer. Mol. Cell. Biol. 15, 3945-3959.

Moon, R. C., & Mehta, R. G. (1989). Chemoprevention of experimental carcinogenesis in animals. Prev. Med. 18, 576-581.

Moon, R. C., Mehta, R. G., & Rao, K. V. N. (1994).Retinoids and cancer in experimental anlimals. In: *The Retinoids: Biology, Chemistry and Medicine*. 2nd Ed. (Sporn, M. B., Roberts, A. B., & Goodman, D. S. eds.), pp. 573-596. Raven Press, New York.

Nagy, L., Thomazy, V. A., Shipley, G. L., Fesus, L., Lamph, W., Heyman, R. A., Chandraratna, R. A., & Davies, P. J. (1995). Activation of retinoid X receptors induces apoptosis in HL-60 cell lines. Mol. Cell. Biol. 15, 3540-3551.

Nakamura, N., Shidoji, Y., Yamada, Y., Hatakeyama, H., Moriwaki, H., & Muto, Y. (1995). Induction of apoptosis by acyclic retinoid in the human hepatoma-derived cell line, HuH-7. Biochem. Biophys. Res. Commun. 207, 382-388.

Nakaoka, H., Perez, D. M., Baek, K. J., Das, T., Husain, A., Misono, K., Im, M. J., & Graham, R. M. (1994). Gh: a GTP-binding protein with transglutaminase activity and receptor signaling function. Science 264, 1593-1596.

Nava, M., Fabrizio, T., & Arioli, N. (1992). Chemoprevention of basal carcinoma by prolonged administration of the synthetic retinoid fenretinide (4-HPR). In: Progress and Perspectives in Chemoprevention. Vol. 79 (De Palo, G., Sporn, M.B., & Veronesi, U., eds.), pp. 161-165. Raven Press, New York.

Nervi, C., Vollberg, T.M., George, M.D., Zelent, A., Chambon, P. & Jetten, A.M. (1991). Expression of nuclear retinoic acid receptors in normal tracheobronchial cells and in lung carcinoma cells. Exp. Cell Res. 195, 163-170.

Nicholson, D. W., Ali, A., Thornberry, N. A., Vaillancourt, J. P., Ding, C. K., Gallant, M., Gareau, Y., Griffin, P. R., Labelle, M., & Lazebnik, Y. A. (1995). Identification and inhibition of the ICE/CED-3 protease necessary for mammalian apoptosis. Nature 376, 37-43.

Noguchi, K., Nakajima, M., Naito, M., & Tsuruo, T. (1995). Inhibition by differentiation-inducing agents of wild-type p53-dependent apoptosis in HL-60 cells. Jpn. J. Cancer Res. 86, 217-223.

Oberhammer, F.A., Hochegger, K., Froschl, G., Tiefenbacher, R., & Pavelka, M. (1994). Chromatin condensation during apoptosis is accompanied by degradation of human lamin A+B, without enhanced activation of cdc2 kinase. J. Cell Biol. 126, 827-837.

Oltvai, Z. N., & Korsmeyer, S. J. (1994). Checkpoints of dueling dimers foil death wishes. Cell 79, 189-192.

Oltvai, Z. N., Milliman, C. L., & Korsmeyer, S. J. (1993). Bcl-2 heterodimerizes in vivo with a conserved homolog, Bax, that accelerates programmed cell death. Cell 74, 609-619.

Onate, S. A., Tsai, S. Y., Tsai, M. J., & O'Malley, B. W. (1995). Sequence and characterization of a coactivator for the steroid hormone receptor superfamily. Science 270, 1354-1371.

Oritani, K., Kaisho, T., Nakajima, K., & Hirano, T. (1992). RA inhibits interleukin-6-induced macrophage differentiation and apoptosis in a murine hematopoietic cell line, Y6. Blood 80, 2298-2305.

Oridate, N., Lotan, D., Mitchell, M. F., Hong, W. K., & Lotan, R. (1995). Induction of apoptosis by retinoids in human cervical carcinoma cell lines. Int. J. Oncol. 7, 433-441.

Oridate, N., Lotan, D., Mitchell, M.F., & Lotan, R. (1995). Inhibition of proliferation and induction of apoptosis in cervical carcinoma cells by retinoids: implications for chemoprevention. J. Cell. Biochem. Suppl. 23, 80-86.

Oridate, N., Lotan, D., Xu, X. C., Hong, W. K., & Lotan, R. (1996). Differential induction of apoptosis by all-trans-RA and N-(4-hydroxyphenyl)retinamide in human head and neck squamous cell carcinoma cell lines. Clin. Cancer Res. in press

Park, J. R., Robertson, K., Hickstein, D. D., Tsai, S., Hockenbery, D. M., & Collins, S. J. (1994). Dysregulated bcl-2 expression inhibits apoptosis but not differentiation of RA-induced HL-60 granulocytes. Blood 84, 440-445.

Pastorino, U., Infante, M., Maioli, M., Chiesa, G., Buyse, M., & Firket, P. (1993). Adjuvant treatment of stage I lung cancer with high dose vitamin A. J. Clinic. Oncol. 11, 1216-1222.

Piacentini, M., Fesus, L., Farrace, M. G., Ghibelli, L., Piredda, L., & Melino, G. (1991). The expression of "tissue" transglutaminase in two human cancer cell lines is related with the programmed cell death (apoptosis). Eur. J. Cell. Biol. 54, 246-254.

Piacentini, M., Annicchiarico-Petruzzelli, M., Oliverio, S., Piredda, L., Biedler, J.L., & Melino, E. (1992). Phenotype-specific "tissue" transglutaminase regulation in human neuroblastoma cells in response to RA: correlation with cell death by apoptosis. Int. J. Cancer. 52, 271-278.

Piacentini, M., Fesus, L., & Melino, G. (1993). Multiple cell cycle access to the apoptotic death program in human neuroblastoma cells. FEBS-Lett. 320, 150-154.

Piacentini, M., Davies, P. J. A., & Fesus, L. (1994). Tissue transglutaminase in cells undergoing apoptosis. In: *Apoptosis II: The Molecular Basis of Apoptosis in Disease* (Tomei, L. D., & Cope, F. O. Eds.), pp. 143-163. Cold Spring Harbor Laboratory Press, New York.

Pienta, K. J., Nguyen, N. M., & Lehr, J. E. (1993). Treatment of prostate cancer in the rat with the synthetic retinoid fenretinide. Cancer Res. 53, 224-226.

Polakowska, R., & Haake, A. R. (1994). Apoptosis: the skin from a new perspective. Cell Death Differ. 1, 19-31.

Pollard, M., Luckert, P. H., & Sporn, M. B. (1991). Prevention of primary prostate cancer by 4-(hydroxyphenyl)retinamide. Cancer Res. 51, 3610-3615.

Ponzoni, M., Bocca, P., Chiesa, V., Decensi, A., Pistoia, V., Raffaghello, L., Rozzo, C., & Montaldo, P. G. (1995). Differential effects of N-(4-hydroxyphenyl)retinamide and RA on neuroblastoma cells: apoptosis versus differentiation. Cancer Res. 55, 853-861.

Roman, S.D., Clarke, C.L., Hall, R.E., Alexander, I.E., & Sutherland, R.L. (1992). Expression and regulation of retinoic acid receptors in human breast cancer cells. Cancer Res. 52, 2236-2242.

Ross, A.C., & Håmmerling, U.G. (1994). Retinoids and the immune system. In: *The Retinoids: Biology, Chemistry, and Medicine*. 2nd Ed.(Sporn, M.B., Roberts, A.B., and Goodman, D.S. Eds), pp. 521-544. Raven Press, New York.

Routmensz, N., De Palo, G., Formelli, F., Costa, A., Marubini, E., Campa, T., Crippa, A., Danesini, G.M., Delle Grottiglie, M., Di Mauro, M.G., Filiberti, A., Gallazzi, M., Guzzon, A., Magni, A., Malone, W., Mariani, L., Palvarini, M., Perloff, M., Pizzichetta, M., & Veronesi, U. (1991). Long term tolerability of fenretinide (4-HPR) in breast cancer patients. Eur. J. Cancer 27, 1127-1131.

Sacks, P. G., Oke, V., Calkins, D. P., Vasey, T., & Terry, N. H. (1990). Effects of β-all-trans-RA on growth, proliferation, and cell death in a multicellular tumor spheroid model for squamous carcinomas. J. Cell Physiol. 144, 237-243.

Schott, A. F., Apel, I. J., Nunez, G., & Clarke, M. F. (1995). Bcl-XL protects cancer cells from p53-mediated apoptosis. Oncogene 11, 1389-1394.

Schwartz, L.M., Smith, S. W., Jones, M. E. E., & Osborne, B. A. (1993). Do all programmed cell deaths occur via apoptosis? Proc. Natl. Acad. Sci. USA. 90, 980-984.

Sedlak, T. W., Oltvai, Z. N., Yang, E., Wang, K., Boise, L. H., Thompson, C. B., & Korsmeyer, S. J. (1995). Multiple Bcl-2 family members demonstrate selective dimerizations with Bax. Proc. Natl. Acad. Sci. U. S. A. 92, 7834-7838.

Seewaldt, V. L., Johnson, B. S., Parker, M. B., Collins, S. J., & Swisshelm, K. (1995). Expression of RA receptor beta mediates RA-induced growth arrest and apoptosis in breast cancer cells. Cell Growth & Diff. 6, 1077-1088.

Seifter, E.G., Rettura, G., & Levenson, S.M. (1981). Decreased resistance of C3H/HeHa mice to C3HBA tumor transplants: increased resistance due to supplemental vitamin A. J. Natl. Cancer Inst. 67, 467-474.

Semba, R. D. (1994). Vitamin A, immunity, and infection. Clin. Infect. Dis. 19, 489-499.

Sen, S. (1992). Programmed cell death: concept, mechanism and control. Biol. Rev. 67, 287-319.

Shan, B., & Lee, W.H. (1994). Deregulated expression of E2F-1 induces S-phase entry and leads to apoptosis. Mol. Cell. Biol. 14, 8166-8173.

Shao, Z. M., Dawson, M. I., Li, X. S., Rishi, A. K., Sheikh, M. S., Han, Q. X., Ordonez, J. V., Shroot, B., & Fontana, J. A. (1995). p53 independent G0/G1 arrest and apoptosis induced by a novel retinoid in human breast cancer cells. Oncogene 11, 493-504.

Sheikh, M.S., Shao, Z.M., Chen, J.C., Hussain, A., Jetten, A.M., & Fontana, J.A. (1993). Estrogen receptor-negative breast cancer cells transfected with the estrogen receptor increase RAR alpha gene expression and sensitivity to growth inhibition by retinoic acid. J. Cell. Biochem. 53, 394-404.

Sheikh, M.S., Shao, Z. M., Li, X. S., Ordones, J. V., Conley, B. A., Wu, S., Dawson, M. I., Han, Q. X., Chao, W. R., Quick, T., & Fontana, J.A. (1995). N-(4-hydroxyphenyl)retinamide (4-HPR)-mediated biological actions involve retinoid receptor-independent pathways in human breast carcinoma. Carcinogenesis 16, 2477-2486.

Shi, Y., Glynn, J M., Guilbert, L .J., Cotter, T. G., Bisonette, R. P., & Green, D. R. Role for c-myc in activation-induced apoptotic cell death in T cell hybridoma. (1992). Science 257, 212-214.

Smith, C. A., Grimes, E. A., McCarthy, N. J., & Williams, G. T. (1994). Multiple gene regulation of apoptosis: significance in immunology and oncology. In: *Apoptosis II: The Molecular Basis of*

Apoptosis in Disease. (Tomei, L. D., & Cope, F. O. Eds.), pp. 43-87. Cold Spring Harbor Laboratory Press, New York.

Steller, H. (1995). Mechanisms and genes of cellular suicide. Science 267, 1445-1449.

Su, I. J., Cheng, A.L., Tsai, T. F., & Lay, J. D. (1993). RA-induced apoptosis and regression of a refractory Epstein-Barr virus-containing T cell lymphoma expressing multidrug-resistance phenotypes. Br. J. Haematol. 85, 826-828.

Swisshelm, K., Ryan, K., Lee., Tsou, H.C., Peacocke, M., & Sager, R. (1994). Down-regulation of retinoic acid receptor β in mammary carcinoma cell lines and its up-regulation in senescing normal mammary epithelial cells. Cell Growth & Diff. 5, 133-141.

Tarcsa, E., Kedei, N., Thomazy, V., & Fesus, L. (1992). An involucrin-like protein in hepatocytes serves as a substrate for tissue transglutaminase during apoptosis. J. Biol. Chem. 267, 25648-25651.

Tate, B. F., Allenby, G., Janocha, R., Kazmer, S., Speck, J., Sturzenbecker, L. J., Abarzua, P., Levin, A. A., & Grippo, J. F. (1994). Mol. Cell. Biol. 14, 2323-2330.

Thacher, S.M., & Rice, R.H. (1985). Keratinocyte-specific transglutaminase of cultured human epidermal cells: relation to cross-linked envelope formation and terminal differentiation. Cell 40, 685-695.

Thompson, E.B. (1994). Apoptosis and steroid hormones. Mol. Endocrinol. 8, 665-673.

Tosi, P., Visani, G., Gibellini, D., Zauli, G., Ottaviani, E., Cenacchi, A., Gamberi, B., Manfroi, S., Marchisio, M., & Tura, S. (1994). All-*trans* RA and induction of apoptosis in acute promyelocytic leukemia cells. Leuk-Lymphom. 14, 503-507.

Truss, M., & Beato, M. (1993). Steroid hormone receptors: interactions with deoxyribonucleic acid and transcriptional factors. Endocrine Rev. 14, 459-479.

Tsai, S., Bartelmez, S., Heyman, R., Damm, K., Evans, R., & Collins, S. J. (1992). A mutated RA receptor-α exhibiting dominant-negative activity alters the lineage development of a multipotent hematopoietic cell line. Genes & Dev. 6, 2258-2269.

Turley, J. M., Funakoshi, S., Ruscetti, F. W., Kasper, J., Murphy, W. J., Longo, D. L., & Birchenall-Roberts, M. C. (1995). Growth inhibition and apoptosis of RL human B lymphoma cells by vitamin E succinate and RA: role for transforming growth factor beta. Cell Growth & Diff. 6, 655-656.

Vaux, D. L. (1993). Toward an understanding of the molecular mechanisms of physiological cell death. Proc. Natl. Acad. Sci. USA 90, 786-789.

Vaux, D. L., Cory, S., & Adams, J. M. (1988). Bcl-2 gene promote haemopoietic cell survival and cooperates with c-myc to immortalize pre-B cells. Nature 335, 440-442.

Veis, D. J., Sorenson, C. M., Shutter, J. R., & Korsmeyer, S. J. (1993). Bcl-2-deficient mice demonstrate fulminant lymphoid apoptosis, polycystic hypopigmented hair. Cell 75, 229-240.

Veronesi, U., De Palo, G., Costa, A., Formelli, Marubini, E., & Del Vecchio, M. (1992). Chemoprevention of breast cancer with retinoids. J. Natl. Cancer Inst. Monogr. 12, 93-98.

van der Burg, B., van der Leede, B.M., Kwakkenbos-Isbrucker, L., Salverda, S., de Laat, S.W., & van der Saag. (1993). Retinoic acid resistance of estradiol-independent breast cancer cells coincides with diminished retinoic acid receptor function. Mol. Cell. Endocrinol. 91, 149-157.

vom Bauer, E., Zechel, C., Heery, D., Heine, M.J., Garnier, J.M., Vivat, V., Le Douarin, B., Gronemeier, H., Chambon, P., & Losson, R. (1996). Differential ligand-dependent interactions between the AF-2 activating domain of nuclear receptors and the putative transcriptional intermediary factors mSUG1 and TIF1. EMBO J. 15, 110-124.

Wallington, L. A., Bunce, C. M., Durham, J., & Brown, G. (1995). Particular combinations of signals, by RA and 1 alpha, 25 dihydroxyvitamin D3, promote apoptosis of HL60 cells. Leukemia 9, 1185-1190.

Wang, Y., Okan, I., Szekely, L., Klein, G., & Wiman, K. G. (1995). bcl-2 inhibits wild-type p53-triggered apoptosis bt not G1 cell cycle arrest and transactivation of WAF1 and bax1. Cell Growth & Diff. 6, 1071-1075.

Welsch, C.W., DeHoog, J.V., & Moon, R.C. (1983). inhibition of mammary tumorigenesis in nulliparous C3H mice by chronic feeding of the synthetic retinoid N-(4-hydroxyphenyl) retinamide. Carcinogenesis 4, 1185-1187.

White, E. (1996). Life, death and the pursuit of apoptosis. Gene & Development 10, 1-15.

Williams, G. T., & Smith, C. A. (1993). Molecular regulation of apoptosis: genetic controls on cell death. Cell 74, 777-779.

Wyllie, A.H., Kerr, J. F. R., & Currie, A. R. (1980). Cell death: the significance of apoptosis. Int. Rev. Cytol. 68, 251-306.

Wyllie, A. H., Morris, R. G., Smith, A. L., & Dunlop, D. (1984). Chromatin cleavage in apoptosis: association with condensed morphology and dependence on macromolecular synthesis. J. Pathol. 142, 67-77.

Yang, Y., Bailey, J., Vacchio, M. S., Yarchoan, R., & Ashwell, J. D. (1995a). Fas and activation-induced Fas ligand mediate apoptosis of T cell hybridomas: inhibition of Fas ligand expression by RA and glucocorticoids. J. Exp. Med. 181, 1673-1682.

Yang, Y., Mercep, M., Ware, C. F., & Ashwell, J. D. (1995b). RA inhibition of ex vivo human immunodeficiency virus-associated apoptosis of peripheral blood cells. Proc. Natl. Acad. Sci. U.S.A. 92, 3051-3055.

Yang, Y, Minucci, S., Ozato, K., Heyman, R. A., & Ashwell, J. D. (1995c). Efficient inhibition of activation-induced Fas ligand up-regulation and T cell apoptosis by retinoids requires occupancy of both retinoid X receptors and RA receptors. J. Biol. Chem. 270, 18672-18677.

Yang, Y., Vacchio, M. S., & Ashwell, J. D. (1993). 9-*Cis*-RA inhibits activation-driven T-cell apoptosis: implications for retinoid X receptor involvement in thymocyte development. Proc. Natl. Acad. Sci. U.S.A. 90, 6170-6174.

Yonish-Rouach, E., Resnitzky, D., lotem, J., Sachs, L., Kimchi, A., & Oren, M. (1991). Wild-type p53 induces apoptosis of myeloid leukaemic cells that is inhibited by interleukin-6. Nature 352, 345-347

Yuan, J. (1995). Molecular control of life and death. Curr. Opin. Cell Biol. 7, 211-214.

Zakeri, Z.F., & Ahuja, H.S. (1994). Apoptotic cell death in the limb and its relationship to pattern formation. Biochem. Cell. Biol. 72, 603-613.

Zauli, G., Visani, G., Vitale, M., Gibellini, D., Bertolaso, L., Capitani, S. (1995). All-*trans* RA shows multiple effects on the survival, proliferation and differentiation of human fetal CD34+ haemopoietic progenitor cells. Br. J. Haematol. 90, 274-282.

Zechel, C., Shen, X. Q., Chambon, P., & Gronemeyer, H. (1994). Dimerization interfaces formed between the DNA binding domains determine the cooperative binding of RXR/RAR and RXR/TR heterodimers to DR5 and DR4 elements. EMBO J. 13, 1414-1424.

Zhang, L. X., & Jetten, A. M. (1995). Regulation of transglutaminases and apoptosis in epithelial cells by specific retinoid receptor signaling pathways. In: *Differentiation Therapy*. Vol. 10 (Kimchi, A, & Waxman, S. Eds.), pp. 311-319. Raven Press, New York.

Zhang, L. X., Mills, K. J., Dawson, M. I., Collins, S. J., & Jetten, A. M. (1995). Evidence for the involvement of RA receptor RARα-dependent signaling pathway in the induction of tissue transglutaminase and apoptosis by retinoids. J. Biol. Chem. 270, 6022-6029.

RETINOIDS IN TUMOR CELL ADHESION, INVASION, AND METASTASIS

Michael Edward

	Abstract	197
I.	Introduction	198
II.	Tumor Invasion and Metastasis	200
III.	Retinoids and Tumor Cell Growth and Differentiation	200
IV.	Effect of Retinoids on Tumor Cell Apoptosis	202
V.	Retinoids and Tumor Angiogenesis	202
VI.	Effect of Retinoids on Tumor Cell Invasion	203
VII.	Retinoids and Tumor Cell Adhesive Interactions	207
VIII.	Effect of Retinoids and Gap Junctional Communication	210
IX.	Retinoid Modulation of Host Immune Response to Tumor Cells	210
X.	Retinoids, Extracellular Matrix Synthesis, and Growth Factor Activities	211
	References	213

ABSTRACT

Retinoids have anti-tumor activity in several pre-malignant and malignant conditions, but considering the research effort and development of new retinoids, they have failed

Advances in Organ Biology
Volume 3, pages 197-217.

ISBN: 0-7623-0285-2

to fulfill their early promise as general anti-cancer agents. There are few cancers that respond well to retinoid monotherapy such as promyelocytic leukemia, but retinoids may be effective in the treatment of several other malignancies in combination with agents such as cytokines. The identification of different nuclear retinoid receptors and several retinoid responsive genes has increased dramatically our understanding of the signaling mechanisms involved, and why some cell lines do not respond to retinoids due to loss of receptors or aberrations in their signaling pathways. In many tumor cell types retinoids induce differentiation, inhibit growth, angiogenesis and invasion, modulate adhesive interactions and host immune responses, or induce apoptosis. This review will focus upon the effects of retinoids on the various steps involved in the development and progression of a tumor from *in situ* to invasive to metastatic lesions.

Further research will undoubtedly increase our understanding of the retinoid signaling pathways, and possibly the interaction of retinoid receptors with other types of receptors, and the identification of retinoid responsive elements on many more genes. Such information may enable the targeting of retinoids to predicted responsive tumors.

I. INTRODUCTION

Although a correlation between vitamin A and cancer was first noted in the 1920s as a result of enhanced premalignant and subsequently malignant lesions in vitamin A deficiency, it was some 40 years later before the use of retinoids as therapeutic agents in malignancy was exploited. This association between vitamin A deficiency and malignancy suggests that physiological levels of retinoids are required to suppress the malignant phenotype, a suggestion that has stimulated considerable effort into the elucidation of the signaling mechanisms involved. Retinoids mediate most of their effects through regulation of gene expression via a signal transducing mechanism involving nuclear retinoic acid receptors that are members of the steroid hormone receptor superfamily. There are two types of receptors, retinoic acid receptors (RARs), which bind all-*trans* retinoic acid and 9-*cis* retinoic acid, and retinoid X receptors (RXRs), which bind only 9-*cis* retinoic acid with each type containing 3 subtypes, α, β and γ. Signaling appears to be mediated by heterodimerization of RARs and RXRs, which then subsequently bind to specific DNA sequences known as retinoic acid response elements (RAREs) that act as ligand-dependent transcriptional regulators for retinoic acid-responsive genes. More comprehensive descriptions of retinoic acid receptors and signaling mechanisms are detailed in chapter 2 of this book, and in recent reviews (Mangelsdorf et al., 1994; Giguere, 1994).

The major route by which retinoids mediate their effects on tumor cell growth and differentiation appears to be through binding to the nuclear retinoic acid receptors, which in turn bind the *cis* regulatory sequences in RAREs, thereby regulating the expression of certain target genes (Love and Gudas, 1994). Defects in such RARs or their signaling mechanisms may contribute to the development of the tumor cell phenotype in a number of cell types. RARβ may play a role in suppression of carcinogenesis as its expression has been shown to be down-regulated in a number of squamous cell carcinomas compared to normal epithelium. Such a reduction in RARβ may enable premalignant cells to escape the anticarcinogenic activity of physiological levels of retinoids. Gene translocations in RARs also appear to be involved in the pathogenesis of certain cancers. In acute promyelocytic leukemia, a number of RARα fusion proteins have been identified as a result of a translocation involving chromosome 15 and 17, which fuses the promyelocytic leukemia-associated gene and the RARα gene. A second type of reciprocal chromosomal translocation involving RARα has now been identified involving chromosome 11 and 17, but while the activity of the resultant fusion proteins may be involved in the block in differentiation in leukemic promyelocytes, the precise mechanism is unknown. However, complete remission in most promyelocytic leukemia patients can be achieved following treatment with all-*trans* retinoic acid, suggesting an involvement of the RAR pathway. It is also likely that the RARγ and RXR retinoid receptors may be aberrently expressed in certain other cancers.

One possible group of genes targeted by retinoid receptors are the homeobox-containing genes, which play important roles in embryonic development, and which have now been shown to be involved in regulating major differentiation-related biological functions of adult cells (Care et al., 1994). Aberrantly activated homeobox-containing genes have been observed in a number of cancers compared to the normal tissue, including those of the kidney, colon, and lung, and with metastatic variants (Cillo, 1994). In view of such observations, it is therefore likely that such homeobox genes are involved in many of the observed effects of retinoids on cellular growth and differentiation.

Murine tumor skin progression to papillomas to squamous cell carcinomas to spindle cell carcinomas is associated with an increased abundance of transcripts for RXRs and decreased presence of RARs, which may favor other nuclear signal tranduction pathways requiring RXRs for heterodimer formation (Darwiche et al., 1995). However, retinoids inhibit tumor progression in the murine experimental carcinogenesis skin model, where retinoic acid applied topically inhibits papilloma formation, and when fed in pharmacological doses blocks the conversion step from papillomas to carcinomas (De Luca et al., 1993). Recent chemopreventive trials have produced some rather disconcerting results in which a high risk population of American smokers and asbestos workers were given supplements of β-carotene plus vitamin A, but the study was terminated early because of increased morbidity and mortality in the supplemented group compared to the control group (Rowe, 1996). A study of Finnish smokers also showed increased lung cancer and mortality with β-carotene supplementation (Alpha-Tocopherol, Beta Carotene Cancer Prevention Study Group, 1994).

The precise mechanisms involved whereby retinoids inhibit either the growth of tumor cells, or the tumor cell's ability to form distant metastases, are poorly understood despite the multitude of reports indicating the effectiveness of retinoids in a number of both *in vitro* and *in vivo* growth, invasion, and experimental metastasis models. The elucidation of such mechanisms has been hindered by the great complexity of the invasive and metastatic process. This review will examine the action of retinoids on a number of steps of the metastatic process.

II. TUMOR INVASION AND METASTASIS

The spread of tumor cells from a primary site and the establishment of metastatic tumors involves many highly selective steps that depend upon a number of complex interactions between the tumor cells and their host environment. Initial growth of a primary tumor beyond approximately 2mm in diameter requires the provision of a blood supply, stimulation of which is a typical characteristic of many growing tumors, and indeed in some, the number of blood vessels may be used as a prognostic marker (Weidner et al., 1992). Stimulation of angiogenesis is mediated by tumor-derived factors and changes in the composition of tumor stroma and surrounding extracellular matrix, such matrix modifications undoubtedly also promoting tumor cell growth and invasion. Invasion of surrounding tissue and traversal of basement membranes by tumor cells is facilitated by changes in both cell-cell and cell-extracellular matrix adhesive interactions, and the production of active proteolytic enzymes such as metalloproteinases. This ability to traverse basement membranes enables tumor cells to enter the vasculature, either as single cells or as cell aggregates, where they must survive the trauma of the vasculature and evade host immune defenses prior to arrest in the capillary bed of the target organ. Specific interactions between the tumor cells and the endothelium and subendothelial basement membrane of the target organ are involved in both the arrest of the tumor cells and in their extravasation, which may ultimately lead to the establishment and growth of a secondary tumor.

For a tumor cell to metastasize, it must be able to complete all of the steps of the metastatic cascade, and therefore would appear to provide a number of potential sites for therapeutic targeting. However, there are a number of redundant factors involved in these steps, and blocking the activity of one factor may merely stimulate the tumor cell to utilize another factor, while most of these steps are normal processes involved in certain biological functions, and general inhibition may lead to serious side effects. This review will focus upon the effect of retinoids on a number of important steps involved in tumor cell dissemination including tumor cell growth, both at the primary and secondary site, angiogenesis, invasion, cell adhesive interactions, and the production of growth factors.

III. RETINOIDS AND TUMOR CELL GROWTH AND DIFFERENTIATION

In recent years there has been a dramatic increase in the number of studies investigating the mechanisms involved in retinoic acid-induced inhibition of tumor cell growth and induction of a differentiated phenotype. Such interest has been prompted by the discovery of retinoic acid receptors through which the retinoids undoubtedly mediate most of their effects, with the diversity of observed effects being related to the expression of RAR/RXR subtypes. Many studies now not only examine the effect of retinoic acid on the tumor cells but also involve an analysis of their RARs. The determination of the tumor cells RAR and RXR subtype profile will enable invaluable information regarding their susceptibility to certain retinoids to be deduced, and correlated with phenotypic changes. Cellular retinoic acid binding proteins (CRABPs) may also play a role in the response of a cell to retinoids, although their primary function appears to be the regulation of intracellular retinoid homeostasis.

Proliferation and differentiation are two processes normally inversely linked during embryonic development and in the maintenance of normal homeostasis. One of the major effects of retinoids is the modulation of growth and differentiation, and indeed many of the therapeutic effects of retinoids in the treatment of both precancerous and cancerous lesions is undoubtedly related to their ability to induce a differentiated phenotype. This is perhaps most apparent in the treatment of acute promyelocytic leukemia with retinoic acid (Degos et al., 1992), and its effects on F9 embryonal carcinoma cells, treatment of which leads to differentiation into primitive endoderm, while combined treatment with retinoic acid and dibutyryl cAMP induces differentiation into parietal endoderm (Strickland et al., 1980). Epithelial cell growth and differentiation is normally under *in vivo* retinoid regulation, with squamous metaplasia developing during vitamin A deficiency, a defect that can be reversed by retinoid supplementation. It is therefore not surprising that retinoids are effective in the treatment of a number of epithelial disorders including both premalignant and malignant lesions.

Retinoids have been shown to inhibit growth and modulate differentiation in a wide range of tumor cell types including melanomas (Edward et al., 1988; Schadendorf et al., 1994), various squamous cell carcinomas (Zou et al., 1994; Sacks et al., 1995), and neuroblastoma (Thiele et al., 1988). Rosewicz and colleagues (Rosewicz et al., 1995) found that all-*trans* retinoic acid treatment of four different ductal pancreatic tumor cell lines resulted in the induction of a more differentiated phenotype characterized by both morphological criteria and by enhanced expression of carbonic anhydrase. They were unable to demonstrate any such effects of retinoic acid on an acinar pancreatic tumor cell line, but did find that this unresponsive cell line lacked RARγ and CRABP II. Schadendorf and coworkers (Schadendorf et al., 1994) found that after screening a number of retinoids for antiproliferative effects on melanoma cells, only those with high RARγ selectivity were active, again suggesting the importance of the retinoid receptors. Melanoma cells have perhaps been studied most in relation to the antiproliferative, antiinvasive, and antimetastatic effects of retinoids, presumably because stages in the progression from *in situ* to invasive to metastatic tumors have been well characterized in this cell type. They therefore provide an ideal tumor system in which to study responses to retinoids by cells that vary in their invasive and metastatic potential. However, the response of melanomas to retinoids varies greatly, both in growth inhibition and production of melanin, which is frequently used as a differentiation marker. Indeed some melanoma cell lines exhibit enhanced growth in the presence of retinoids while others may upregulate melanin synthesis in the absence of growth inhibition. These variable responses are probably related to differences in RAR expression and to the heterogeneous nature of the tumor in which cells at various stages of differentiation may be present. Indeed Kim and colleagues (Kim et al., 1995) have recently shown that despite functional retinoid receptors in malignant human bronchial epithelial cells, these cells remained relatively resistant to all-*trans* retinoic acid, while growth in their normal counterparts was inhibited.

IV. EFFECT OF RETINOIDS ON TUMOR CELL APOPTOSIS

In addition to inhibiting growth and inducing differentiation in a number of tumor cell types, retinoids have been found to induce apoptosis in some cell lines. The synthetic retinoid N-(4-hydroxyphenyl) retinamide induces apoptosis in neuroblastoma cells, while in contrast, retinoic acid fails to induce apoptosis but inhibits growth by differentiation (Di Vinci et al., 1994). The acyclic retinoid all-*trans* 3,7,11,15, tetramethyl-2,4,6,10,14-hexadecapentaenoic acid induces apoptosis in a human hepatoma-derived cell line as early as 8hr after addition of the retinoid, while all-*trans* retinoic acid and 9-*cis* retinoic acid fail to induce apoptosis (Nakamura et al., 1995). Retinoic acid-induced growth arrest and apoptosis in breast cancer cells has also been associated with the expression of RARβ (Seewaldt et al., 1995). While RARβ appears to be absent or downregulated in most breast cancer cell lines, RARβ-negative MCF-7 cells, when transfected with the RARβ gene, become susceptible to growth inhibition by 1μm all-*trans* retinoic acid, and undergo apoptosis after four to six days treatment. Exposure of an ovarian adenocarcinoma cell line to all-*trans* retinoic acid results in the induction of substantial cell death, which has been correlated with upregulation of c-myc transcript levels (Krupitza et al., 1995). However, the mechanisms involved in c-myc upregulation appear to be via RAR-independent pathways involving membrane-associated phospholipases.

V. RETINOIDS AND TUMOR ANGIOGENESIS

For a solid tumor to grow beyond approximately 2mm in diameter, the development of new blood vessels is essential, as passive diffusion of nutrients into or waste products out of the tumor becomes inadequate. The development of a well-developed blood supply may also facilitate the spread of the tumor cells to metastatic sites without significant local invasion by the primary tumor. Tumor angiogenesis is the directional sprouting, migration, and proliferation of endothelial cells from capillaries to form new vessels that grow toward the solid tumor, and are likely to be stimulated by factors released by the tumor cells, modulation of the peritumoral extracellular matrix, and possibly by host inflammatory cells attracted to the tumor site. Inhibition of angiogenesis by retinoids may be mediated by an effect on the production of tumor-derived factors or by a direct effect on the capillary endothelial cells. Angiogenesis in the CAM of the six-day chick embryo is inhibited by 1μg of retinoic acid, while vitamin A injected intramuscularly in rabbits inhibits angiogenesis in the V2 carcinoma implanted intracorneally.

All-*trans*-, 13-*cis*-, and 9-*cis* retinoic acid exhibit antiangiogenic effects in an experimental model of cutaneous angiogenesis, induced by intradermal injection of human breast and vulval carcinoma cells into immunosuppressed mice (Majewski et al., 1995). The antiangiogenic effects of the retinoids were observed both by exposure of the tumor cells to the retinoids prior to injection, or by treatment of the animals intraperitoneally for five days before the angiogenesis assay. The inhibitory effects of the retinoids could be counteracted by a selective antagonist of RARα, suggesting an involvement of RARα in the mechanism of angiogenesis inhibition. The effect of the retinoids is enhanced by combination with interferon α or 1,25-dihydroxyvitamin D_3, the latter compounds synergistic effect possibly being mediated by the formation of heterodimers between the corresponding receptors. Retinoids may also inhibit angiogenesis by a direct effect on the endothelial cells, either by affecting growth or migration, a process that involves the production of certain proteolytic enzymes and adhesive interactions, both of which will be discussed in subsequent sections. A number of *in vitro* studies have demonstrated that metalloproteinases play an important role in angiogenesis by facilitating extracellular matrix degradation and promoting endothelial cell migration, while *in situ* detection of metalloproteinases suggests that it is interstitial collagenase that is produced in significant amounts in tumor blood vessels (Karelina et al., 1995).

VI. EFFECT OF RETINOIDS ON TUMOR CELL INVASION

One of the major characteristics of malignant tumor cells is their ability to invade tissues and to form secondary tumors at distant sites. Tumor invasion is a highly complex process and is facilitated by modulation of the surrounding matrix and several adhesive interactions that will be discussed in the following section. Modification of the matrix surrounding a tumor is mediated by both compositional changes and by the action of a number of proteolytic enzymes. Liotta (1986) has proposed a three-step theory of invasion through basement membranes that still holds true today, in which the tumor cell attaches to laminin via laminin receptors, followed by degradation of type IV collagen, and cell migration. Ligation of certain cell surface adhesion molecules may result in signal transduction culminating in the release of proteases. This theory is supported by the discovery that laminin, or more effectively laminin fragments induce the release of type IV collagenase by a number of metastatic cell lines. That collagenases and indeed other classes of degradative enzymes are important in mediating tumor cell invasion is supported by the correlation between invasive/metastatic capacity of certain cell lines and the production of a number of proteolytic enzymes such as type IV collagenase. However, the production of proteolytic enzymes is not sufficient for lytic activity as most of the proteases involved in the invasive process are released in an inactive form that requires subsequent activation, but that may then be inactivated by various classes of inhibitors. Invasive tumor cells may also stimulate surrounding normal stromal cells to synthesize and release large amounts of proteolytic enzymes including interstitial collagenase, gelatinases, and stromelysins (Ito et al., 1995). However, these enzymes appear to degrade the extracellular matrix in a controlled manner, and there is increasing evidence that they bind to the surface of tumor cells where they may be activated (Chen, 1992), and indeed such activities have been localized to the invading front or invadopodia of tumor cells (Monsky et al., 1994).

Tumor cell invasion and degradation of extracellular matrix has been associated with increased production of a number of classes of proteolytic enzymes including matrix metalloproteinases (MMPs) such as the collagenases, serine proteases such as plasminogen activators, lysosomal cysteine proteases including cathepsin B and L, and lysosomal aspartic proteases such as cathepsin D.

The MMPs all contain a zinc ion at their active site, exhibit sequence homology, are almost all secreted in a latent form and are inhibited by specific tissue inhibitors of metalloproteinases (TIMPs). The MMPs are grouped into three main subclasses of interstitial collagenases (MMP-1), type IV collagenases (MMP-2 or 72 Kd gelatinase A, and MMP-9 or 92 Kd gelatinase B), and stromelysins. The interstitial collagenase, MMP-1, cleaves fibrillar collagen types I, II, III, VII, and X at a single site and thus destroys the higher order structure of the collagen, rendering it susceptible to other proteases. A number of studies have demonstrated suppression of MMP-1 by retinoic acid, while the stimulated synthesis of MMP-1 by cytokines and growth factors such as interleukin-1α and -1β (IL-1α and -β), Tumor necrosis factor-α (TNF-α), epidermal growth factor (EGF) and platelet derived growth factor (PDGF), and phorbol esters is suppressed by retinoic acid (Woessner, 1991). A major source of MMP-1 utilized in tumor invasion is likely to be synthesized by normal fibroblasts associated with the tumor stroma, and therefore the action of retinoids in suppressing MMP-1 synthesis may be either upon production of tumor-derived factors or a direct effect on the stromal cells (Ito et al., 1995). In fibroblasts, Pan and Brinckerhoff (1994) have shown that RXR-α mRNA is constitutively expressed and not regulated by retinoids while mRNAs for RAR-α, -β, and -γ are constitutively low, but are inducible with retinoids. It appears that RAR/RXR heterodimers are involved in complex formation with certain sequences mediating suppression of the collagenase gene and undoubtedly involves the activator-1 (AP-1) site, although there is no direct evidence that the RAR/RXRs bind directly to AP-1 sites, but may involve interactions with *fos* and *jun*. Fisher and colleagues (Fisher et al., 1996) have recently shown that exposure of human skin to UVB light results in the rapid induction of MMPs and upregulation of transcription factors AP-1 and NF-κB, known stimulators of MMPs, and that retinoic acid pretreatment of the skin substantially reduces AP-1 and collagenase induction. Although it is now generally accepted that the AP-1 site plays an important role both in the induction and suppression of MMP gene expression, there is increasing evidence for the involvement of additional sites.

A number of tumor cell types have been shown to express gelatinases A and B (MMP-2 and -9) and indeed enhanced expression has frequently been correlated with tumor invasive and metastatic potential. One of the major substrates for gelatinase A and B is type IV collagen, suggesting that they are likely to play a major role in invasion and metastasis by facilitating tumor cell traversal of basement membranes. *In vitro* invasive capacity is usually assessed by invasion through filters coated with a reconstituted basement membrane gel, and in many instances, invasion into the gel can be inhibited by treatment of the tumor cells with retinoids and correlated with a reduced expression of type IV collagenases (Hendrix et al., 1990). *In vivo* assessment is by utilizing an experimental metastasis model, in which pretreated cells are injected into the tail veil of mice and lung tumors counted after approximately three weeks. A number of studies have demonstrated that pretreatment of tumor cells with retinoids blocks their ability to form lung tumors in experimental metastasis (Edward and MacKie, 1989), but the precise mechanism involved is difficult to assess, and may involve a number of cell-cell and cell-extracellular matrix interactions as well as suppression of type IV collagenase production.

Stromelysin 3 is a recently characterized MMP that appears to be differentially expressed in certain benign and malignant tumors, and in normal tissues that undergo extensive remodelling such as the placenta, uterus, post-lactation mammary glands, and during wound healing. Unlike other MMPs, it is secreted as an active enzyme and has a unique substrate specificity, with the serine protease inhibitor α1-protease inhibitor (serpin) being a major substrate while types I and IV collagen, fibronectin, laminin, and gelatin are not degraded. A major source of stromelysin 3 appears to be stromal cells adjacent to growing and invading tumors, presumably stimulated by tumor-derived factors (Basset et al., 1990). Retinoic acid has been shown to inhibit the induction of stromelysin 3 in normal fibroblasts by 12-O-tetradecanoylphorbol-13-acetate and certain growth factors (Anderson et al., 1995). Such an inhibition may have an effect on tumor development as α1-protease inhibitor has a suppressive effect on tumor cell growth and invasion. Another stromelysin, matrilysin (MMP-7), has been shown to be associated with colon cancer cell invasion *in vitro*, and to be suppressed by treatment with retinoic acid (Yamamoto et al., 1995).

While retinoids suppress the synthesis of MMPs in a number of cell types, the activity of the MMPs may also be modified by retinoids through their effects on the production of specific inhibitors, the so-called tissue inhibitors of metalloproteinases (TIMPs). Three TIMPs have currently been identified that inhibit active MMPs, binding in a 1:1 molar ratio to form tight noncovalent complexes. TIMP 1 and 2 also have been found to form tight-binding complexes with the latent forms of MMP-2 (gelatinase B) and MMP-9 (gelatinase A) respectively. Retinoids have been found to upregulate the expression of TIMPs while simultaneously inhibiting the synthesis of MMPs in various cell types, but in addition, retinoic acid-induced stimulation of TIMPs in fibroblasts has been shown to be synergistically enhanced with basic fibroblast growth factor (bFGF) and EGF (Bigg and Cawston, 1995).

The serine proteases plasminogen activators catalyze the conversion of plasminogen into plasmin, a very broad specificity protease. Of the two types of plasminogen activators most commonly studied, tissue-plasminogen activator (t-PA) and urokinase-plasminogen activator (u-PA), it is u-PA that is most frequently implicated in tumor invasion. The expression of free plasmin activity is efficiently blocked by inhibition of plasminogen activation by plasminogen activator inhibitors (PAI-1, PAI-2, and PAI-3) and by the plasma proteinase inhibitors α_2-antiplasmin and α_2-macroglobulin, which inhibit plasmin in solution. Both u-PA and plasminogen bind to many cell types, and the concomitant localization of both u-PA and plasminogen on the cell surface ensures focal activation will be localized to invadopodia, similar to that observed with MMPs. Activation of plasminogen is by the cleavage of a single bond to generate the broad specificity protease plasmin, which can degrade a number of extracellular matrix molecules including fibronectin and laminin, but it can also function as an activator of latent MMPs.

In a number of tumor cell types, tumor infiltrates contain more u-PA positive cells than normal tissues, but the staining is predominantly confined to fibroblast-like cells and endothelial cells within the tumor stroma. The u-PA receptor (u-PAR), however, is detected in malignant epithelial cells, and therefore, like the MMPs, the tumor cells appear to stimulate normal stromal cells to produce u-PA, which the tumor cells then recruit to their cell surface via their u-PARs (Ossowski et al., 1991). Retinoids may therefore affect the activity of plasmin by either inhibiting the synthesis of PAs and cell surface u-PARs, or by upregulating the expression of PAIs. Treatment of human prostate adenocarcinoma cell lines with N-(4-hydroxyphenyl)-retinamide results in a small reduction in u-PA activity despite a substantial increase in the total activatable u-PA, presumably due to an observed upregulation of PAI-1 production (Kim et al., 1995). Waghray and Webber (1995) have also demonstrated a retinoic acid-induced reduction in u-PA activity in prostatic carcinoma cells and have correlated such a reduction with inhibition of extracellular matrix degradation and invasion into reconstituted basement membrane. In contrast, other groups have demonstrated increases in receptor-bound u-PA in human fibrosarcoma cell lines following exposure to retinoic acid (Miwa et al., 1995). Such apparent inconsistent effects of retinoids on PA expression underline the importance of determining the expression of not only u-PA, but also PAIs and RAR expression.

In addition to the important role of certain classes of proteases in facilitating tumor invasion, the expression of glycosidases and in particular, heparanase may also be involved in facilitating tumor cell invasion. The activity of heparanase has been directly correlated with the lung colonization potential of human melanoma cells (Nakajima et al., 1986), and has also been found in the sera of patients with metastatic disease. Heparan sulphate proteoglycans form an important part of the extracellular matrix, regulating both the assembly and integrity of the matrix, but also provide an extracellular matrix reservoir of certain heparin-binding growth factors. They are major components of basement membranes where they may function as selective filters due to their high anionic charge. Blocking of heparanase activity in highly metastatic B16 melanoma cells with structural analogues of heparin inhibits experimental metastasis (Vlodavsky et al., 1994), but there is currently little evidence to suggest that retinoids modulate heparanase activity. It is tempting to speculate that heparanase activity would be an effective target for retinoids in blocking tumor cell traversal of basement membranes and thus metastasis.

VII. RETINOIDS AND TUMOR CELL ADHESIVE INTERACTIONS

The progression from *in situ* to invasive to metastatic tumors involves a highly complex sequence of both cell-cell and cell-extracellular matrix interactions (Edward and MacKie, 1993). Such interactions by the tumor cells require the expression of a vast array of adhesion molecules that are involved in the release of tumor cells from the primary tumor and local invasion, entry into the vasculature and interactions with certain blood cells, arrest in target organs by means of interactions with capillary endothelial cells, and extravasion and invasion into the target organ tissue. These adhesion molecules may be highly specific for a particular ligand or capable of binding multiple ligands such as certain integrins, but the expression of such a repertoire of adhesion molecules may not only determine tumor cell adhesive properties, but also metastatic potential (Edward, 1995a).

Retinoic acid has been shown to reduce the adhesion of a human intestinal epithelial cell line to plastic, while also reducing the cells' ability to grow in clusters, suggesting that both cell-substratum and cell-cell interactions were affected. Retinoic acid also reduces cell-cell adhesion in BHK21/C13 cells, but unlike the intestinal epithelial cells, cell-substratum adhesion is increased (Kamei, 1983). The attachment of 3T3 fibroblast-like cells to type IV collagen and laminin is enhanced by retinoic acid, while virally transformed 3T3 cells, which adhere readily to these substrates, are unaffected by retinoic acid pretreatment (Kato and De Luca, 1987). We have found that metastatic B16 melanoma cell adhesion to fibronectin is unaffected by retinoic acid pretreatment, but that adhesion to type IV collagen and laminin is significantly reduced (Edward et al., 1989), while the melanoma cells' ability to form, or maintain the integrity of tumor spheroids, is inhibited by retinoic acid (Edward et al., 1992a). There have been many studies involving adhesion receptor expression in melanoma cells primarily because changes in adhesion receptor expression have been correlated with the stage of tumor progression (Albelda, 1993), and perhaps the most studied family of adhesion molecules are the integrins.

The integrins constitute a large family of heterodimeric transmembrane receptors that mediate both cell-cell and cell-extracellular matrix interactions, each integrin receptor being composed of an α and a β subunit. There are currently at least 20 α and 8 β subunits identified that can associate to form a number of receptors with distinct binding specificities. Integrins are the major group of adhesion molecules responsible for mediating cell-extracellular matrix interactions, and changes in the expression of certain of these molecules have been identified with the development of the malignant phenotype and with tumor progression. That such integrins are important in tumor cell dissemination can be demonstrated by inhibition of experimental metastasis following preincubation of the tumor cells with peptides containing the arginine-glycine-aspartic acid (RGD) sequence, which is recognized by many of the integrins, while antibodies against certain integrin subunits may also block experimental metastasis, adhesion, and invasion (Humphries et al., 1986; Melchiori et al., 1995). Care has to be taken in selecting antibodies, as some may not block the adhesive interaction of the integrin but activate the integrin and actually potentiate its affinity for its ligand.

Retinoic acid treatment of Lewis lung carcinoma cells induces a marked reduction in the amount of β4 integrin subunit, which is associated with an inhibition of the malignant phenotype (Gaetano et al., 1994). There is also a quantitative switch in integrin expression accompanying differentiation of F9 embryonal carcinoma cells treated with retinoic acid, in which a 10-25 fold increase in the amount of $\alpha 3\beta 1$ integrin protein was observed with a reduction of $\alpha 5\beta 1$ expression (Burdsal et al., 1994). Santos and colleagues (Santos et al., 1994) found that retinoic acid treatment of MeWo melanoma cells enhanced their adhesion to vitronectin, which accompanied increased levels of the αv integrin subunit. It is also likely that retinoic acid may affect not only the expression of certain adhesion molecules, but their activation state and their signal-transducing ability, such as the induction of type IV collagenase following ligation of the laminin receptor or certain integrins (Seftor et al., 1993).

The cadherin family of adhesion molecules are calcium ion-dependent cell-cell adhesion molecules that have an inhibitory effect on cell invasion. E-cadherin has an antimetastatic and antiinvasive role in epithelial cells. When MDCK cells are treated with anti-E-cadherin antibodies, the cells acquire an invasive phenotype, while highly invasive clones transfected with E-cadherin plasmids lose their invasive capacity (Uleminckx et al., 1991), suggesting an invasion suppressor role for E-cadherin. Vermeulen and colleagues (Vermeulen et al., 1995) have demonstrated the enhanced aggregation of MCF-7/6 breast cancer cells following exposure to retinoic acid, and have shown that the aggregation was mediated via E-cadherin as assessed by antibodies functionally blocking the adhesion molecule.

The highly glycosylated transmembrane protein, CD44, was originally described as a homing receptor for lymphocytes but is now found to be expressed by a number of different cell types. Several variants of the standard CD44 have been identified, with certain variant forms being associated with tumor cell aggressiveness. The exact function of CD44 is as yet unclear, although its ability to bind hyaluronate would suggest a possible role in mediating cell interactions with, and migration into hyaluronate-rich matrices. There are few reports of retinoic acid modulating CD44 expression, but Gross and coworkers (Gross et al., 1995) recently demonstrated a retinoic acid-induced increase in neuroblastoma cell CD44 positive cells and increased density of cell surface CD44. In most tumor cell models, overexpression of CD44, and in particular its variant isoforms, is associated with progression and metastasis, whereas its presence on neuroblastoma cells appears to suppress tumorigenic and metastatic properties, and is associated with a retinoic acid-induced neuronal differentiated phenotype.

Intercellular adhesion molecule-1 (ICAM-1) is a cell surface glycoprotein and a member of the immunoglobulin superfamily whose expression is regulated in a wide variety of cells by various proinflammatory cytokines such as interleukin-1, interferon-γ, and tumor necrosis factor-α. ICAM-1 is the counter-receptor for the β2 integrins lymphocyte function associated antigen-1 (LFA-1/CD11a) and Vc_{3bi} complement receptor (MAC-1). ICAM-1 interacts with LFA-1, expressed preferentially on leukocytes modulating a variety of biological responses such as the recruitment of activated lymphocytes and macrophages to inflammatory sites, and immune surveilance of tumors by cytotoxic T lymphocytes. Although constitutively expressed by relatively few cell types, ICAM-1 is readily induced in many cell types by proinflammatory cytokines. ICAM-1 can be induced by retinoic acid in a number of different tumor cell lines (Bouillon et al., 1991) and in a neuroblastoma cell line both ICAM-1 and RAR-β are elevated. The promoter region of the ICAM-1 gene has been identified as containing a RARE necessary for the retinoic acid-mediated induction of ICAM-1, while cotransfection experiments with different retinoic acid receptors suggest that the ICAM-1 gene is regulated by retinoic acid in a RAR-β/RXR-α-dependent fashion (Aoudjit et al., 1995).

ICAM-1 expression in a number of melanoma cell lines is induced by exposure to retinoic acid, and Cilenti and colleagues (Cilenti et al., 1995) have shown that the ICAM-1 promoter is retinoic acid-inducible and contains a RARα complemented with RXRs, and that mutation of the RARE abrogates promoter responsiveness to retinoic acid. As enhanced expression of ICAM-1 on the tumor cell surface should render tumor cells more susceptible to cell lysis by immune effector cells, and indeed agents that enhance ICAM-1 expression have been used in cancer therapy (Webb et al., 1991), it is somewhat surprising that ICAM-1 expression may be a positive correlate of high-grade metastatic potential in melanoma. Altomonte and coworkers (Altomonte et al., 1993) however, have demonstrated that soluble

ICAM-1 isolated from the sera of melanoma patients will inhibit effector cell lysis of melanoma cells, presumably by binding to LFA-1 on the effector cells. It is tempting to speculate that the increased expression of ICAM-1 associated with highly metastatic melanoma cells, and its subsequent release from the cell surface in a functional soluble form, may block tumor cell-effector cell interactions and hence tumor cell lysis. It is therefore possible that retinoic acid may not only stimulate ICAM-1 expression, but reduce its release from the cell surface in a soluble form.

VIII. EFFECT OF RETINOIDS ON GAP JUNCTIONAL COMMUNICATION

Gap junctional communication (GJC) is believed to play an important role in the maintenance of tissue homeostasis and in the regulation of cellular proliferation. Cells may communicate directly via specialized intercellular channels called gap junctions by exchange of low molecular weight molecules such as metabolites, ions, and various intracellular second messengers. Each gap junctional complex is composed of a channel formed by the association of two hemichannels provided by the two adjacent cells, where each hemichannel is composed of six proteins, the connexins, forming a central pore. Many tumor cells appear to exhibit decreased intercellular communication, particularly with surrounding normal cells, and it is this heterologous communication that may lead to tumor suppression. A number of tumor promoters such as phorbol esters block GJC (Klaunig and Ruch, 1990), which correlates with their tumor-promoting activities *in vivo*.

Retinoic acid is able to inhibit the transformation of C3H10T1/2 cells at physiological concentrations, an effect that has been correlated with an increase in GJC (Mehta and Loewstein, 1991), while in rat liver epithelium, retinoic acid induces enhanced GJC and enhanced expression of connexin-43 (Bex et al., 1995). Gap junctional formation is regulated by cell adhesion molecules, and therefore some of the observed effects of retinoic acid in enhancing GJC may be mediated in part by upregulation of certain adhesion molecules, and also glycosaminoglycans, two classes of molecules known to be modulated by retinoids. As tumors progress from in situ to invasive to metastatic cells, GJC may play vital roles at different stages of the metastatic process. GJC, in conjunction with cell-cell adhesive interactions, will be lost when cells are released from the primary tumor, but significantly increased adhesion and communication levels between metastatic cancer cells and endothelium of the target organ implies that GJC is important in facilitating extravasation (Elsabban and Pauli, 1994). Just how retinoids would affect such a step is currently unclear.

IX. RETINOID MODULATION OF HOST IMMUNE RESPONSE TO TUMOR CELLS

The growth of a primary tumor and the subsequent successful establishment of a metastasis depends upon the tumor cells' ability to evade immune surveilance.

While retinoids may affect ICAM-1 expression on tumor cells, rendering them more susceptible to effector cell lysis, retinoids may also enhance the activity of the immune response. A number of studies have demonstrated the ability of retinoids to modulate host antitumor immune responses, but the exact mechanisms involved have yet to be elucidated. Some of the possible mechanisms may involve the suppression of shedding of tumor cell plasma membrane fragments that suppress T-lymphocyte activation (Jiang et al., 1990), and modulation of cytokine responses. That retinoids do affect T-lymphocytes has been demonstrated by the inhibition of growth and metastasis of transplantable tumors in syngeneic mice, but not in nude mice (Eccles et al., 1985).

Lymphokine killer activity, which may be derived from both T-lymphocytes and natural killer cells, is significantly enhanced by retinoic acid. This enhanced activity appears to be mediated via increased production of tumor necrosis factor and interferon gamma, which in turn enhances interleukin-2 receptor expression (Fegan et al., 1995). N-(4-hydroxyphenyl)-retinamide has been shown to reduce the incidence of experimental tumors in mice, but a recent study of natural killer cell activity in a group of mastectomised breast cancer patients treated with the retinamide showed that natural killer cell activity was augmented (Villa et al., 1993). The retinoid does not, however, appear to act directly upon the natural killer cells, as incubation of peripheral blood mononuclear cells from blood donors with the retinoid does not modify their natural cytotoxicity.

X. RETINOIDS, EXTRACELLULAR MATRIX SYNTHESIS, AND GROWTH FACTOR ACTIVITIES

Retinoids may affect tumor cell growth and invasion by modulating the extracellular matrix composition in the tumor stroma and thereby affect tumor cell proliferation, invasion and angiogenesis. Retinoids undoubtedly modify the extracellular matrix by influencing the activity of a number of degradative enzymes such as the MMPs and PAs, but they may also affect the synthesis of a number of extracellular matrix components. Many growing and invading tumors exhibit a modified peritumoral extracellular matrix, which in many instances is similar to that observed during wound healing, and is characterized by an increased production of hyaluronate, modified collagen synthesis, and enhanced cellular proliferation and migration. Such changes in the tumor stroma may facilitate tumor growth, invasion, and angiogenesis, and as with MMPs, much of this enhanced extracellular matrix synthesis appears to be derived from normal peritumoral stromal cells, apparently stimulated by tumor cell-derived factors. Any effect of retinoic acid upon peritumoral extracellular matrix synthesis may therefore be due to a direct effect upon either the tumor cells or the surrounding stromal cells, or by a modulation of the activity of tumor cell-derived factors. We have demonstrated a dramatic increase in hyaluronate synthesis by normal skin fibroblasts exposed to melanoma cell-conditioned medium (Edward et al., 1992b; Godden et al., 1996). Exposure of the tumor cells to retinoic acid during medium conditioning or fibroblasts exposed to the tumor cell-conditioned medium failed to block the stimulation in hyaluronate synthesis (unpublished observation). Indeed in one melanoma cell line exposed to retinoic acid, a further stimulation in hyaluronate was observed, particularly at low (10^{-9}M-10^{-8}M) retinoic acid concentrations.

Retinoids have been shown to counteract the tumor promoter-stimulated release of fibronectin from a number of cell lines (Zerlauth and Wolf, 1985), to decrease collagen synthesis in both normal and keloid fibroblasts, and to inhibit the production of thrombospondin and fibronectin in human epidermis (Varani et al., 1989). Retinoic acid does however stimulate collagen synthesis in ultraviolet-B irradiated mouse skin while having no effect on nonirradiated skin (Schwartz et al., 1989). Hyaluronate synthesis is inhibited by retinoic acid in cultured skin fibroblasts (Edward, 1995b), but is stimulated in human epidermal skin organ cultures (Tammi et al., 1989) and during retinoic acid-induced differentiation of teratocarcinoma cells (Prehm, 1980). The effect of retinoic acid on extracellular matrix synthesis is therefore dependent not only upon the cell type, but their proliferative and differentiated state. We have demonstrated a retinoic acid-induced inhibition of synthesis and changes in the properties of sulphated glycosaminoglycans of metastatic B16 melanoma cells (Edward and MacKie, 1989), but just how such changes may affect the tumor cells' ability to metastasize is unclear, as the glycosaminoglycans may be involved in cellular adhesive interactions, gap junction formation, and the regulation of growth factor activities.

The interaction of certain growth factors with extracellular matrix components may modulate their activity and bioavailability, and therefore any change in the extracellular matrix composition may affect their interaction with cell receptors.The binding of fibroblast growth factors (FGFs) to heparin and heparan sulphate appears to protect them from degradation and therefore may provide an extracellular matrix or cell surface bound reservoir of the growth factor. As basic FGF requires prior binding to heparan sulphate for interaction with its receptor, tumor cell modulation of the extracellular matrix by either proteolytic degradation of heparan sulphate proteoglycan, or limited heparanase activity may affect the availability and activity of bFGF, and hence the growth of tumors, many of which utilize FGF as an autocrine growth factor. FGF is also a potent angiogenic factor, therefore any retinoid-induced modulation of the extracellular matrix that affects the availability of bFGF may modulate both tumor cell growth and angiogenesis. Transforming growth factor-β (TGF-β) is also found associated with the extracellular matrix, frequently bound to the proteoglycan decorin, an interaction that results in inactivation of the TGF-β.

Retinoic acid may modulate the response of a tumor cell to certain growth factors by affecting their bioavailability, synthesis, activation status, or the expression of their receptors. TGF-β may be induced in a number of normal and tumor cell types by retinoic acid, coincident with decreased growth rate and/or differentiation (Danielpour, 1996; Falk et al., 1991), while Cohen and colleagues (Cohen et al.,

1995) have recently demonstrated a retinoic acid-induced expression of TGF-β receptors in addition to TGF-β in human neuroblastoma cells. Modulation of TGF-β activity will in itself result in changes in cellular growth and extracellular matrix as TGF-β can inhibit the expression of various proteases including collagenases, while stimulating the synthesis of a number of matrix molecules and protease inhibitors in many cell types. Retinoids may therefore mediate some of their observed effects on tumor cells indirectly through modulation of certain growth factor activities.

REFERENCES

Albelda, S.M. (1993). Biology of disease: role of integrins and other cell adhesion molecules in tumor progression and metastasis. Lab. Invest. 68, 4-17.

Alpha-Tocopherol, Beta-Carotene Cancer Prevention Study Group. (1994). The effect of Vitamin E and beta carotene on the incidence of lung cancer and other cancers in male smokers. N. Eng. J. Med. 330, 1029-1035.

Altomonte, M., Gloghini, A., Bertola, G., Gasparollo, A., Carbone, A., Ferrone, S., & Maio, M. (1993). Differential expression of cell adhesion molecules CD54/CD11a and CD58/CD2 by human melanoma cells and functional role in their interaction with cytotoxic cells. Cancer Res. 53, 3343-3348.

Anderson, I.C., Sugarbaker, D.J., Ganju, R.K., Tsarwhas, D.G., Richards, W.G., Sunday, M., Kobzik, L., & Shipp, M.A. (1995). Stromelysin-3 is overexpressed by stromal elements in primary non-small cell lung cancers and regulated by retinoic acid in pulmonary fibroblasts. Cancer Res. 55, 4120-4126.

Aoudjit, F., Brochu, N., Morin, N., Poulin, G., Stratowa, C., & Audette, M. (1995). Heterodimeric retinoic acid receptor-β and retinoid X receptor-α complexes stimulate expression of the intercellular adhesion molecule-1 gene. Cell Growth Differ. 6, 515-521.

Basset, P., Bellocq, J.P., Wolf, C., Stoll, I., Hutin, P., Limacher, J.M., Podhajcer, C.L., Chenard, M.P., Rio, M.C., & Chambon, P. (1990). A novel matrix metalloprotease gene specifically expressed in stromal cells of breast carcinoma. Nature (Lond.) 348, 699-704.

Bex, V., Mercier, T., Chaumontet, C., Gaillard-Sanchez, I., Flechon, B., Mazet, F., Traub, O., & Martel, P. (1995). Retinoic acid enhances connexin43 expression at the post-transcriptional level in rat liver epithelial cells. Cell Biochem. Funct. 13, 69-77.

Bigg, H.F., & Cawston, T.E. (1995). All-*trans*-retinoic acid interacts synergistically with basic fibroblast growth factor and epidermal growth factor to stimulate the production of tissue inhibitor of metalloproteinases from fibroblasts. Arch. Biochem. Biophys. 319, 74-83.

Bouillon, M., Tessier, P., Boulianne, R., Destrempe, R., & Audette, M. (1991). Regulation by retinoic acid of ICAM-1 expression on human tumor cell lines. Biochim. Biophys. Acta. 1097, 95-102.

Burdsal, C.A., Lotz, M.M., Miller, J., & McClay, D.A. (1994). Quantitative switch in integrin expression accompanies differentiation of F9 cells treated with retinoic acid. Devel. Dynamics 201, 344-353.

Care, A., Testa, U., Bassani, A., Tritarelli, E., Montesoro, E., Samoggia, P., Cianetti, L., and Peschle, C. (1994). Coordinate expression and proliferative role of HoxB genes in activated adult T lymphocytes. Mol. Cell. Biol. 14, 4872-4877.

Chen, W-T. (1992). Membrane proteases: roles in tissue remodelling and tumor invasion. Curr. Opin. Cell Biol. 4, 802-809.

Cilenti, L., Toniato, E., Ruggiero, P., Fusco, C., Farina, A.R., Tiberio, A., Hayday, A.C., Gulino, A., Frati, L., & Martinotti, S. (1995). Transcriptional modulation of the human intercellular adhesion molecule gene 1 (ICAM-1) by retinoic acid in melanoma cells. Exp. Cell Res. 218, 263-270.

Cillo, C. (1994). HOX genes in human cancer. Invasion Metast. 95, 38-49.

Cohen, P.S., Letterio, J.J., Gaetano, C., Chan, J., Matsumoto, K., Sporn, M.B., & Thiele, C.J. (1995). Induction of transforming growth factor β1 and its receptors during all-*trans*-retinoic acid (RA) treatment of RA-responsive human neuroblastoma cell lines. Cancer Res. 2380-2386.

Danielpour, D. (1996). Induction of transforming growth factor-β autocrine activity by all-trans-retinoic acid and 1α,25-dihydroxyvitamin D3 in NRP-152 rat prostatic epithelial cells. J. Cell. Physiol. 166, 231-239.

Darwiche, N., Celli, G., Tennenbaum, T., Glick, A.B, Yuspa, S.H., & De Luca, L.M., (1995). Mouse skin tumor progession results in differential expression of retinoic acid and retinoid X receptors. Cancer Res. 55, 2774-2782.

Degos, L. (1992). Retinoic acid in acute promyelocytic leukemia: a model for differentiation therapy. Curr. Opin. Oncology 2, 45-52.

De Luca, L.M., Sly, L., Jones, C.S., & Chen, L.C. (1993). Effects of dietary retinoic acid on skin papilloma and carcinoma formation in female SENCAR mice. Carcinogenesis 14, 539-542.

Di Vinci, A., Geido, E., Infusini, E., & Giaretti, W. (1994). Neuroblastoma cell apoptosis induced by the synthetic retinoid N-(4-hydroxyphenyl)retinamide. Int. J. Cancer 59, 422-426.

Eccles, S.A., Purvies, H.P., Barnett, S.C., & Alexander, P. (1985). Inhibition of growth and metastasis of syngeneic transplantable tumors by an aromatic retinoic acid analogue. 2. T-cell dependence of retinoid effects *in vivo*. Cancer Immunol. Immunother. 19, 115-120.

Edward, M. (1995a). Integrins and other adhesion molecules involved in melanocytic tumor progression. Curr. Opin. Oncology 7, 185-191.

Edward, M. (1995b). Effect of retinoids on glycosaminoglycan synthesis by human skin fibroblasts grown as monolayers and within contracted collagen lattices. Br. J. Dermatol. 133, 223-230.

Edward, M., Gold, J.A., & MacKie, R.M. (1988). Different susceptibilities of melanoma cells to retinoic acid-induced changes in melanotic expression. Biochem. Biophys. Res. Commun. 155, 773-778.

Edward, M., Gold, J.A., & MacKie, R.M. (1989). Modulation of melanoma cell adhesion to basement membrane components by retinoic acid. J. Cell Sci. 93, 155-161.

Edward, M., Gold, J.A., & MacKie, R.M. (1992a). Retinoic acid-induced inhibition of metastatic melanoma cell lung colonisation and adhesion to endothelium and subendothelial extracellular matrix. Clin. Exp. Metast. 10, 61-67.

Edward, M., Grant, A.W., & MacKie, R.M. (1992b). Human melanoma cell-derived factor(s) stimulate fibroblast glycosaminoglycan synthesis. Int. J. Cancer 52, 499-503.

Edward, M., & MacKie, R.M. (1989). Retinoic acid-induced inhibition of lung colonisation and changes in the synthesis and properties of glycosaminoglycans of metastatic B16 melanoma cells. J. Cell Sci. 94, 537-543.

Edward, M., & MacKie, R.M. (1993). Cell-cell and cell-extracellular matrix interactions during melanoma cell invasion and metastasis. Melanoma Res. 3, 227-234.

Elsabban, M.E., & Pauli, B.U. (1994). Adhesion mediated gap junctional communication between lung metastatic cancer cells and endothelium. Invasion Metast. 14, 164-176.

Falk, L.A., De Benedetti, F., Lohrey, N., Birchenall-Roberts, M.C., Ellingsworth, L.W., Faltynek, C.R., & Ruscetti, F.W. (1991). Induction of TGF-β receptor expression and TGF-β protein production in retinoic acid-treated HL-60 cells: possible TGF-β-mediated autocrine inhibition. Blood 77, 1248-1255.

Fegan, C., Bailey-Wood, R., Coleman, S., Neale, S.A., Hoy, T., & Whittaker, J.A. (1995). All *trans* retinoic acid enhances human LAK activity. Eur. J. Haematol. 54, 95-100.

Fisher, G.J., Datta, S.C., Talwar, H.S., Wang, Z-Q., Varani, J., Kang, S., & Voorhees, J.J. (1996). Molecular basis of sun-induced premature skin aging and retinoid antagonism. Nature (Lond) 379, 335-339.

Gaetano, C., Melchiori, A., Albini, A., Benelli, R., Falcioni, R., Modesti, A., Modica, A., Scarpa, S., & Sacchi, A. (1994). Retinoic acid negatively regulates β4 integrin expression and suppresses the malignant phenotype in a Lewis lung carcinoma cell line. Clin. Exp. Metastasis 12, 63-72.

Giguere, V. (1994). Retinoic acid receptors and cellular retinoic acid binding proteins: complex interplay in retinoid signalling. Endocrine Rev. 15, 61-79.

Godden, J.L., Edward, M., Grant, A.W., & MacKie, R.M. (1996). Partial characterisation of human melanoma cell-derived factors that stimulate fibroblast glycosaminoglycan synthesis. Pathobiology 64, 262-268.

Gross, N., Beck, D., Beretta, C., Jackson, D., & Perruisseau, G. (1995). CD44 expression and modulation on human neuroblastoma tumors and cell lines. Eur. J. Cancer 31A, 471-475.

Hendrix, M.J.C., Wood, R., Seftor, E.A., Lotan, D., Nakajima, M., Misiorowski, R.L., Seftor, R.E.B., Stetler-Stevenson, W.G., Bevacqua, S.J., Liotta, L.A., Sobel, M.E., Raz, A., & Lotan, R. (1990). Retinoic acid inhibition of human melanoma cell invasion through a reconstituted basement membrane and its relation to decreases in the expression of proteolytic enzymes and motility factor receptor. Cancer Res. 50, 4121-4130.

Humphries, M.J., Olden, K., and Yamada, K.M. (1986). A synthetic peptide from fibronectin inhibits experimental metastasis of murine melanoma cells. Science 233, 467-470.

Ito, A., Nakajima, S., Sasaguri, Y., Nagase, H., & Mori, Y. (1995). Co-culture of human breast carcinoma MCF-7 cells and human dermal fibroblasts enhances the production of matrix metalloproteinases 1, 2 and 3 in fibroblasts. Br. J. Cancer 71, 1039-1045.

Jiang, C.G., Taylor, D.D., & Black, P.H. (1990). Effect of retinoic acid on tumor-mediated immunologic alterations in mice bearing a variant of the B16 melanoma. Int. J. Cancer 46, 1054-1058.

Kamei, H. (1983). Effect of retinoic acid on cell-cell adhesiveness in cloned BHK21/C13 cells which form piling-up colonies. Exp. Cell Res. 148, 11-20.

Karelina, T., Goldberg, G.I., & Eisen, A.Z. (1994). Matrix metalloproteinases in blood vessel development in human fetal skin and in cutaneous tumors. J. Invest. Dermatol. 105, 411-417.

Kato, S., & De Luca, L.M. (1987). Retinoic acid modulates attachment of mouse fibroblasts to laminin substrates. Exp. Cell Res. 173, 450-462.

Kim, Y-H., Dohi, D.F., Han, G.R., Zou, C-P., Oridate, N., Walsh, G.L., Nesbitt, J.C., Xu, X-C., Hong, W.K., Lotan, R., & Kurie, J.M. (1995). Retinoid refractoriness occurs during lung carcinogenesis despite functional retinoid receptors. Cancer Res. 55, 5603-5610.

Kim, J.H., Tanabe, T., Chodak, G.W., & Rukstalis, D.B. (1995). *In vitro* anti-invasive effects of N-(4-hydroxyphenyl)-retinamide on human prostatic adenocarcinoma. Anticancer Res. 15, 1429-1434.

Klaunig, J.E., & Ruch, R.J. (1990). Role inhibition of intercellular communication in carcinogenesis. Lab. Inves. 62, 135-146.

Krupitza, G., Hulla, W., Harant, H., Dittrich, E., Kallay, E., Huber, H., Grunt, T., & Dittrich, C. (1995). Retinoic acid induced death of ovarian carcinoma cells correlates with *c-myc* stimulation. Int. J. Cancer 61, 649-657.

Liotta, L.A. (1986). Tumor invasion and metastasis—role of the extracellular matrix. Cancer Res. 46, 1-7.

Love, J.M., & Gudas, L.J. (1994). Vitamin A, differentiation and cancer. Curr. Opin. Cell Biol. 6, 825-831.

Majewski, S., Marczak, M., Szmurlo, A., Jablonska, S., & Bollag, W. (1995). Retinoids, interferon α, 1,25-dihydroxyvitamin D_3 and their combination inhibit angiogenesis induced by non-HPV-harboring tumor cell lines. RARα mediates the antiangiogenic effect of retinoids. Cancer lett. 89, 117-124.

Mangelsdorf, K.J., Umesono, K., & Evans, R.M. (1994). The retinoid receptors. In: *The Retinoids: Biology, Chemistry and Medicine* (Sporn, M.B., Roberts, A.B., & Goodman, D.S., eds.) pp 319-350. Raven Press, New York.

Mehta, P., & Loewenstein, W.R. (1991). Differential regulation of communication by retinoic acid in homologous and heterologous junctions between normal and transformed cells. J. Cell Biol. 113, 371-379.

Melchiori, A., Mortarini, R., Carlone, S., Mrchisio, P.C., Anichini, A., Noonan, D.M., & Albini, A. (1995). The $\alpha 3\beta 1$ integrin is involved in melanoma cell migration and invasion. Exptl. Cell Res. 219, 233-242.

Miwa, K., Yamada, C., Kono, T., & Osada, H. (1995). Retinoic acid enhances plasminogen activation on the cell surface. Thrombosis Res. 80, 47-56.

Monsky, W.L., Lin, C-Y., Aoyama, A., Kelly, T., Akiyama, S.K., Mueller, S.C., & Chen, W-T. (1994). A potential marker protease of invasiveness, seprase, is localised on invadopodia of human malignant melanoma cells. Cancer Res. 54, 5702-5710.

Nakajima, M., Irimura, T., & Nicolson, G.L. (1986). Tumor metastasis-associated heparanase (heparan sulphate endoglycosidase) activity in human melanoma cells. Cancer Lett. 31, 277-283.

Nakamura, N., Shidoji, Y., Yamada, Y., Hatakeyama, H., Moriwaki, H., & Muto, Y. (1995). Induction of apoptosis by acyclic retinoid in the human hepatoma-derived cell line, HuH-7. Biochem. Biophys. Res. Commun. 207, 382-388.

Ossowski, L., Clunie, G., Masucci, M-T., & Balsi, F. (1991). *In vivo* paracrine interactions between urokinase and its receptor—effect on tumor cell invasion. J. Cell Biol. 115, 1107-1112.

Pan, L., & Brinckerhoff, C.E. (1994). Inhibition of collagenase gene expression in synovial fibroblasts by all-*trans* and 9-*cis* retinoic acid. Ann. N.Y. Acad. Sci. 732, 335-347.

Prehm, P. (1980). Induction of hyaluronic acid synthesis in teratocarcinoma stem cells by retinoic acid. 111, 295-298.

Rosewicz, S., Stier, U., Brembeck, F., Kaiser, A., Papadimitriou, C.A., Berdel, W.E., Wiedenmann, B., & Riecken, E-O. (1995). Retinoids: effects on growth, differentiation, and nuclear receptor expression in human pancreatic carcinoma cell lines. Gastroenterology 109, 1646-1660.

Rowe, P.M. (1996). Beta carotene takes a collective beating. Lancet 347, 249.

Sacks, P.G., Harris, D., & Chou, T-C. (1995). Modulation of growth and proliferation in squamous cell carcinoma by retinoic acid: a rationale for combination therapy with chemotherapeutic agents. Int. J. Cancer 61, 409-415.

Santos, C.L.S., Giorgi, R.R., Frochtengarten, F., Elias, M.C.Q.B., Chammas, R., & Brentani, R.R. (1994). Regulation of vitronectin receptor expression by retinoic acid on human melanoma cells. Int. J. Clin. Lab. Res. 24, 148-153.

Seewaldt, V.L., Johnson, B.S., Parker, M.B., Collins, S.J., & Swisshelm, K. (1995). Expression of retinoic acid receptor β mediates retinoic acid-induced growth arrest and apoptosis in breast cancer cells. Cell Growth Differ. 6, 1077-1088.

Schadendorf, D., Worm, M., Jurgovsky, K., Dippel, E., Michel, S., Charpentier, B., Bernardon, J-M., Reichert, U., & Czarnetzki, B.M. (1994). Retinoic acid receptor-γ-selective retinoids exert antiproliferative effects on human melanoma cell growth *in vitro*. Int. J. Oncol. 5, 1325-1331.

Schwartz, E., Cruikshank, F.A., Perlish, J.S., & Fleischmajer, R. (1989). Alterations in dermal collagen in ultraviolet irradiated hairless mice. J. Invest. Dermatol. 93, 142-146.

Seftor, R.E.B., Seftor, E.A., Stetler-Stevenson, W.G., & Hendrix, M.J.C. (1993). The 72kDa type IV collagenase is modulated via differential expression of $\alpha v\beta 3$ and $\alpha 5\beta 1$ integrins during human melanoma cell invasion. Cancer Res. 53, 3411-3415.

Strickland, S., Smith, K.K., & Marotti, K.R. (1980). The hormonal induction of differentiation in teratocarcinoma stem cells: generation of parietal endoderm by retinoic acid and dibutyryl cyclic AMP. Cell 21, 347-355.

Tammi, R., Ripellino, J.A., Margolis, R.U., Maibach, H.I., & Tammi, M. (1989). Hyaluronate accumulation in human epidermis treated with retinoic acid in skin organ culture. J. Invest. Dermatol. 92, 326-332.

Thiele, C.J., Deutsch, L.A., & Israel, M.A. (1988). The expression of multiple proto-oncogenes is differentially regulated during retinoic acid induced maturation of human neuroblastoma cell lines. Oncogene 3, 281-288.

Uleminckx, K., Vackat, L., Mareel, M., Fiers, W., & Van Roy, F.V. (1991). Genetic manipulation of E-cadherin expression by epithelial tumor cells reveals an invasion suppressor role. Cell 66, 107-119.

Varani, J., Nickloff, B.J., Dixit, V.M., Mitra, R.S., & Voorhees, J.J. (1989). All-trans retinoic acid stimulates growth of adult human keratinocytes cultured in growth factor-deficient medium, inhibits production of thrombospondin and fibronectin, and reduces adhesion. J. Invest. Dermatol. 93, 49-454.

Vermeulen, S., Bruyneel, E., Van Roy, F., Mareel, M., & Bracke, M.(1995). Activation of the E-cadherin/catenin complex in human MCF-7 breast cancer cells by all-trans-retinoic acid. Br. J. Cancer 72, 1447-1453.

Villa, M.L., Ferrario, E., Trabattoni, D., Formelli, F., De Palo, G., Magni, A., Veronesi, U., & Clerici, E. (1993). Retinoids, breast cancer and NK cells. Br. J. Cancer 68, 845-850.

Vlodavsky, I., Mohsen, M., Lider, O., Svahn, C.M., Ekre, H.P., Vigoda, M., Ishaimichaeli, R., & Peretz, T. (1994). Inhibition of tumor metastasis by heparanase inhibiting species of heparin. Invasion Metast. 14, 290-302.

Waghray, A., & Webber, M.M. (1995). Retinoic acid modulates extracellular urokinase-type plasminogen activator activity in DU-145 human prostatic carcinoma cells. Clin. Cancer Res. 1, 747-753.

Webb, D.S.A., Mostowski, H.S., & Gerrard, T.L. (1991). Cytokine-induced enhancement of ICAM-1 expression results in increased vulnerability of tumor cells to monocyte-mediated lysis. J. Immunol. 146, 3682-3686.

Weidner, N., Folkman, J., & Pozza, F. (1992). Tumor angiogenesis: a new significant and independent prognostic indicator in early-stage breast carcinoma. J. Natl. Cancer Inst. 84, 1875-1887.

Woessner, J.J. (1991). Matrix metalloproteinases and their inhibitors in matrix remodelling. FASEB J. 5, 2145-2154.

Yamamoto, H., Itoh, F., Hinoda, Y., & Imai, K. (1995). Suppression of matrilysin inhibits colon cancer cell invasion *in vitro*. Int. J. Cancer 61, 218-222.

Zerlauth, G., & Wolf, G. (1985). Studies on the tumor promoter-induced release of fibronectin from human lung fibroblasts, and its counteraction by retinoic acid. Carcinogenesis 6, 531-534.

Zou, C-P., Clifford, J.L., Xu, X-C., Sacks, P.G., Chambon, P., Hong, W.K., & Lotan, R. (1994). Modulation by retinoic acid (RA) of squamous cell differentiation, cellular RA-binding proteins, and nuclear RA receptors in human head and neck squamous cell carcinoma cell lines. Cancer Res. 54, 5479-5487.

RETINOID RECEPTORS AND CANCER

Joseph A. Fontana and Arun K. Rishi

	Abstract	219
I.	Introduction	220
II.	Hematopoietic Cells	221
III.	Lung Cancer	223
IV.	Mammary Carcinoma	224
	References	226

ABSTRACT

Retinoids display therapeutic efficacy in a number of premalignant and malignant diseases. Retinoids modulate cellular phenotypes by binding to a number of retinoic acid nuclear receptors (RARα, β, or γ) or retinoic X nuclear receptors (RXRα, β or γ). Most cells express more than one RAR and RXR receptor. Various RAR and RXR subtypes activate different and distinct genes by binding to specific retinoid response elements located in the regulatory regions of target genes. Modulation of the expression of these receptors has a profound effect on the physiology of the cells and their acquisition of a malignant phenotype. RARα appears to regulate hematopoietic differentiation and its loss of mutation results in aberrant growth. RARβ is expressed in both normal lung and breast tissue while RARβ expression is lost in their malignant

Advances in Organ Biology
Volume 3, pages 219-230.

ISBN: 0-7623-0285-2

counterparts. The mechanism(s) involved in the loss of RARβ in these tissues is unclear. Finally, RARα expression in breast carcinoma is regulated by the estrogen receptor and its presence is necessary for the retinoic acid-mediated inhibition of growth.

I. INTRODUCTION

Retinoids are a unique class of compounds that have demonstrated therapeutic benefit in a number of premalignant and malignant diseases (Huang et al., 1988; Hong et al., 1986; 1990; Kraemer et al., 1988; Veronesi and Cosra, 1988). Retinoids mediate the phenotype of a variety of cell types through their ability to modulate the expression of a large number of genes. Retinoids accomplish this by binding to a number of nuclear receptors. Retinoic acid nuclear receptors (RARα, β and γ) and retinoid X receptors (RXRα, β and γ) have been found (Evans, 1988; O'Malley, 1990; Gudas, 1992; Lohnes et al., 1992). Each of the RARs and RXRs are the products of separate genes. Each receptor subtype has at least two isoforms that arise due to the differential splicing and promoter usage. The RARs and RXRs mediate gene transcription through a variety of mechanisms. These nuclear receptors have been found to bind to specific DNA sequences termed consensus sequences. These *cis*-DNA sequences, also called RAREs or RXREs, are located in the regulatory region of retinoid target genes (Gudas, 1992; Lohnes et al., 1992). Retinoid nuclear receptor binding to the RAREs preferentially occurs through heterodimer formation between the RAR and RXR (Leid et al., 1992; Zhang et al., 1992). In contrast, RXRs can mediate gene transcription through both homodimer formation as well as by heterodimer formation with the RARs, depending upon the specific consensus sequence involved (Leid et al., 1992; Zhang et al., 1992; Kliewer et al., 1992). In addition, the RXRs modulate the expression of a number of other genes through their ability to form heterodimers with the vitamin D and thyroid hormone nuclear receptors as well as a number of orphan receptors (Leid et al., 1992; Zhang et al., 1992; Kliewer et al., 1992). The retinoid response elements have usually been found to be direct repeats (RAREs, 5' AGTTCA 3'; RXREs, 5' AGGTCA 3') in which the half sites are separated by a certain number of base pair spacers. In general, RARs bind to consensus sequences in which the direct repeats are separated by 5 or 2 base pair spacers, while RXR homodimer formation occurs on consensus sequences in which the direct repeats are separated by a one base pair spacer. RARs can also mediate gene activity through their modulation of AP-1 mediated gene transcription. AP-1 is an important mediator of transcription for a variety of genes (Pfahl, 1993). RAR and RXR complexed to their ligands inhibit c-*jun*/c-*fos* binding to the AP-1 consensus sequence present in the promoter of several genes and thus subsequent activation of these genes is prevented (Pfahl, 1993). RAR and RXR inhibition of AP-1 mediated gene transcription does not require RAR or RXR binding to DNA; nevertheless, the DNA binding portion of these receptor is still required for this inhibition (Schüle et al., 1991).

Most cell types express more than one RAR and RXR receptor. While homologous recombination studies have suggested that functional redundancy exists among the RARs, other studies have indicated that the various receptor subtypes may possess separate functions (Nagpal et al., 1992; Boylan et al., 1995). These results would suggest that the various RAR and RXR subtypes activate different and distinct genes. Further support for the concept that these RAR subtypes modulate the transcription of unique genes is derived from the studies of Boylan and colleagues (Boylan et al., 1993). Utilizing RARγ or RARα null F9 terato-carcinoma cells, these investigators have demonstrated that the inducibility of a number of target genes is differentially affected by disruption of either the RARα or RARγ gene, suggesting these receptors perform specific functions in these cells (Boylan et al., 1995; 1993). Interestingly, the loss of one receptor in cells with a subsequent defect in retinoid mediated phenotype can often be overcome by the overexpression of another receptor (Robertson et al., 1992a; Taneja et al., 1995). Whether this is simply due to marked overexpression of the second receptor to levels over those noted physiologically, thus enabling it to assume functions not usually mediated by this receptor, remains to be determined (Kastner et al., 1995). Loss of specific RARs has been found in a number of malignancies and in certain situations their loss has been found to play a major role in the malignant process.

I. HEMATOPOIETIC CELLS

The expression of the retinoid nuclear receptors has been extensively studied in hematopoietic cells. The vast majority of these studies have been conducted with the human myeloblastic leukemia cell line HL-60 (Nagy et al., 1995; Robertson et al., 1991; Nervi et al., 1989; Hashimoto et al., 1989; Gallagher et al., 1989). A number of studies have demonstrated the presence of a variety of retinoid nuclear receptors in these cells. Numerous investigators have demonstrated the expression of RARα while several investigators have found both RARα and RARβ expression in these cells (Nagy et al., 1995; Robertson et al., 1991; Nervi et al., 1989; Hashimoto et al., 1989; Gallagher et al., 1989). Nagy and colleagues (Nagy et al., 1995), utilizing RNase protection assays and a variety of HL-60 cell lines, demonstrated that these cells possessed RARα and did not express RARβ or RARγ. In addition, these investigators found that these cells expressed RXRα and RXRβ while RXRγ could not be found.

That RARα plays a pivotal role in hematopoietic cell differentiation has been emphasized by a number of investigators (Collins et al., 1990; Tsai and Collins, 1993; Onodera et al., 1995; Blondi et al., 1991). Deletion of the terminal fifty amino acids of the ligand binding domain of RARα receptor gene in HL-60 cells results in a nuclear receptor displaying marked reduction in affinity for all-*trans* retinoic acid (Robertson et al., 1992b). Cells possessing this receptor no longer display sensitivity to retinoic acid-induced differentiation (Robertson et al., 1992). Introduction of the wild-type RARα gene into these cells results in the re-acquisition of sensitivity to all-*trans*-retinoic acid-induced differentiation along the myelocytic pathway (Collins et al., 1990). Tsai and Collins (1993) showed that granulocyte-macrophage-colony stimulating factor dependent promyelocytic cell lines could be established by the transfer of a dominant negative RARα expression vector into bone marrow cells. The expression of this dominant negative mutant arrested cells at the promyelocyte stage and blocked any further differentiation along the myeloid pathway (Tsai and Collins, 1993). Postulating that RARα regulates myeloid differentiation through its heterodimerization with an RXR, Onodera and coworkers (Onodera et al., 1995) transfected an RARα expression vector into murine bone marrow cells and examined RARα over expression on myeloid maturation and C-kit expressing cells. RARα overexpression resulted in the maturation arrest of the myeloid cell linage and a marked increase in the number of cells expressing C-kit (Onodera et al., 1995). These investigators speculate that overexpression of RARα results in the formation of RARα homodimers at the expense of RARα-RXR heterodimer formation, which is necessary for normal myeloid differentiation (Onodera et al., 1995). Thus, RARα appears to play a major role in the normal maturation of myeloid cells.

Acute promyelocytic leukemia (APL) is a hematopoietic malignancy, representing approximately 6 percent of myeloid leukemia. APL is characterized by the presence of myeloid precursors blocked at the promyelocytic proliferative stage as well as by the specific reciprocal chromosomal translocation t(15:17) (Mitleman, 1988). This translocation juxtaposes the RARα gene located on chromosome 17 to a zinc finger gene, PML, located on chromosome 15 (de Thé et al., 1990a; Borrow et al., 1990; Alcalay et al., 1991). Two reciprocal fusion genes are then generated, namely, PML-RARα located on chromosome 15 and RARα-PML located on chromosome 17 with corresponding mRNA transcripts and protein products (de The et al., 1991; Kakizula et al., 1991; Pandolfi et al., 1991; Goddard et al., 1991; Kastner et al., 1992). Numerous investigators have now demonstrated that the PML-RARα hybrid appears to be exerting dominant negative inhibitory effects on normal myeloid differentiation (Dycu et al., 1994; Weis et al., 1994). Recent immunohistochemistry investigations suggests that PML is part of a unique macromolecular organelle termed as PML oncogenic domains (Dycu et al., 1994; Weis et al., 1994). The presence of the PML-RARα fusion product results in the inhibition of this organelle structure. Whether this inhibition of PML oncogenic domain structure by the PML-RARα fusion product or the ability of the PML-RARα fusion product to inhibit RARα mediated gene transcription as demonstrated by other investigators (Kastner et al., 1992) is responsible for the block in myeloid differentiation is still unclarified. In addition, other investigators have found that PML-RARα fusion product cooperates with c-*jun* and c-*fos* to enhance the transcription of both synthetic and natural genes containing AP-1 consensus sequences (Doucas et al., 1993). This is in marked contradiction to the RARα receptor that inhibits AP-1 mediated gene transcription in the presence of ligand

(Pfahl, 1993). The addition of all-trans-retinoic acid to these cells results in the generation of normal-appearing PML oncogenic domains, normal RARα mediated gene transcription, RARα inhibition of AP-1 mediated gene transcription, as well as normal differentiation of these cells both in culture as well as in patients (Kastner et al., 1992; Dycu et al., 1994; Weis et al., 1994; Doucas et al., 1993).

III. LUNG CANCER

Lung epithelial cells express RARα, β, and γ while RARβ is frequently deleted in lung tumor cells (Stanbridge, 1990). The human RARβ gene is localized on chromosome 3 (Mattei et al., 1988) and expression of RARβ is often strongly induced in many tissues by treatment with RA. A retinoic acid response element (RARE) has been found in the promoter regions of both the human and murine RARβ gene (de Thé et al., 1990b; Sucov et al., 1996; Hoffmann et al., 1990). RAR expression has been examined in a large number of cell lines derived from the four major histological types of lung tumors (small cell, epidermoid carcinoma, adeno-carcinoma, and large cell carcinoma). These studies have revealed that although RARβ expression can be readily found in normal lung tissue, RARβ expression was weak or undetectable in a large number of these cell lines (Houle et al., 1991; Gerbert et al., 1991; Nervi et al., 1991). This was only true for RARβ since RARα and RARγ could be easily detected in both normal and malignant lung tissue (Houle et al., 1994). Gerbert and colleagues (Gerbert et al., 1991) examined the expression of RARβ in normal lung, 33 lung cancer cell lines, and nine primary lung tumors. Approximately 50 percent of the cell lines, and approximately 30 percent of the tumor samples revealed altered RARβ expression and/or inducibility by retinoic acid. RARβ is normally expressed as two transcripts of 3.1 Kb and 2.8 Kb differing at their 5' ends. Many of the tumor cell lines and samples displayed the specific loss of one transcript (Gerbert et al., 1991). Loss of RARβ expression could be subsequent to mutations in the RARβ gene or to alterations in *trans*-factors required for RARβ transcription. The possibility of mutations in the RARβ gene was markedly enhanced by the observation that 3p region is commonly deleted in lung cancer (Stanbridge, 1990). Gerbert and colleagues (Gerbert et al., 1991) were able to detect gross changes in the RARβ gene in only a minority (10%) of the cell lines not expressing RARβ and in nonc of the tumor samples displaying loss or undetectable RARβ expression. These results suggested that RARβ expression may be altered due to aberrations in a required *trans* factor.

RARβ expression is often dependent on the presence of RA, which regulates RARβ expression through a β_2RARE present in the RARβ (isoform 2) promoter (de Thé et al., 1990b; Sucov et al., 1996; Hoffmann et al., 1990). The inability to detect RARβ expression both in the absence and presence of RA is secondary to a defect in the transcriptional regulation of the RARβ promoter. Zhang and coworkers (Zhang et al., 1994) tested this hypothesis by transfecting various constructs of the RARβ promoter linked to the *CAT* gene into human lung carcinoma cell lines with varying degrees of RARβ mRNA expression following exposure to RA. These investigators found that the RARβ promoter was no longer RA responsive in a number of lung carcinoma cells displaying loss of RARβ expression (Zhang et al., 1994). These results strongly suggested that the RA non-responsiveness of the RARβ gene may be the result of a defect in the RARβ promoter (Zhang et al., 1994). To examine whether the decrease in RARβ promoter activity was due to a defect in RARβ RARE activation, these investigators transfected a CAT reporter construct containing the β RARE linked to the heterologous (thymidine kinase) tk promoter (Hoffmann et al., 1990) into the lung carcinoma cells possessing either RA-responsive or nonresponsive *RAR*β genes. Surprisingly, strong activation of CAT activity was noted in all of the lung carcinoma cell lines examined (Zhang et al., 1994). Thus, the βRARE can confer retinoid responsiveness in all of these cell lines in the presence of the tk promoter, suggesting that the loss of RARβ expression is RARβ promoter specific.

Moghal and Neel (1995) further examined this issue in a number of human lung carcinoma cell lines. RARβ2 is the most prominent RARβ isoform expressed in human lung carcinoma cell lines (Houle et al., 1994). In order to distinguish between *cis* and *trans* mechanisms resulting in the marked decrease in RARβ expression, a 5 Kb sequence upstream of the RARβ2 transcription start site was fused to the luciferase reporter gene and transfected into RARβ-positive and -negative lung carcinoma cell lines. Failure of RA to induce luciferase activity was noted in a number of cell lines indicating the presence of *trans*-acting transcriptional defects in the RARβ2 promoter. Surprisingly, as opposed to the findings of Zhang and colleagues (Zhang et al., 1994), these investigators found that the minimal promoter containing the β2RARE linked to a luciferase reporter gene also displayed minimal activity in lung carcinoma cells displaying loss of RARβ2 expression. Utilizing a number of constructs of the RARα receptor in which chimeras were constructed between DNA and ligand binding regions of the thyroid nuclear receptor and that of the RARα receptors, these investigators found that the most likely explanation for the loss of RARβ expression is the loss of transactivation factors (e.g., activating factors 2; AF-1/ AF-2) that specifically interact with the ligand binding portion of the thyroid and retinoid nuclear receptors and enhance transactivation through this portion of the nuclear receptor (Moghal and Neel, 1995). These factors appear to be frequently inactivated in human lung cancer and their identities remain unknown.

Restoration of RARβ expression in human lung carcinoma cells has profound effects on the biology of these cells. Overexpression of RARβ in the human lung carcinoma cell lines H157 and Calu I markedly inhibits the anchorage independent growth of these cells as well as the growth of these cells in nude mice (Houle et al., 1993). In addition, the restoration of RARβ expression in HeLa cells results in RA-dependent suppression of proliferation (Frangioni et al., 1994). These results strongly suggest that the loss of RARβ expression may play a major role in the development of certain human lung carcinomas.

IV. MAMMARY CARCINOMA

Normal breast epithelial cells express RARα, β, and γ as well as RXRα and β mRNAs. Numerous studies have now found that the loss of a number of these receptors occurs in breast carcinoma (Roman et al., 1992; Titcomb et al., 1994; Shao et al., 1994). As previously described in lung carcinoma cell, loss of RARβ has been found in the majority of breast carcinoma cell lines (Shao et al., 1994; Swisshelm et al., 1994; Van der Burg et al., 1993). Interestingly, other investigators have found that immortalization of certain breast epithelial cells also may result in the loss of RARβ mRNA expression while others maintain RARβ mRNA expression (Swisshelm et al., 1994; Van der Burg et al., 1993; Li et al., 1995). The ability of retinoic acid to induce RARβ mRNA was maintained in the normal mammary carcinoma cells but appeared to be uniformally lost in malignant cells (Swisshelm et al., 1994; Van der Burg et al., 1993; Li et al., 1995). As previously described for human lung carcinoma cells, the mechanism by which RARβ expression is lost is complex. The human estrogen receptor positive cell line MCF-7 was found to display significant retinoic acid-inducible activity from a heterologous β_2-RARE (Swisshelm et al., 1994). Interestingly, the inducible activity was approximately 2-fold higher than that noted in normal mammary cells. When a variety of constructs of the normal RARβ promoter was transfected into MCF-7 cells, significantly less luciferase activity (12-26 fold) were noted when compared to normal mammary cells. These results strongly supported the contention that either the loss of a necessary accessory factor(s) or the presence of inhibitory factor(s) was responsible for the marked decrease in RARβ promoter activity in the mammary carcinoma cells (Swisshelm et al., 1994). Utilizing electromobility gel shift assays, Swisshelm and colleagues (Swisshelm et al., 1994) demonstrated that nuclear extracts obtained from immortalized and malignant breast cells displayed markedly different complex formation with a synthetic probe containing the β-RARE than that noted when nuclear extracts derived from normal mammary carcinoma cells were utilized. The nuclear extracts obtained from malignant cells displayed a higher molecular weight complex as well as additional complexes. What these additional factors are that bind to regulatory regions in the RARβ promoter and their role/s in inhibiting RARβ transcriptional activity remains to be defined.

The role of RARβ in the regulation of breast carcinoma proliferation has been examined by a number of groups (Swisshelm et al, 1994; Li et al, 1995; Seewald et al, 1995). The results from a number of studies strongly suggest that RARβ plays a major role as a negative regulator of growth in these cells. Swisshelm and colleagues (Swisshelm et al., 1994) have found that expression of RARβ mRNA markedly increases during normal mammary cell senescence. Transfection of RARβ into ER-negative breast carcinoma cells markedly inhibits nonadherent growth of these cells and results in their growth now being inhibited in the presence of retinoids (Li et al, 1995). Seewald and coworkers (Seewald et al., 1995) have demonstrated that the stable transfection of RARβ into MCF-7 cells results in RA-mediated growth arrest and apoptosis in these cells.

Numerous investigators have demonstrated the loss of RARα in ER-negative breast carcinoma cells (Roman et al., 1992; Titcomb et al., 1994; Shao et al., 1994). The ability of the ER to up-regulate RARα expression in ER-positive breast carcinoma cells was found by a number of investigators (Roman et al., 1993; Sheikh et al., 1994). The addition of estradiol to estradiol depleted medium results in a 2- to 3-fold increase in RARα mRNA expression in ER-positive cells. That this up-regulation of RARα by ER-estradiol complex is physiologically important was further demonstrated by the observation that the stable transfection of ER-cDNA into ER-negative MDA-MB-231 cells results in expression of ER, up-regulation of RARα expression through a unique mechanism and RA inhibition of growth (Sheikh et al., 1994). Utilizing deletional analysis, Rishi and colleagues (Rishi et al., 1995) demonstrated that a promoter sequence of 42 base pairs located approximately 100 bases upstream of the transcriptional start site contains elements responsible for the estradiol-mediated enhancement of RARα expression. The ligand bound ER up-regulates RARα mRNA expression through a weak interaction with the SP-1 transcription factor utilizing an SP-1 motif and an imperfect half-palindromic estrogen response motif (5' GGTGA 3') located ten bases from the SP-1 site (Rishi et al., 1995).

REFERENCES

Alcalay, M., Zangrilli, D., Pandolfi, P.P., Wongo, L., Mencarelli, A., Giacommucci, A., Rocchi, M., Biondi, A., Rambaldi, A., Lo Coco, F., Diverio, D., Donti, E., Grignani, F., & Pleicci, P.G. (1991). Translocation breakpoint of acute promyelocytic leukemia lies within the retinoic acid receptor α locus. Proc. Natl. Acad. Sci. USA 88, 1977-1981.

Blondi, A., Rambaldi, A., Alcalay, M., Pandolri, P-P., LoCoco, F., Diverio, D., Rossi, V., Mencarelli, A., Longo, L., Zangrilli, D., Masera, G., Barbui, T., Mandelli, F., Grignani, F., & Pelicci, P.G. (1991). *RARα* gene rearrangement as a genetic marker for diagnosis and monitoring in acute promyelocityic leukemia. Blood 77, 1418-1422.

Borrow, J., Goddard, A.D., Sheer, D., & Solomon, E. (1990). Molecular analysis of acute promyelocytic leukemia breakpoint cluster region on chromosome 17. Science 249, 1577-1580.

Boylan, J.F., Lohnes, D., Taneja, R., Chambon. P., & Gudas, L.J. (1993). Loss of retinoic acid receptor γ function in F9 cells by gene disruption in aberrant Huxa-1 expression and differentiation upon retinoic acid treatment. Proc. Natl. Acad. Sci. USA 90, 9061-9065.

Boylan, J.F., Lufkin, T., Achkar, C.C., Taneja, R., Chambon, P., & Gudas, L.J. (1995). Targeted disruption of retinoic acid receptor α (RARα) and RARγ results in receptor-specific alterations in retinoic acid mediated differentiation and retinoic acid metabolism. Mol. Cell Biol. 15, 843-851.

Collins, S.J., Robertson, K.A., & Mueller, L. (1990). Retinoic acid-induced granulocytic differentiation of HL-60 myeloid leukemia cells is mediated directly through the retinoic acid receptor (RARα). Mol. Cell. Biol. 10, 2154-2163.

de Thé, H., Chomienne, C., Wanotte, M., Degos, L., & Dejean, A. (1990a). The t(15:17) translocation of acute promyelocytic leukemia fuses the retinoic acid receptor α gene to a novel transcribed locus. Nature 347, 538-561.

de Thé, H., Vivanco-Ruiz, M.M., Tiollais, P., Stunnenberg, H., & Dejean, A. (1990b). Identification of a retinoic acid responsive element in the mouse retinoic acid receptor type β gene. Nature 343, 177-180.

de Thé, H., Lavau, C., Marchio, A., Chomienne, C., Degos, L., & Dejean, A. (1991). The PML-RARα fusion mRNA generated by the t(15:17) translocation in acute promyelocytic leukemia encodes a functionally altered RAR. Cell 66, 675-684.

Doucas, V., Brocues, J.P., Yaniv, M., de Thé, H., & Dejean, A. (1993). The PML retinoic acid receptor α translocation converts the receptor from an inhibitor to a retinoic acid-dependent activator of transcription factor AP-1. Proc. Natl. Acad. Sci. USA 90, 9345-9349.

Dycu, J.A., Mauli, G.G., Miller, W.H., Chen, D.J., Kanizvka, A., & Evans, R.M. (1994). A novel macromolecular structure is a target of the promyelocyte—retinoic acid receptor oncoprotein. Cell 76, 333-343.

Evans, R.M. (1988). The steroid and thyroid hormone receptor family. Science 240, 889-895.

Frangioni, J.V., Moghal, N., Stuart-Tilley, A., Neel, B.G., & Alper, S.L. (1994). The DNA binding domain of retinoic acid receptor β is required for ligand-dependent suppression of proliferation. J. Cell Sci. 107, 827-838.

Gallagher, R.E., Said, F., Papenhausen, P.R., Paietta, T., & Wiernik, P.H. (1989). Expression of retinoic acid receptor-α mRNA in human leukemic cells with variable responsiveness to retinoic acid. Leukemia 3, 789-795.

Gerbert, J.F., Moghal, N., Frangioni, J.V., Sugarbaker, D.J., & Neel, B.G. (1991). High frequency of retinoic acid receptor β abnormalities in human lung cancer. Oncogene 6, 1859-1868.

Goddard, A.D., Borrow, J., Freemont, P-S., & Solomon, E. (1991). Characterization of a zinc finger gene disrupted by the t(15:17) in acute promyelocytic leukemia. Science 254, 1371-1374.

Gudas, L.J. (1992). Retinoids, retinoid-responsive genes, cell differentiation and cancer. Cell Growth Differ. 3, 655-662.

Hashimoto, Y., Petkovich, M., Gaub, M.P., Kagechika, H., Shudo, K., & Chambon, P. (1989). The retinoic acid receptors α and β are expressed in human promyelocytic leukemia cell line HL-60. Mol. Endocrinol. 3, 1046-1052.

Hoffmann, B., Wehmann, J.M., Zhang, X.-K., Hermann, T., Graupner, G., & Pfahl, M. (1990). A retinoic acid receptor specific element controls the retinoic acid receptor-β promoter. Mol. Endocrinol. 4, 1734-1743.

Hong, W.K., Endicott, J., Itri, L.M., Doos, W., Batsakis, J.G., Bell, R., Fofonoff, S., Byers, R., Atkinson, E.N., Vaughan, C., Toth, B.B., Kramer, A., Dimery, I.W., Skipper, P., & Strong, S. (1986). 13-*Cis*-retinoic acid in the treatment of oral leukoplakia. N. Engl. J. Med. 315, 1501-1505.

Hong, W.K., Lippman, S.M., Itri, L.M., Karp, D.D., Lee, J.S., Byers, R.M., Schantz, S.P., Kramer, A.M., Lotan, R., Peters, L.J., Dimery, I.W., Brown, B.W., & Goepfert, H. (1990). Prevention of second tumors with isoretinoin in squamous-cell carcinoma of the head and neck. N. Engl. J. Med. 323, 825-827.

Houle, B.C., Leduc, F., & Bradley, W.E.C. (1991). Implication of RARβ in epidermoid (squamous) lung cancer. Genes Chromosomes Cancer 3, 358-366.

Houle, B.C., Pelletier, M., Wu, J., Goodyear, C., & Bradley, W.E.C. (1994). Fetal isoform of human retinoic acid receptor β expressed in small cell lung cancer lines. Cancer Res. 54, 365-369.

Houle, B.C., Rochette-Egly, C., & Bradley, W.E.C. (1993). Tumor-suppressive effect of the retinoic acid receptor β in human epidermoid lung cancer cells. Proc. Natl. Acad. Sci. USA 90, 985-989.

Huang, M., Ye, Y.C., Chen, S.R., Chai, J.R., Zhoa, L., Gu, L.-J., & Wang, Z. (1988). Use of all-*trans* retinoic acid in the treatment of acute leukemia. Blood 72, 567-572.

Kakizuka, A., Miller, W.H. Jr., Frankel, S.R., V.V.V.S., Dmitrousuu, E., & Evans, R.M. (1991). Chromosomal translocation t(15:17) in human acute promyelocytic leukemia fuses RARa with a novel putative transcription factor PML. Cell 66, 663-674.

Kastner, P., Mark, M., & Chambon, P. (1995). Nonsteroidal nuclear receptors: What are genetic studies telling us about their role in real life? Cell, 83, 859-869.

Kastner, P., Perez, A., Lutz, Y., Rochette-Egly, C., Gaub, M.P., Durand, B., Lanotte, M., Berger R., & Chambon, P. (1992). Structure, localization and transcriptional properties of two classes of retinoic acid receptor a fusion proteins in acute promyelocytic leukemia (APL): structural similarities with a new family of oncoproteins. EMBO J. 11, 629-642.

Kliewer, S.A., Umesono, K., Mangelsdorf, D.J., & Evans, R.M. (1992). Retinoic X receptor interacts with nuclear receptors in retinoid acid, thyroid hormone and vitamin d_3 signalling. Nature 355, 446-449.

Kraemer, K.H., Digiovanna, J.J., Moshell, A.N., Tarone, R.E., & Peck, G.L. (1988). Prevention of skin cancer in xeroderma pigmentosum with the use of oral isoretinoin. N. Engl. J. Med. 318, 1633-1637.

Leid, M., Kastner, P., Lyons, R., Nakshatri, H., Saunders, M., Zacharewshi, T., Chen, J.Y., Staub, A., Garnier, J.M., Mader, S., & Chambon, P. (1992). Purification, cloning, and RXR identity of the HeLa cell factor with which RAR or TR heterodimerizes to bind target sequences efficiently. Cell 68, 377-395.

Li, X.-S., Shao, Z.-M., Sheikh, M.S., Eiseman, J.L., Sentz, D., Jetten, A.M, Chen, J.-C., Dawson, M.I., Aisner, S., Rishi, A.K., Gutierrez, P., Schnapper, L., & Fontana, J.A. (1995). Retinoic acid nuclear receptor β inhibits breast carcinoma anchorage independent growth. J. Cell. Physiol. 165, 449-4581.

Lohnes, D., Dierich, A., Ghyselinck, N., Kastner, P., Lampron, C., LeMeur, M., Lufkin, T., Mendelsohn, C., Nakshatri, H., & Chambon, P. (1992). Retinoid receptors and binding proteins. Cell Sci. 16(Suppl), 69-76.

Mattei, M.-G., de Thé, H., Mattei, J-F., Marchio, A., Tiollais, P., & Dejean, A. (1988). Assignment of the human hap retinoic acid receptor *RAR*β gene to the p24 band of chromosome 3. Human Genet. 80, 189-190.

Mitleman, F. (1988). *Catalog Of Chromosome Aberrations In Cancer*. Alan R. Liss, New York.

Moghal, N., & Neel, B.G. (1995). Evidence for impaired retinoic acid receptor-thyroid hormone receptor AF-2 cofactor activity in human lung cancer. Mol. Cell Biol. 15, 3945-3959.

Nagpal, S., Saunders, M., Kastner, P., Durand, B., Nakshatri, H., & Chambon, P. (1992). Promoter context- and response element-dependent specificity of the transcriptional activation and modulating function of the retinoic acid receptor. Cell 70, 1007-1019.

Nagy, L., Thomazy, V.A., Shipley, G.L., Fejus, L., Lamph, W., Heyman, R.A., Chandraratna, R.A.S., & Davies, P.J.A. (1995). Activation of retinoid X receptors induces apoptosis in HL-60 cell lines. Mol. Cell. Biol. 15, 3540-3551.

Nervi, C., Grippo, J.F., Sherman, M.I., George, M.D., & Jetten, A.M. (1989). Identification and characterization of nuclear retinoic acid-binding activity in human myeloblastic leukemia HL-60 cells. Proc. Natl. Acad. Sci. USA 86, 5854-5858.

Nervi, C., Voolberg, T.M., George, M.D., Zelent, A., Chambon, P., & Jetten, A.M. (1991). Expression of nuclear retinoic acid receptors in normal tracheobronchial cells and in lung carcinoma cells. Exp. Cell Res. 195, 163-170.

O'Malley, B.W. (1990). The steroid superfamily: More excitement predicted for the future. Mol. Endocrinol., 4, 363-369.

Onodera, M., Kunisada, T., Nishikawa, S., Sakiyama, Y., Matsumoto, S., & Nishikawa, S-I. (1995). Overexpression of retinoic acid receptor α suppresses myeloid cell differentiation at the promyelocyte stage. Oncogene 11, 1291-1298.

Pandolfi, P.P., Grigani, F., Alcalay, M., Mencarelli, A., Biondi, A., Loco, F., Grignani, F., & Pelicci, P.G. (1991). Structure and origin of the acute promyelocytic leukemia PML/RARa cDNA and characterization of its retinoid binding and transactivation properties. Oncogene 5, 1285-1292.

Pfahl, M. (1993). Nuclear receptor/AP-I interactions. Endocrine Reviews 14, 651-658.

Rishi, A.K., Shao, Z.-M., Baumann, R.G., Li, X.-S., Sheikh, M.S., Kimura, S., Bashirelahi, N., & Fontana, J.A. (1995). Estradiol regulation of the human retinoic acid receptor α gene in human breast carcinoma cells is mediated via an imperfect half-palindromic estrogen response element and SP1 motifs. Cancer Res. 55, 4999-5006.

Robertson, K.A., Emami, B., Mueller, L., & Collins, S.J. (1992a). Multiple members of the retinoic acid receptor family are capable of mediating the granulocytic differentiation of HL-60 cells. Mol. Cell. Biol. 12, 3743-3749.

Robertson, K.A., Emami, B., & Collins, S.J. (1992b). Retinoic acid-resistant HL-60 R cells harbor a point mutation in the retinoic acid receptor ligand-binding domain that offers dominant negative activity. Blood 80, 1885-1889.

Robertson, K.A., Mueller, L., & Collins, S.J. (1991). Retinoic acid receptors in myeloid leukemia: Characterization of receptors in retinoic acid-resistant K-562 cells. Blood 77, 340-347.

Roman, S.D., Clarke, C.L., Hall, R.E., Alexander, I.E., & Sutherland, R.L. (1992). Expression and regulation of retinoic acid receptors in human breast cancer cells. Cancer Res. 52, 2236-2242.

Roman, S.D., Ormandy, C.J., Manning, D.L., Blamey, R.W., Nicholson, R.I., Sutherland, R.L., & Clarke, C.L. (1993). Estradiol induction of retinoic acid receptors in human breast cancer cells. Cancer Res. 53, 5940-5945.

Schüle, R., Rangarajan, P., Yang, N., Kliewer, S., Ransone, L., Bolando, J., Verma, I.M., & Evans, R.M. (1991). Retinoic acid is a negative regulator of AP-I-responsive genes. Proc. Natl. Acad. Sci. USA 6092-6096.

Seewald, V.L., Johnson, B.S., Parker, M.B., Collins, S.J., & Swisshelm, K. (1995). Expression of retinoic acid receptor β mediates retinoic acid-induced growth arrest and apoptosis in breast cancer cells. Cell Growth Differ. 6, 1077-1088.

Shao, Z.M., Sheikh, M.S., Chen, J.-C., Kute, T., Aisner, S., Schnaper, L., & Fontana, J.A. (1994). Expression of the retinoic acid nuclear receptors (RARs) and retinoid X receptor positive and negative breast cancer. Int. J. Oncol. 4, 859-863.

Sheikh, M.S., Shao, Z.-M., Chen, J.-C., Hussain, A., Jetten, A.M., & Fontana, J.A. (1993). Estrogen receptor-negative breast cancer cells transfected with the estrogen receptor exhibit increased RARα gene expression and sensitivity to growth inhibition by retinoic acid. J. Cell. Biochem. 53, 394-404.

Sheikh, M.S., Shao, Z.M., Li, X.-S., Dawson, M.I., Jetten, A.M., Wu, S., Conley, B.A., Garcia, M., Rochefort, H., & Fontana, J.A. (1994). Retinoid-resistant estrogen receptor negative human breast carcinoma cells transfected with retinoic acid receptor a acquire sensitivity to growth inhibition by retinoids. J. Biol. Chem. 269, 21440-21447.

Stanbridge, E. (1990). Human tumor suppressor genes: Ann. Rev. Genet. 24, 615-657.

Sucov, H.M., Murakami, K.K., & Evans, R.M. (1996). Characterization of an autoregulated response element in the mouse retinoic acid receptor type β gene. Proc. Natl. Acad. Sci. 87, 5392-5396.

Swisshelm, K., Ryan, K., Lee, X., Tsou, H.C., Peacocue, M., & Sager, R. (1994). Down-regulation of retinoic acid β in mammary carcinoma cell lines and its up-regulation in sequencing normal mammary epithelial cells. Cell Growth Differ. 5, 133-141.

Taneja, R., Bouillet, P., Boylan, J.F., Gaub, M.-P., Roy, B., Gudas, L.J., & Chambon, P. (1995). Reexpression of retinoic acid receptor (RAR)γ or overexpression of RARα or RARβ in RARγ-null F9 cells reveals a partial functional redundancy between the three RAR types. Proc. Natl. Acad. Sci. USA 92, 7854-7858.

Titcomb, M.W., Gottardis, M.M., Pine, J.W., & Allegretto, E.A. (1994). Sensitive and specific detection of retinoid receptor subtype protein in cultured cell and tumor extracts. Mol. Endocrinol. 8, 870-877.

Tsai, S., & Collins, S.J. (1993). A dominant negative retinoic acid receptor blocks neutrophil differentiation at the promyelocytic stage. Proc. Natl. Acad. Sci. USA 90, 7153-7157.

Van der Burg, B., van der Leede, B.-J. M., Kwakkenbos-IsGrücker, L., Salverda, S., de Loat, S.W., & Van der Saag, P.T. (1993). Retinoic acid resistance of estradiol-independent breast cancer cells coincides with diminished retinoic acid receptor function. Mol. Cell Endocrinol. 91, 149-157.

Veronesi, U., & Costa, A. (1988). Chemoprevention of contralateral breast cancer with the synthetic retinoid Fenretinamide. Cancer Invest. 6, 639-642.

Weis, K., Rambaud, S., Lavau, C., Jansen, J., Carvalho, T., Carmo-Fonseca, M., Lamond, A., & Dejean, A. (1994). Retinoic acid regulates abberant nuclear localization of PML-RARα with acute promyelocytic leukemia. Cell 76, 345-356.

Zhang, X-K., Hoffmann, B., Tran, P.B.-V., Graupner, G., & Pfahl, M. (1992). Retinoid X receptor is an auxiliary protein for thyroid hormone and retinoic acid receptors. Nature 355, 441-446.

Zhang, X-K, Liu, Y., Lee, M.-O., & Pfahl, M. (1994). A specific defect in the retinoic acid response associated with human lung cancer cell lines. Cancer Res. 54, 5663-5669.

RETINOIDS IN THE MANAGEMENT OF CENTRAL NERVOUS SYSTEM (CNS) TUMORS

M. E. Westarp

	Abstract	232
I.	Clinically Prevalent CNS Tumors	232
II.	Therapeutic Situation in Malignant Glioma	233
III.	Rationale for Use of Retinoids	235
IV.	Selection of Oral Retinoic Acid Derivatives	237
V.	Oral Inhibitors of Retinoic Acid Catabolism	238
VI.	Clinical Application of Retinoic Acid Plus Liarozole	241
VII.	Tolerance of Retinoic Acid Plus Liarozole	245
VIII.	Retinoids and Radiotherapy	246
IX.	Retinoids in Non-glial CNS Neoplasms	249
X.	Intratumoral Medication—HSV-TK, Retinoids, and More	250
XI.	Future Adjuvant Medical Options	251
XII.	Summary—Retinoids in CNS Tumors	252
	Acknowledgments	253
	References	253

Advances in Organ Biology
Volume 3, pages 231-260.

ISBN: 0-7623-0285-2

ABSTRACT

We have analyzed and investigated with clinical means the potential use of retinoids in primary CNS (brain) tumors. Established to be effective in ectodermal skin disorders, retinoids also effectively alter the growth of neuro-epithelial tissue, i.e. cells of neuro-ectodermal origin. 13-*cis* Retinoic acid and inhibitor of RA catabolism, liarozole, can both be given orally and as additional, adjuvant medication to standard treatment protocols. Particularly in conjunction with liarozol to avoid decreasing plasma and tissue concentrations, 13cRA seems to be safe and promising in the therapy of intracranial tumors. The combined medication did not lead to increased intracranial pressure, was well tolerated, and may be able to induce tumor cell differentiation, slow de-differentiation and improve anti-tumoral responses. Retinoids are compatible with all other treatment modalities, including radiotherapy, anti-edematous Boswellia acids and intra-tumoral approaches such as herpes-simplex thymidine kinase/ganciclovir insertion, and may prove useful even in lower-grade astrocytoma or other neuro-epithelial, e.g. spinal, neoplasia.

I. CLINICALLY PREVALENT CNS TUMORS

Depending on the population examined and the diagnostic methods used, the most prevalent central nervous system (CNS = brain and spinal cord) tumors are glioma, meningioma, or metastatic spread from lung, breast, prostate, laryngeal cancer and lymphoma. Primary CNS neoplasia is different from systemic malignancies in that

1. metastasis practically does not occur to sites outside the CNS;
2. the skull early exerts a constraint of volume, that is, tumor growth means compression of brain;
3. biopsies from deeper CNS cannot be taken without damage to functional tissue;
4. neurons are all post-mitotic, that is, cannot divide or replace themselves; and
5. brain and spinal cord are relatively radio-sensitive, that is, prone to radiation-induced dysfunction; tumor relapses after initial radiotherapy regularly cannot be irradiated again.

Glioma is the most common adult intracranial tumor and the most frequent primary brain malignancy. Anaplastic astrocytoma and glioblastoma multiforme are together addressed as malignant glioma. They have a higher median age of incidence (58 years) than more benign tumors, and will therefore become more prevalent as the population pyramid is aging. No convincing risk factors have been identified, with childhood head irradiation (Ron et al., 1988), electromagnetic field window effects (Cleary et al., 1990), and chemicals (Barbone et al., 1994) still being discussed. Epidermal growth factor receptor (EGFR) amplification is a characteristic feature of more aggressive glioma, and EGFR upregulation plus loss of

Table 1. Frequent Primary Central Nervous System Tumors

Tumor Designation	*Frequency**	*Comment*
Meningioma	17 %	benign tumors (not meningosarcoma)
Glioblastoma	15 %	neuro-epithelial malignancy
Anaplastic astrocytoma	2 %	neuro-epithelial malignancy
Astrocytoma	13 %	neuro-epithelial malignancy
Neurinoma, neurofibroma	7 %	nerve-sheath tumors
Hypophysary adenoma	6 %	neuro-epithelial tumors
Oligodendroglioma	5 %	neuro-epithelial malignancy
Spongioblastoma	5 %	neuro-epithelial malignancy
Medulloblastoma	4 %	neuro-epithelial malignancy
[CNS metastases	12.5 %	bronchial, mammary carcinoma etc.]

Note: * Adjusted according to Zülch, 1960; Tönnies & Peiffer, 1983; Mahaley et al., 1989.

heterozygosity on chromosome 10 has been demonstrated in 26 percent of high-grade astrocytoma (Bello et al., 1994). Loss of alleles at chromosome-10p-specific loci significantly correlates with poor prognosis in younger glioma patients (Diedrich et al., 1995). Mutation of the p53 tumor suppressor gene may be one of the earliest identified genetic lesions during malignant progression of human astrocytoma (Van Meir et al., 1995), although in other systemic cancers p53 abnormalities seem to be a late event in tumor progression. We thus may not yet perceive the earlier events in the development of malignant glioma at all.

II. THERAPEUTIC SITUATION IN MALIGNANT GLIOMA

The incidence of both glioma and astrocytoma increases with age up to 70-74 years of age (Burger and Green, 1987; Vecht et al., 1990; Polednak and Flannery, 1995), and survival seems negatively correlated with age. The most frequent primary brain malignancy is malignant glioma, that is, anaplastic glioma (WHO grade III) and glioblastoma multiforme (WHO grade IV). Median survival in glioblastoma is 44 to 48 weeks (Hildebrand, 1994). Five-year-survival remains around 6 percent in malignant astrocytoma (Salcman et al., 1994) and 4 percent in glioblastoma (Burger and Green 1987; Vecht et al., 1990). Neurosurgery, radiotherapy, plus cytotoxic chemotherapy have not prolonged mean glioblastoma survival beyond 48 weeks (Hildebrand, 1994). Even testing of large numbers of immortalized or primary tumor cell lines has not resulted in an efficient chemotherapy. Younger patients with SPECT/PET tumor perfusion superior to that of surrounding normal brain may benefit from 4 x 4 infusion series with 80 $mg/m^2/d$ ACNU = nimustine, and trials are continuing including other nitroso-ureas or carboplatin 3 x 600 mg i.v. A therapeutic stand-still in malignant glioma is at present coinciding with an increasing number of patients facing a grim prognosis.

In the most malignant glioma, glioblastoma, both radiotherapy and cytotoxic chemotherapy simultaneously face a notoriously heterogeneous tumor cell population (glioblastoma *multiforme*) with variable degrees of neovascularization, marked infiltrations into healthy brain tissue, reduced oxygenation in partially necrotic areas, variably long cell cycles and a broad spectrum of preexisting cellular multidrug resistances. Multiple-drug resistance gene *mdr*-1 and its product P-glycoprotein have been implicated in acquired resistance to cytotoxic drugs to which the tumors were never exposed, mediated by an increase of drug efflux (Riordan and Ling, 1985). Expression of P-glycoprotein mRNA can spontaneously be found in untreated cell cultures (Schlüsener, 1990). To overcome this situation, large numbers of substances have been examined for reversion of multidrug resistance and inhibition of celluar p-glycoprotein. Until now, reasonable *in vitro* results were either restricted to concentrations not obtainable *in vivo* (Westarp et al., 1993b) or have not been reproduced in clinical trials. In several human neuroblastoma lines, 10^{-6} M all-*trans* retinoic acid (RA) enhanced the uptake of vinblastine, vincristine, or actinomycin (Bates et al., 1989) via P-glycoprotein functions. RA may synergistically complement N-alkyl aromatic modulators of multiple drug resistance during conventional chemotherapy such as trifluoperazine and verapamil (Riordan & Ling, 1985), though an increase in chemosensitivity of normal cells may limit this approach (Pommerenke et al., 1992).

In malignant glioma standard basic therapies are resection and irradiation (Table 2). Comparing these classical tools, neurosurgical resection has a persisting morbidity of 5-10 percent, brain biopsy of 1.2 percent, and interstitial radiotherapy of 3.1 percent (Ostertag and Kreth, 1992:25). Medical approaches are much more controversial in malignant glioma than surgery or radiotherapy.

Anti-invasive metalloprotease inhibitors such as Timp-2 (Lokeshwar et al., 1992) or human anti-epidermal growth factor receptor monoclonal antibodies (Brady et al., 1992) have not sufficiently proven their usefulness. Although, human malignant gliomas tend to be rather heterogeneous in EGFR expression even within one tumor (Strommer et al., 1990), retinoids inhibit human glioma cell growth at least in part by modulation of the EGF receptor phosphotyrosine kinase activity (Young et al., 1989). Latest novel approaches including Boswellia acids (3 × 1200mg/d) for the treatment of perifocal edema (D.K. Böker, Gießen, pers. comm. 1996), or high-dose tamoxifen (120-400 mg/d) aiming at the proteinkinase-C of the tumor (H.-D. Herrmann, Hamburg-Eppendorf, pers. comm. 1996) have not become part of standard protocols yet. The use of combined adjuvant medications to improve clinical outcome via different modes of action seems therefore justified. Differentiation-inducers, such as

Table 2. Basic Approach to Classical Malignant Glioma Therapy

Clinical State [1]	*Under Age 70 Years*	*Above Age 70 Years*
Karnofsky index > 70%	resection[2] radiotherapy	diagnostic biopsy radiotherapy
Karnofsky index < 70%	radiotherapy (diagnostic biopsy)	diagnostic biopsy

Note: [1] Clinical morbidity and functional state according to the Karnofsky scale (100% = unimpaired, 10% = fully dependent on external help).

[2] No resection if tumor diffuse without mass effect and within eloquent location.

retinoids that are established in dermatology, may be of particular interest for malignant glioma as neuro-epithelial glioma, that is, cancer of similar embryonic origin as the skin (ectoderm).

III. RATIONALE FOR USE OF RETINOIDS

Retinoids are derivatives with the typical structure of vitamin A (Biesalski, 1989), inhibiting neoplastic transformation by blocking the G_0 to G_1 transition in response to promoting factors (Mordan, 1989). Intracellularly, retinol is partially and irreversibly converted to retinoic acid by alcohol dehydrogenase (EC 1.1.1.1) in the presence of NAD^+ (Favennec, 1988). Retinoic acid (RA) has been postulated to be the main retinoid morphogen (Biesalski 1989; Benjamin et al., 1989), and more than 20 nuclear receptors have been characterized (Giguere et al., 1987; Mangelsdorf et al., 1990). Retinoic acid is a potent differentiation inducer (Bollag, 1983), and in frogs it may lead to duplicate limb redevelopment after amputation (Summerbell and Maden, 1990). It is vitamin A's main epidermal tissue derivative and is involved in immune functions (see part XI; Harant et al., 1995). Nuclear retinoic acid receptors of the RAR-alpha type are predominantly found in haematopoietic cells, and RAR type-β prevail in spinal cord, cerebellum, brain cortex, and glia cells (de Thé et al., 1989; 1990).

13-*cis*-retinoic acid (isotretinoin) reverses premalignant lesions in man (Gold et al., 1983), and is being tested as adjuvant medication in a number of human cancers (Boyd, 1989; Hong and Itri, 1994). In oropharyngeal carcinoma, significantly fewer relapses occurred during 12 months of 50 - 100 $mg/m^2/d$ *cis*-retinoic acid (Hong and Lippman, 1991). Retinoic acid potently induces differentiation in human astrocytoma (Rutka et al., 1988), neuroblastoma (Rutka et al., 1988), and glioblastoma cells (Yung et al., 1989; Steck et al., 1990). Retinoic acid induces differentiation in a majority of human glioma cell culture subpopulations (Yung, 1989), while re-dedifferentiation occurs after termination of exposure (Fischer et al., 1987). Both cutaneous side effects and *in vitro* proto-oncogene down-regulation develop simultaneously within one week of exposure. In glioma cells, both the all-*trans* and the 13-*cis* stereoisomer up-regulate cell adhesion (Reboul et al., 1990). In a tumorigenic glioblastoma cell line (SNB-19), retinoic acid induces pronounced changes in fucose-containing glycoprotein expression (van der Meulen et al., 1994). Macrophageal production of tumor necrosis factor (TNF) and of nitric oxide, both pivotal mediators potentially involved in brain edema, may be attenuated by retinoic acid (Mehta et al., 1994). Judged by both mRNA and protein levels, 10^{-9} M of RA are inducing intracellular retinol-binding protein *before* affecting the intracellular retinoic acid-binding protein (CRABP), possibly enhancing intracellular retinol usage (Wei et al., 1989).

One rationale for their use in neuroendothelial-neuroectodermal cancer is that retinoic acids clinically induce differentiation in embryonically related tissue, that is, skin disease. Two synthetic derivatives of RA, isotretinoin and etretinate, have been approved for dermatological indications, including basal cell nevus syndrome, a disease associated with intracranial calcifications and neuroectodermal medulloblastoma.

Synthetic retinoids have to be RA competitors to be pharmacologically active in human epidermis (Siegenthaler and Saurat, 1989) and to retain differentiating activity as RA derivatives (Frey et al., 1991). Exposure to retinoic acid for more than one cell cycle induces changes in inositol phospholipid metabolism (Lanciotti and Ponzoni, 1988) and is accompanied by a decrease in c-myc/n-myc proto-oncogene expression (Nakamura and Hart, 1989). In glioblastoma RA alters epidermal growth factor receptor associated glycoconjugates (Steck et al., 1990) as well as phosphotyrosine kinase activity (Yung et al., 1989). The polyamine spermidine has been reported to be elevated in proliferative phases of glioblastoma tumors (Ernestus et al., 1992); in SK-N-BE neuroblastoma cells, RA precipitates a large spermine catabolism, interfering with ornithine decarboxylase (Melino, 1988).

Most adult brain malignancies, notably glioblastoma, astrocytoma, and spongioblastoma, are of neuro-epithelial origin. Malignant glioma derive from glial cells that regularly express nuclear retinoic acid receptors of the beta type (RAR-β). Via these response elements, retinol derivatives seem to control target cell differentiation. Free retinol *in vivo* circulates at more than a 100 times lower concentrations than RBP-bound retinol (Creek et al., 1989). Unlike vitamin D and vitamin D-binding protein, retinol carrier RBP is physiologically regulated by its ligand (Cooke and Haddad, 1989). RBP binding sites have been demonstrated on choroidal epithelium, which is suggestive of retinol transport across the blood brain barrier (MacDonald et al., 1990). Neuro-ectodermal tumors harbor the potential to respond to retinoids by virtue of shared embryonic origins with an established target of retinoic acid therapy, the skin. A disruption of a retinoic acid-receptor gene as in

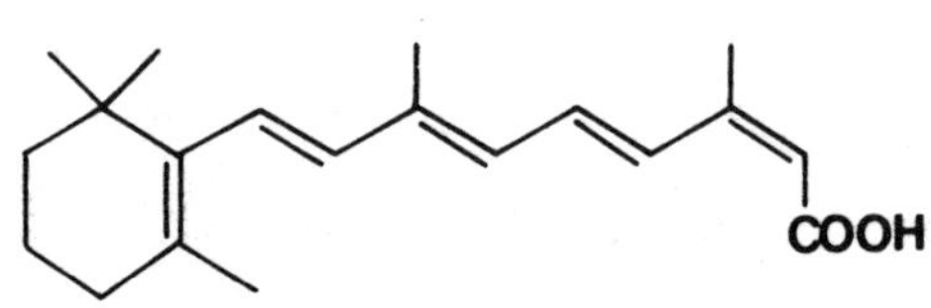

Isotretinoin
13-cis RETINOIC ACID Roaccutan®/Accutane®

Figure 1. Chemical structure of 13cRA.

retinoic acid-responsive acute promyelocytic leukemia (Lo et al., 1991) has not been identified in CNS tumors.

IV. SELECTION OF ORAL RETINOIC ACID DERIVATIVES

Intracellular retinol derivatives approved for human use are 13cRA, ATRA, and tretinoin. In cultured human amnion cells, RA visibly increases cell-to-cell communication at physiological concentrations (Figure 2). Both 0.5 hours and 24 hours after micro-injection of lucifer yellow, the dye expands to more neighboring cells in the presence of 10^{-7}M all-*trans* retinoic acid.

ATRA may be a more potent dysmorphogen than 13cRA (Creech-Kraft and Juchau, 1992), and therefore a more powerful differentiation inducer. However, 13cRA has a longer plasma half-life, additional metabolites, and dramatically less effect on intracranial pressure compared to all-*trans*-RA (McElwee et al., 1991). Unlike retinol, synthetic cis-retinoic acid does not accumulate in the liver (Meyskens et al., 1985). Despite high protein binding, significant amounts of serum retinoic acids can be expected to cross the blood brain barrier (Meyskens et al., 1985), particularly in patients with an impaired barrier function.

On two established human glioblastoma cell line cultures ("M" and "N," kindly provided by Prof. Dr. D. Stavrou, Neuropathology UKE Hamburg), we have tested the following substances by quantitating β decay radioactivity of ^{3}H-thymidine one microcurie per well, added at 75 percent confluence for 16-24 hours incubation:

Nimustine [ACNU = (1-4-amino-2-methyl-5-pyrimidinyl)-methyl-3-(2-chloroethyl)-3-nitrosourea, mw 309.2];

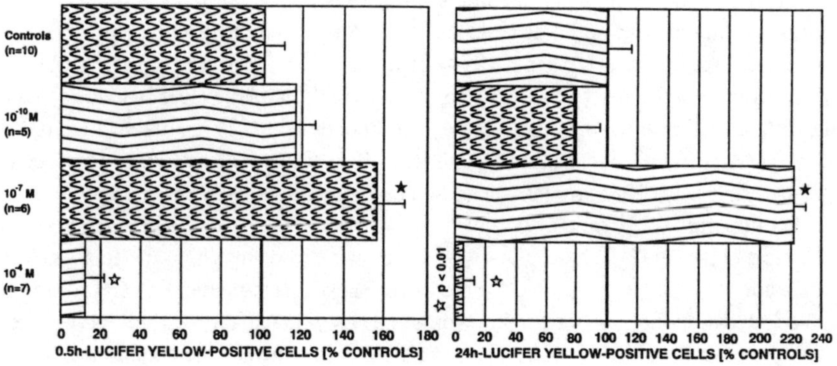

Figure 2. All-trans retinoic acid on human amnion cell monolayers increases lucifer-yellow exchange between cells. Both 0.5 hours and 24 hours after micro-injection of lucifer yellow, the dye expands to more neighboring cells in the presence of 10^{-7}M all-trans retinoic acid.

13cRA = Ro4-3780 [(2Z,4E,6E,8E)-3,7-dimethyl-9-(2,6,6,-trimethyl-1-cyclohexen-1-y1)-2,4,6,8-nonatetraenoic acid] mw 300.4;

Ro13-7410 [p-((E)-2-(5,6,7,8-Tetrahydro-5,5,8,8,-tetramethyl-2-naphthyl)-1-propenyl)benzoic acid] mw 348.5;

acitretin = Ro10-1670 [all-trans-9-(4-methoxy-2,3,6-trimethylphenyl)-3,7-dimethyl-2,4,6,8-nonatetraenoic acid] mw 326.4;

Ro13-6307 [(all-E)-3-methyl-7-(5,6,7,8-tetrahydro-5,5,8,8,-tetramethyl-2-naphthyl)-2,4,5-octatrienoic acid] mw 338.5;

Ro14-6113 [temarotene hydroxic metabolite] mw 320.5;

Ro22-6595 [13cis-4-oxo-retinoic acid] mw 314.4, and interferon-α_{2a} at 10^{-7} to 10^{-9} final molar concentrations.

Thymidine incorporation was reduced by 13cRA somewhat better than by ATRA. In addition in a 3-weeks-on and 1-week-off therapeutic scheme, 13/37 glioma patients remained stable under high-dose isotretinoin (Yung et al., 1993). Considering the importance of less headache and/or intracranial pressure in brain tumor patients we chose to focus on 13cRA for further clinical consideration.

V. ORAL INHIBITORS OF RETINOIC ACID CATABOLISM

The imidazole-derived steroid synthesis modulator liarozole (5-[(3-chlorophenyl)(1H-imidazol-1-yl)methyl]-1H-benzimidazole monohydrochloride ($C_{17}H_{13}Cl\,N_4$.HCl, R75251) effectively inhibits retinoic acid degradation (End et al., 1991; Van Wauwe et al., 1994) and reaches brain tissue when given orally (De Coster et al., 1991). Liarozole (R75251 and R85246, mw 345.2) interferes with cytochrome P_{450}-dependent degradation of retinoic acid and prevents a decrease of plasma levels over time (Miller et al., 1994). It modulates human steroid synthesis, and *in vitro* both liarozole and 13cRA may reduce glioblastoma thymidine incorporation (Westarp et al., 1994). In a phase I and pharmacokinetic study we further combined adjuvant 13-*cis* retinoic acid with liarozole (R75251) to increase the differentiation-inducing retinoid effect (Muindi et al., 1992). Liarozole, filed under R75251 as hydrochloride (Figure 3), and R85246 as fumarate, was kindly provided in 150 mg tablets (R75251) by Janssen Research Foundation, B-2340 Beerse.

Though liarozole would also inhibit the all-*trans* retinoic acid (ATRA) catabolism (Wouters et al., 1992), ATRA gives a headache in 90 percent within the first week of medication, that is, is unsuitable for the treatment of space-demanding CNS tumors. Given one additional oxo-metabolite plus a longer serum half-life of 13cRA compared to ATRA, we consider 13cRA plus liarozole a safe adjuvant therapeutic option for neuro-ectodermal, neuro-epithelial malignancies. Other inhibitors of retinoid degradation such as ketoconazole have severe (gastro-intestinal) side-effects (Mahler et al., 1993) that render them unsuitable for experimental adjuvant application.

Three eligible patients did not consent to enter the study, and two patients discontinued medication for mucocutaneous side effects. In the remaining patients

Liarozole

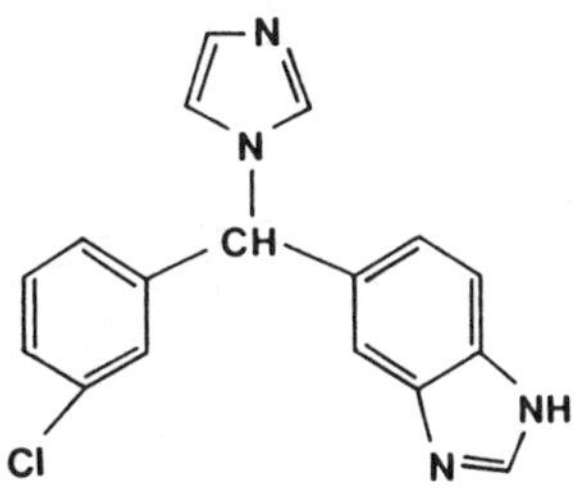

5-[(3-chlorophenyl)(1H-imidazol-1-yl)methyl]-1H-benzimidazole monohydrochloride

Figure 3. Chemical structure of liarozole (R75251).

we started medication with 13cRA to guarantee the patients' tolerance, that is, retinoid concentrations under liarozole alone could not be evaluated separately. We have thus far treated ten patients for more than six months or until death with 2.5-20 mg/d 13-cis retinoic acid and 75-300 mg/d liarozole p.o, covered by liability agreements and the Ulm University Ethics Commission (Table 3). Patients with therapeutical retinol, estrogen, progesterone, tetracycline or iron application, malabsorption, maldigestion, cholestasis, cholestyramine medication, hepatic or renal insufficiency, chronic pancreatitis or pancreatic cancer, known skeletal hyperostosis (Periquet et al., 1991), and patients with severe acute or subacute skin rashes were excluded unless otherwise indicated. Toxicity was assessed by weekly monitoring of red and white blood cell count, hepatic transaminases, serum bilirubin, creatinine, urea, uric acid, Na, K, Cl, cholesterol, triglycerides, retinol, retinol-binding protein, erythrocyte sedimentation rate for the first five weeks, continuous documentation of any reported side effects, and followed by regular quarterly controls plus intermittent determinations as required by intercurrent disease, side effects or course of the malignancy. Clinical assessment was done by monitoring of activities of daily living, Karnofsky index, regular outpatient reports from family and physicians, and quarterly neurological examinations.

In conjuction with 10 mg/m^2 13cRA, daily maintenance doses of 50 mg/m^2 liarozole p.o. were sufficient to regularly increase plasma 13cRA from traces < 2.5 ng/ml to more than 70 ng/ml, and 4-oxo-13cRA from < 7.5 ng/ml to more than 150 ng/ml, while tretinoin remained below 10 percent of 13cRA. In 16/17 untreated malignant brain tumor patients, retinoic acids were not quantifiable in fasting serum; in a single brain tumor patient we measured 2,5 ng/ml 13-*cis*-retinoic acid and 7,5 ng/ml 4-oxo-13-*cis*-retinoic acid. Combined with 20 mg retinoic acid, 150 mg liarozole was as effective as 300 mg in increasing and maintaining retinoid

Table 3. Brain Tumor Patient Characteristics

	Age[1]	*Sex*	*Time*[2] to Tx	*Time*[3] *to radiation*	*Duration*[4] *of medication*	*Time*[5] *to neurosurgery*	*Survival*[6]	*Tumor site*[7]	*Diagnosis*[8]
1+	75	F	2°	5X	5*	1	53	precentr.	MM
2	50	F	9	5	10*	unres.	14	br. stem	GM
3	44	F	1.5	1	5*	unres.	4.5	3 cysts	GM
4	48	M	1.5	6X	3*	1	7	bihemi.	GM
5	51	M	3	19	2	1	22	temp. par.	GM
6+	42	F	3	24	0	refused	24	frontop.	03+
7	68	F	5	6	3	0.5 part	11	par. temp.	GM
8	65	M	3	9	2	1; 7	12	cent. par.	GM
9	24	M	2	12	2	1	28	parietal	A3
10	71	F	6	5	5	4 part	12	lumbar	GM+
11	49	M	5	3	3	5 ster.	20	frontbas.	A4
12	62	M	3	5*	5*	unres.	6	thalam.	A4GM

Notes: +) patient still alive, (1) age at beginning of retinoid therapy; (2) time from onset of disease to retinoid therapy; (3) months from onset of disease to external radiotherapy; *= in five patients, retinoid medication was instituted before radiotherapy; (4) months of retinoid medication; (5) months from onset of disease to neurosurgery; unres.= unresectable/ tumor not treatable by neurosurgery, refused= no consent for surgery, part.= partial resection, ster.= stereotactic surgery only; (6) months of survival from clinical signs or symptoms; (7) tumor site/ localization: precentr.= precentro-temporo-parietal, 3 cysts= three cystic glioblastomata, bihemi.= bihemispheric frontoparietal "butterfly glioma," br.stem= brain stem and thalamus, temp.par.= temporo-parietal, frontop.= fronto-parietal, par.temp.= parieto-temporal, cent.par.= centro-parietal, thalam.= thalamic; (8) diagnosis: GM= glioblastoma multiforme, A3/A4= astrocytoma grade III/IV, A4GM= astrocytoma grade IV with glioblastomatous components, MM= "malignant meningioma" of sarcomatous character, unusual survival according to (Forbes and Goldberg, 1984) O3= oligodendroglioma grade III+, o) relapsing "malignant sarco-meningioma," liarozole refused, intermittently treated with 20 mg/d 13cRA p.o. for six weeks each time during radiotherapy and three, six and 18 months after surgery, ^{x}) 3 x 80 mg/m^{2}d ACNU (nimustine) i.v. after neurosurgery and radiotherapy.

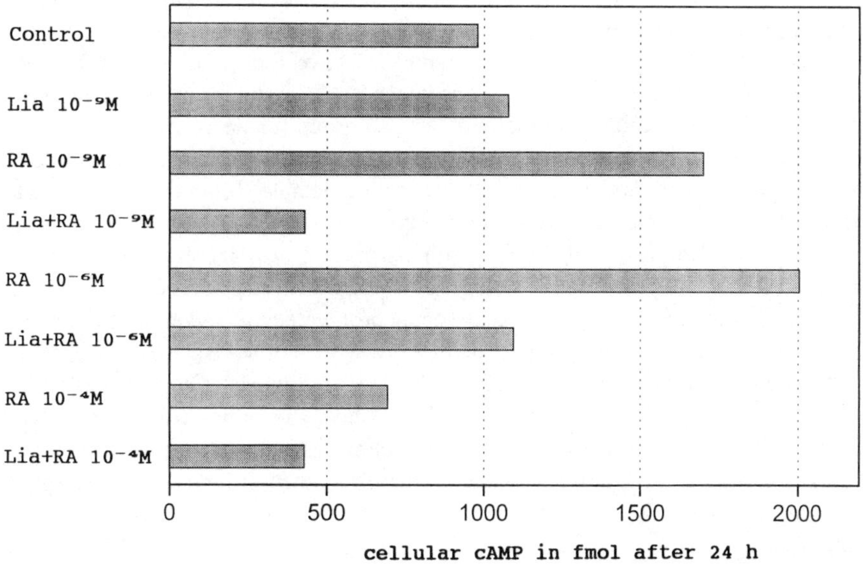

Figure 4. At various concentrations liarozole modulates the effect of all-trans retinoic acid on intracellular cAMP levels. At as little as 10^{-9} M, ATRA increases cAMP, and liarozole 10^{-9} M does not affect cAMP at all in transformed 3T3 cell monolayers treated for 24 h. In combination, ATRA plus liarozole cause a net reduction of intracellular cAMP to values below controls (human 3T3 fibroblast cells).

serum levels (mean retinoic acid 115 and 116 ng/ml, 4-oxo-13cRA 289 and 334 ng/ml, tretinoin 4.2 and 2.6 ng/ml).

At 10^{-7}M, 13cRA also demonstrates an *in vitro* synergism with both arotinoid Ro13-7410 and liarozole. Interferon-α_{2a} plus 13cRA have been successfully combined in squamous cell skin carcinoma (Lippman et al., 1992). In our dose range tested, both IFN-α_{2a} and IFN-α_{2a} plus 13cRA (10^{-7}M to 10^{-9}M) were equally effective in reducing glioblastoma nucleoside turnover.

At as little as 10^{-9} M, ATRA increases cAMP, and liarozole 10^{-9} M does not affect cAMP at all in transformed 3T3 monolayers treated for 24 h. In combination, ATRA plus liarozole cause a net reduction of intracellular cAMP to values below controls (human 3T3 fibroblast cells).

VI. CLINICAL APPLICATION OF RETINOIC ACID PLUS LIAROZOLE

From 1990 to 1993, we treated 12 patients (5 women, 7 men) aged 42 to 76 years and suffering from CNS malignancies in an open therapeutic trial with 13cRA and liarozole. All of them had tumors graded III to IV after Kernohan, and seven of eight primary glioblastoma multiforme (4 men, 4 women), and a total of eight of 12 patients with neuro-epithelial malignancies have died until now. Effective duration of treatment was 1 to 20 months (mean 8 + 5.7 months), and survival from diagnosis 4.5 to 22 months in the glioblastoma group (mean 11.75 months). For the four non-glioblastoma patients still alive, mean survival now amounts to 30 months. In five patients, including all three "unresectable" brain tumors, retinoid medication was instituted before radiotherapy

We have combined oral 13cRA with liarozole to (a) increase intracellular retinoid effects, (b) prevent a physiologic downregulation of plasma concentrations as described in retinoic-acid responsive acute promyelocytic leukemia (Muindi et al., 1992), and (c) allow out-patient therapy with a low risk of headache or pseudotumor cerebri. We gave 2.5-20 mg/d 13-*cis* retinoic acid and 75-300 mg/d liarozole p.o. with the approval of the Ulm University ethics committee—five bioptically defined glioblastoma multiforme patients, six other malignant gliomata, and one sarcomatous "malignant meningioma" (5 women, 7 men; age at first signs or symptoms 42 to 76 years, mean 57.5 + 12 years) for more than six months or until death.

Effective duration of retinoid treatment was 1 to 24 months (mean 8.5 months + 7 SD, n=12). In five patients, including all three "unresectable" brain tumors, retinoid medication was instituted before radiotherapy. One out of eight primary glioblastoma multiforme (4 women, 4 men including patient #12 classified as anaplastic astrocytoma with glioblastiform neuroradiologic changes) suffered from a rare spinal glioblastoma. Survival from diagnosis in the cranial glioblastoma group was 4.5 to 22 months (mean 11.8 months).

Retinoids were quantitated as described (Biesalski and Weise, 1990; Westarp et al., 1993b) from sera stored at -85°C and protected from light and oxygen. Retinyl esters in six patients were tested by Prof. Biesalki, Stuttgart. Samples were sealed and covered with aluminum foil from bedside to HPLC to minimize oxidation and degradation. Retinol and retinol binding protein are stable under these conditions, and slight hemolysis does not interfere with the results obtained.

We have treated 11 patients with primary CNS glioma (grade III to IV) with 13cRA plus liarozole^{INN}, and one highly malignant sarco-meningioma with 13cRA alone (Table 3). High retinoid plasma levels can be achieved by 13cRA medication alone, yet may become downregulated over time (Runde and Aul 1992; Westarp et al., 1991; 1992, 1993b). Liarozole > 70 mg/m^2/d given orally in conjunction with > 5 mg/m^2/d 13cRA are sufficient to increase and maintain plasma concentrations of 13cRA and 4-oxo-13cRA that correspond to a 10^{-7} molar steady state. At 10^{-7} M, both 13cRA and liarozole *in vitro* reduce glioblastoma thymidine incorporation in two cell lines (Yung et al., 1989; Westarp et al., 1994; 1993b).

In 424 patients from four studies, the survival of irradiated glioblastoma multiforme has been calculated to be 46 weeks with narrow 95 percent-confidence intervals of 44-48 weeks (Hildebrand, 1994; Szuwart et al., 1989). Mean postdiag-

Table 4. Medication Calculated Per m^2 Body Surface and Sum of Plasma 13cRA Derivatives

Pat No.	*Body weight*	*Height in cm*	*Surface m^2***	*13cRA/ liarozole [mg/d]*	*13cRA/ liarozole $mg/m^2/d$*	*Plasma level in ng/mL of total 13cRA-derivatives*
1	62 kg	166	1.68	20/300	12/180	368.4
2	81 kg	164	1.87	20/300	11/160	337.5
3	61 kg	158	1.62	10/150	6.90	454.4
4	74 kg	178	1.91	20/300	10/160	1073
				10/300	5/160	6395
				10/150	5/80	243.0
5	50 kg	186	1.67	20/300	12/180	407.3
				20/150	12/90	381.8
				10/75	6/45	299.4
	(one year later)			7.5/150	4.5/90	171.3
8	73 kg	175	1,88	15/150	8/80	33.2
		173		5/150	2.7/80	230.5
9	73 kg		1.86	10/600	5/325	259.8
	(one month later)			10/600	5/325	177.0

Notes: * fasting mean plasma steady state level of 13cRA + 4-oxo-13cRA + tretinoin (ng/mL) available 14 hours after last medication (5, 10, 15 or 20 mg/d 13cRA p.o., plus 75, 150, 300 or 600 mg/d liarozole p.o.) without interference from other therapies; for calculation of mean values, "traces" were defined as := 1 ng/mL.

** Body surface according to DuBois

** = $kg^{0.425} \times cm^{0.725} \times 71.84$

Source: Adopted from Westarp et al. 1994.

nostic survival in the 11 brain glioblastoma patients treated with 13cRA plus liarozole was eleven months (n.s.).

Even if different methods should lead to incomparable plasma levels, the amount of retinoic acid and its catabolic inhibitor were indeed correlated with fasting 13cRA and 4-oxo-13cRA (Table 4, Figures 5 and 6).

VII. TOLERANCE OF RETINOIC ACID PLUS LIAROZOLE

All patients treated developed slight cutaneous reactions, dry skin, mild cheilitis, and a temporary nonitching rash that promptly resolved upon dose reduction down to 5 mg/d. Leukocyte counts, liver enzymes did not change significantly, erythrocyte sedimentation rate remained unaffected, plasma retinyl esters measured in six tumor patients were not elevated, and even after two years of continuous combined adjuvant medication, plasma vitamin A and retinol-binding protein remained within normal ranges. In every case dry lips, peeling skin on upper arms and thighs, as well as an aseptic conjunctival injection corresponded to total plasma retinoic acid concentrations above 100 ng/ml. We adusted dosages avoiding painful lips, generalized facial erythema or extensive livid dermal efflorescences, and neither ob-

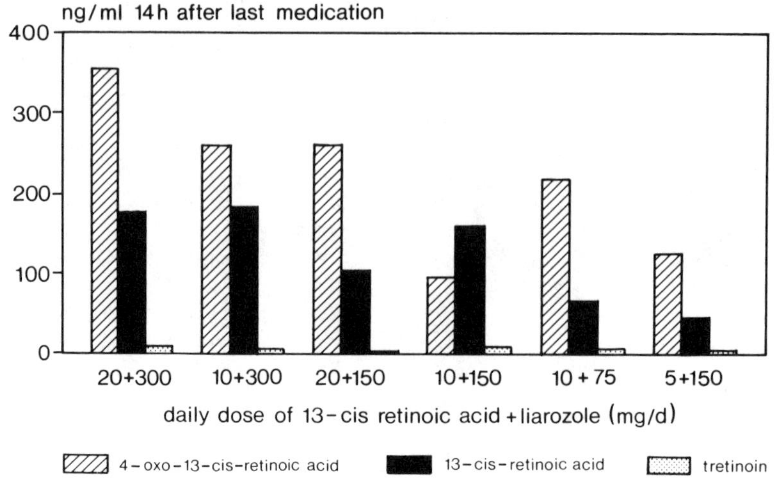

Figure 5. The amount of oral 13cRA and liarozole correlates with steady state plasma concentrations of 13cRA and 4-oxo-13cRA.

served fever as in typical "retinoic acid syndrome," nor musculoskeletal side effects.

Though liarozole upregulates progesterone and reduces testosterone, estradiol, aldosterone, and to some extent plasma cortisol, patients did not show signs or symptoms of hormonal dysfunctions (Westarp et al., 1993a), despite the fact that 13cRA itself may decrease plasma testosterone by interfering with 5-α -reductase activity (Rademaker et al., 1991). Cutaneous side effects occurred after one to two weeks of retinoid therapy and promptly responded to dosage adjustments. While successful differentiation therapy of acute promyelocytic leukemia with all-*trans* retinoic acid (> 45 mg/m^2) has regularly been accompanied by headache and increased intracranial pressure, fatigue was the only central nervous side effect we have seen up until now in ten patients treated with 13-*cis* retinoic acid plus liarozole.

Side effects (from dry skin, mild cheilitis, or a temporary non itching rash promptly resolving upon dose reduction to a facial erythema and/or peeling proximal extremities) and plasma levels were equilibrated best in adults medicated with 75 to 150 mg/d liarozole plus 5 to 15 mg/d 13-*cis* retinoic acid p.o. Plasma retinyl esters measured in six CNS tumor patients were not elevated in any case; one transiently nondetectable retinyl palmitate concentration was measured after seven months of liarozole plus 13cRA (pat. #5). During the observation period of up to two years potential long-term side effects such as generalized hyperostosis could not be observed.

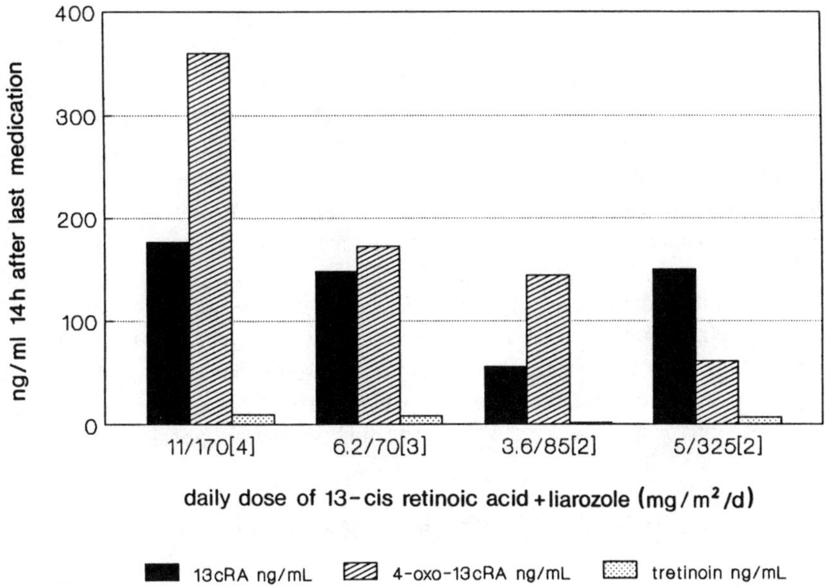

Figure 6. Of all single parameters, liarozole/m^2 correlates best with sum of plasma levels of 13cRA and derived retinoids. Mean relative dose of 13cRA and liarozole [number of patients] and corresponding mean fasting plasma retinoid levels (n=11)

Unlike retinol, 13cRA does not accumulate in the liver (Meyskens et al., 1985) and is not eliminated more slowly in obese recipients (Chien et al., 1992). Maximum plasma levels occur 1-4 hours after oral application (Meyskens et al., 1985), with a biphasic mean elimination half-life of 17 hours for 4-oxo-13cRA. Significant amounts approximating 20 percent of plasma retinoic acids can be expected to cross the blood brain barrier (Meyskens et al., 1985), particularly in brain tumor patients with impaired barrier function. Though 13cRA alters hepatic methionine catabolism and transmethylation (Schalinske and Steele 1991), and accumulates intracellularly due to a liarozole-mediated inhibition of 5-alpha hydroxylase and cytochrome P_{450}, plasma parameters indicate a good hepatic tolerance. A lack of headache under combined retinoid medication argues against a space-demanding growth during terminal differentiation. Fatigue was the only central nervous side effect we have observed. Cutano-conjunctival side effects occurred after one to two weeks of therapy and promptly responded to dosage adjustments. Retinoic acid side effects may be attenuated by vitamin E supplementation (Drott et al., 1993).

Further studies will see whether liarozole influences intracellular CRABP as a retinoic acid-absorbing "sink compartment" or alters retinoid glucuronidation.

While oral vitamin A increases plasma 13cRA (Eckhoff and Nau, 1990), oral 13cRA plus liarozole leave plasma retinol unaffected. This could be decisive for lymphocytic retinoid metabolism (Moriguchi et al., 1985), potentially impaired by synthetic retinoids that down-regulate plasma retinol (Buck et al., 1991; Clerici et al., 1992). Cutaneous tolerance and pharmacologic concentrations came out best in patients with 75 to 150 mg/d liarozole plus 5 to 15 mg/d 13-*cis* retinoic acid p.o.

In our CNS tumor series, extracellular retinol, retinol binding protein (RBP), and plasma carotene were not markedly altered in patients without radiotherapy (Figure 7). This may even be of importance for lymphocyte retinoid metabolism (Moriguchi et al., 1985, Buck et al., 1991). In this respect, 13cRA plus liarozole contrast with synthetic retinoids such as fenretinide. Fenretinide has been reported to reduce plasma retinol in breast cancer patients by 38-71 percent (Formelli et al., 1993). Liarozole may upregulate plasma progesteron and reduce testosteron, estradiol, aldosteron, and to some extent plasma cortisol. Nevertheless, patients did not show signs or symptoms of hormonal dysfunction (Westarp et al., 1993b; 1994).

VIII. RETINOIDS AND RADIOTHERAPY

In human glioblastoma cell cultures, retinoic acid increased rather than diminished gamma ray sensitivity (W.K.A. Yung, personal communication, Hamburg 1990), arguing against a negative effect of a differentiating agent on glioblastoma radiosensitivity. Radiosensitivity is restricted in grade IV brain tumors, and does not correlate with degree of differentiation (Walter and Müller 1959; Zülch, 1960). In other cancers, both chemotherapy and radiotherapy have profited from retinoid supplementation (Rutz and Little, 1989).

All study patients had been started on retinoid medication either after relapse or because of an unresectable tumor mass. Nevertheless, a clinical benefit of adjuvant retinoids remains to be proven. Four of the subjectively more favorable cases reported had received retinoid differentiation therapy during whole brain irradiation, supporting a safe adjuvant use of differentiation-inducing retinoids.

Retinol-binding protein (RBP) markedly varied under radiotherapy of the skull, a phenomenon not yet explained (Okuno et al., 1992). RBP increased after approximately 20 of 60 Gy total brain irradiation (Figure 8). Should an increased liberation from the liver be a direct result of cranial radiotherapy? Very low EQ values < 0.50 were regularly seen after radiotherapy without changes in renal retention or hepatic enzymes. As far as we know relevant liver radiodoses can be excluded as cause for the marked changes in mean plasma RPB during radiotherapy of the skull (Figure 8). We cannot yet explain why RBP should increase after the first third of radiotherapy (tumor + 2 cm, 60 Gy whole brain dose).

A novel cytoreductive therapy for malignant tumors in general and non-metastasizing brain tumors in particular is the transduction of rapidly dividing tumor cells

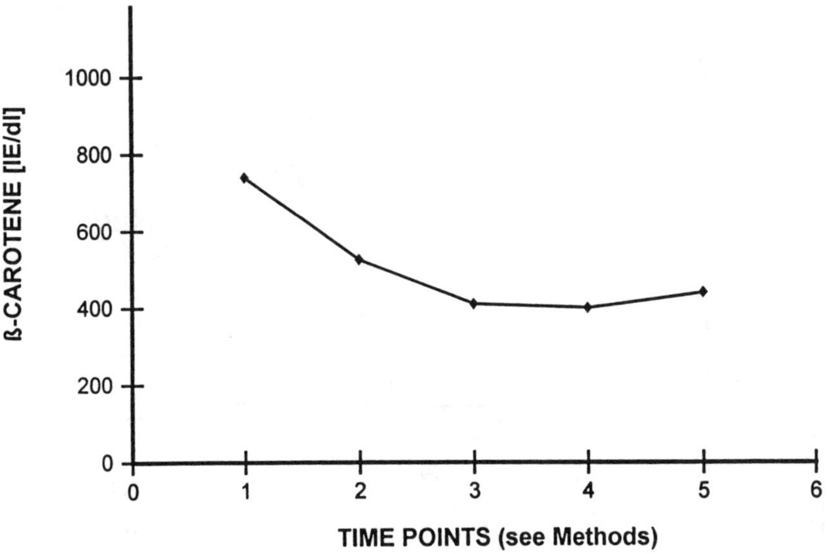

Figure 7. Plasma β-carotene seems to decline within references after initiation of oral retinoid therapy. Though there is no significant change demonstrable due to variability and small sample size, β-carotene seems to decrease during early medication with 13cRA and liarozole.

with herpes simplex virus thymidine kinase (HSV-TK) plus subsequent therapy with ganciclovir (Shimizu et al., 1995; Chambers et al., 1995). This approach may also be exploited to sensitize glioma cells to radiation damage. Pretreatment with the antiviral pyrimidine analogue 5-(2-bromovinyl)-2'-deoxyuridine (BVdU) has enhanced the radiation-induced cytotoxicity in human glioma (Kim et al., 1994).

IX. RETINOIDS IN NON-GLIAL CNS NEOPLASMS

Childhood neuroblastoma may spontaneously regress or differentiate into less malignant ganglioneurinoma, and there have been reports on therapeutic retinoid effects (Thiele et al., 1985). Spongioblastoma are of neuro-epithelial/neuro-ectodermal origin as well, and an adjuvant retinoid therapy may be warranted particularly in spinal tumors without further interventional option, including the recurrence of spinal ependymoma. Retinoids do not interfere with coumarin anticoagulation or nonsteroidal anti-inflammatory medication both given to reduce post-irradiation myelopathy in patients with spinal radiotherapy (Glantz et al., 1994). Retinoids also do not interfere with cytotoxic chemotherapies or experimental treatments such as high-dose tamoxifen.

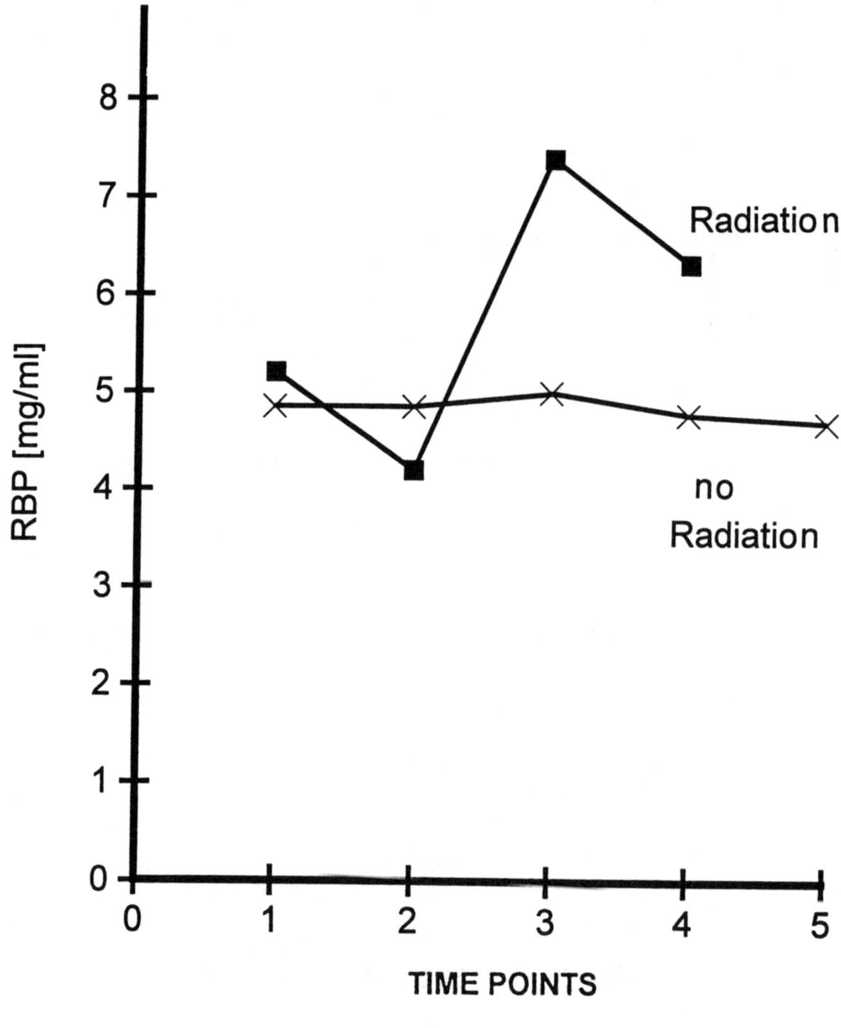

Figure 8. Does plasma retinol binding protein (RBP) increase during brain radiotherapy (up to 60 Gy)? We cannot yet explain, why RBP should increase after the first third of radiotherapy (tumor + 2 cm, 60 Gy whole brain dose).

At present, there is little indication for the use of retinoids in management of benign meningioma (Meyskens et al., 1982). For large and basal meningioma, which are difficult to resect in-toto, both preoperative balloon occlusion and tumor embolization with polyvinyl alcohol particles have been achieved (Tymianski et al., 1994; Dean et al., 1994; Morimura et al., 1994).

Table 5. Low Serum Vitamin-A Increases Under 13cRA Medication

Patient	*Clinical Description*	*RBP/Retinol*	*μg/ml*[2]
PEm *1935	metastatic colon cancer, craniotomy	< 12/none	EQ= -
	begin of 13cRA[1] medication	< 12/.163	EQ= -
	(dexamethasone continued)	< 12.161	EQ= -
	β-carotene = 418 μg/dl	< 12/none	EQ= -
	slightly dry skin	13/none	EQ= -
	β-carotene = 417 μg/dl	< 12/none	EQ= -
	β-carotene = 521 μg/dl	17/.201	EQ= -1.28
	β-carotene =1025 μg/dl	15/.301	EQ= -0.75
	4 weeks prior to death	10/.342	EQ= -0.84

Notes: [1] 13-cis retinoic acid 10 mg- 0- 10 mg daily p.o. after meals.
[2] = EQ equimolar ratio of retinol to retinol-binding protein.

Metastases are frequent intracranial space-demanding malignant lesions. Even if they, depending on the primary tumor, respond to palliative radiotherapy, their edema and growth causes headache and multiple neurologic deficits. To control intracranial tumor growth at least initially by adding an adjuvant retinoid medication is worthwhile testing for both medical and humane reasons.

There is no information available yet on the potential effects of retinoids on tumor spread within the CNS. Personally, however, I have not seen a metastasizing glioma yet, although up to 14 percent of malignant glioma may show leptomeningeal dissemination (Higuchi et al., 1994), and 4 percent are reported to occur multilocally. Indeed we have seen multiple glioblastoma in individual patients, with rapidly fatal outcome. In all these clinical cases known to us, multiple glioblastoma lesions appeared to have originated simultaneously rather than by tumor spread.

X. INTRATUMORAL MEDICATION—HSV-TK, RETINOIDS, AND MORE

Glioblastoma multiforme is a primary brain disease that very rarely metastasizes. The first and probably most important step of tumor invasion is a block of tumor cell differentiation at the invasion front. So intracavitary treatment of primary brain tumors may offer therapeutic windows not yet completely opened. Both a diagnostically necessary biopsy in conjunction with rapid histomorphologic evaluation, and an open resection offer the opportunity to administer anti-tumor agents "on site." A particularly attractive intralesional intervention model is the insertion of herpes simplex virus thymidine kinase (HSV-TK) into tumor cells via injection of cells into the tumor that produce retroviral constructs containing HSV-TK. HSV-TK then activates administered ganciclovir and the tumor cells die (Culver et al., 1992; Ram et al., 1993). Replication-defective adenovirus (ADV) carrying the HSV-TK gene controlled by the Rous sarcoma virus promoter can be injected into the tumor either directly (Chen et al., 1994) or via fibroblasts producing retroviral vectors carrying HSV-TK (Barba et al., 1994). Not all tumor cells have to be transduced since there is said to be a remarkable bystander killing of neighboring cells that do not express HSV-TK (Wu et al., 1994; Perez-Cruet, 1994).

Intralesional recombinant interleukin-2 (rIL-2) plus lymphokine activated killer (LAK) cells have not proven clear benefits in malignant brain tumors. Chromosome-9 deletions including interferon α and β genes and loss of loci on chromosome 10 were predominantly found in highly malignant glioma (Venter and Thomas, 1991). Combined intratumoral injection of IL-2 activated killer cells, IFN-β and ACNU led to a casuistic regression of relapsing glioblastoma in a 26-year-old patient (Naganuma et al., 1989). Intratumoral interferon and IL-2/LAK cell combinations have been employed intracavitarily (Barba et al., 1988), while interferon therapy prior to or during radiotherapy had disadvantages due to cell cycle arrest in human glioma models (Nakamura, 1986). As single agents, intracavitary IFN or IL-2 given through Omaya reservoirs did not improve treatment (Fetell et al., 1987), and ten percent of patients developed sterile abscesses at the injeciton site (Atkinson, 1989).

Intra-operative wounding may liberate transforming growth factor TGF-β like wound factors stimulating tumor growth (Fürstenberger et al., 1989) and contribute to a further weakening of thc anti-cancer host defenses (Pfeilschifter, 1990). 10^{-8} M RA inhibit tumor growth in the presence of TGF-β in myc-oncogene transfected fibroblasts (Roberts et al., 1985). Hypothetically, intracavitary retinoic acid could block TGF-β associated effects, besides inducing differentiation and enhancing cellular anti-tumor immunity. For retinoic acid, hydrophobic encapsulation into a water soluble β-cyclodextrin complex—already available for retinol (Sigma R0635, methyl-beta-cyclodextrin)—might facilitate intralesional administration.

Altered lipid compositions are reported more frequently in nervous system tumors (Gulaya et al., 1990). Sufficient amounts of long chain n-6 and n-3 fatty acids are essential for human fetal brain maturation (Crawford et al., 1989). A blocked conversion of linoleic acid to less saturated gamma linolenic acid (GLA) or dihomogammalinolenic acid may lead to impaired prostaglandin PGE_1 and cyclic adenosine monophosphate (cAMP) synthesis, necessary for control of a differentiated state (Horrobin, 1990). In neuroblastoma cell lines, retinoic acid prevents the desensitization of the PGE_1-mediated cAMP elevation (Yu et al., 1988). While linolenic acid deficiency alters brain membrane fluidity (Bourre, 1989), gamma-linolenic acid (GLA) can alter specific lipases controlling the arachidonate pathway. Lithium ions strongly interact with primary polyphosphoinositide metabolism (Busa, 1988), a potential final pathway of several growth-stimulating systems (Wahl and Carpenter, 1988). Lithium-gamma linolenic acid (GLA-Li^+) could therefore exert specific anti-proliferative effects in dedifferentiated neuro-ectodermal tumors—and serve as lipophil vehicle for intralesional retinoids. While the continuous instillation of IL-2 or interferons has been accompanied by severe side effects such as brain infection associated with the delivery systems used, sterile lipophilic or micro-encapsuled retinoids could be left at the lesion site without further need for a catheter.

XI. FUTURE ADJUVANT MEDICAL OPTIONS

Both retinoic acid and vitamin D_3 recognize a human osteocalcin gene vitamin D response element (Schule et al., 1990). Active vitamin D_3, which binds to a nuclear human vitamin D receptor homologous to steroid hormone receptors, has been shown to inhibit numerous carcinoma cells. Active vitamine D_3 can be secreted extra-renally by macrophages, and dexamethasone inhibits $1\alpha,25(OH)_2D_3$ production by activated macrophages (Reichel, 1990). Vitamin D_3 is therefore another candidate adjuvant. Differentiation induction with retinoic acid plus vitamin D_3 is being successfully tested in myelodysplastic syndromes (Blaszek et al., 1990). The mechanisms of retinoid therapy in primary central nervous system cancer are incompletely understood. They may be different from haematological diseases, including a dual regulatory role as transcriptional activators and suppressors in analogy to thyroid hormone receptors (Ross, 1993). Parallel to the combined adjuvant use of retinoic acids and a specific catabolic inhibitor for glioma treatment (Yung et al., 1993), potential synergisms between different retinoic acid derivatives should be considered systematically.

Interferon-α_2 exerts an antiproliferative effect on human glioblastoma cells (Hamada et al., 1986). Since bystander killing and an enhanced anti-tumoral immunity are observed under HSV-TK/ganciclovir treatment (Barba et al., 1994), interferons could act as useful amplifiers of intralesional therapy.

Phenylacetate, a non toxic differentiation inducer, can enhance the efficacy of retinoids and interferon-α at nontoxic, pharmacologic concentrations. Again, the effects of phenylacetate or phenylbutyrate seem to be less pronounced in poorly differentiated cells (Liu et al., 1994). Data on the combination of differentiation inducers with unorthodox angiogenesis inhibitors such as thalidomide (D'Amato et al., 1994) are not available at present.

Tumor suppressor genes code for DNA binding or transcription-regulating proteins participating in the control of cell gene expression. Proto-oncogenes may code for growth or maintenance factors or their receptors, and their untimely activation or over-expression may contribute to malignant CNS tumor growth. Proto-oncogene function can principally be antagonized by transfected anti-sense oligonucleotides as has been evidenced in a syngeneic rat model for IGF_1 (Trojan et al., 1993).

Beneficial retinoic acid effects may involve RA-dependent immune mechanisms (Micksche et al., 1977; Carman et al., 1989; McGarry et al., 1989). In non-CNS cancer, retinoic acid has been shown to ameliorate depressed cutaneous T-recall antigen reactivity (Serrou and Cupissol, 1980), while in CNS malignancies, cell-mediated immunity is known to be impaired (Thomas et al., 1975); retinol does not enhance skin delayed hypersensitivity (Rosales et al., 1992). Retinoic acids modulate B-cell function (Bollag and Peck, 1991) and stimulate natural killer cell activity (Clerici et al., 1992). Oral retinoic acid also has a potential to enhance cell mediated immunity in tumor patients (Dennert and Lotan, 1978; Serrou and Cupissol 1980; Moriguchi et al., 1985; Harant et al., 1995). Induction of differentiation plus enhancement of antitumor immunity by retinoids (Wolf, 1984; Moriguchi et al., 1985) are particularly attractive complementary approaches (Anstey, 1992) compatible with further therapies.

The availability of a complete stem cell biology, a good recovery potential of the cellular body, and combination of therapeutic agents have finally resulted in a well-established and successful regimen in childhood leukemia. Progress in cellular and molecular biology as well as clinical efforts including intralesional therapies now present a realistic chance to offer additional therapeutic options to malignant glioma patients and to overcome the therapeutic stand-still in CNS malignancies in general and malignant glioma in particular.

XII. SUMMARY—RETINOIDS IN CNS TUMORS

Vitamin A's intracellular metabolite retinoic acid induces differentiation in glioma cultures and recognizes retinoic acid receptors on human glial cells. Retinoic acid degradation can be inhibited by the novel steroid synthesis modulator liarozole (5-((3-chlorophenyl)(1H-imidazol-1-yl)methyl)-1H-benzimidazole HCl). Measured by ^{3}H-thymidine incorporation, liarozole itself has an intrinsic attenuating effect on glioblastoma growth *in vitro*. In contrast to a number of other retinoids, pharmacotherapeutic doses of 13cRA do not cause headache. When retinoic acid is given alone, plasma levels are known to decrease over time despite continued intake. Liarozole is an effective catabolic inhibitor for 13cRA, and both can be given orally as long-term adjuvant in potential synergy. Due to a dose-dependent enhancement of ectodermal differentiation plasma retinoid concentrations can be titrated by inspection. Oral liarozole approximating 75 mg/m^2 plus oral 13cRA approximating 10 mg/m^2 have been well tolerated and have maintained permanently high plasma concentrations of 13cRA and 4-oxo-13cRA. We have neither observed hormonal dysfunction, a "retinoic acid syndrome," elevated intracranial pressure, nor major headache. The most frequent side effects, for example, cheilitis, conjunctivitis, and desquamation, have promptly responded to dose reduction; they may be attenuable by vitamin E. Liver enzymes and plasma retinol and retinol-binding protein did not change significantly, and serum retinol palmitate was not elevated. As long as liarozole awaits approvement by the FDA or European drug administration, 13-*cis* retinoic acid can be prescribed as adjuvant medication 10 mg capsules b.i.d.

In malignant glioma retinoic acid modulates the expression of growth signaling molecules such as EGFR. Of particular interest will be whether glioma-associated oncogenes such as c-sis (Diedrich, 1992), HA/N-ras and c-myc (Orian et al., 1992) can be down-regulated by long-term retinoid medication. Retinoid therapy should be examined intralesionally, alone or in conjunction with HSV-TK/ganciclovir and anti-oedematous Boswellia acids treatment in primary malignant central nervous system tumors. In our opinion, all these strategies are well compatible with subsequent respectively overlapping percutaneous or interstitial radiotherapy.

ACKNOWLEDGMENTS

We thank Prof. Hans-Peter Richter, PD., Dr. Ulrich Kunz, and Prof. Peter Krauseneck, Bamberg, for punctual discussion, and J.-C. Stein and all former collaborators for inclusion of data.

REFERENCES

Anstey, A. (1992). Retinoids. In: *Principles of Pharmacology. Basic Concepts And Clinical Applications.* (Munson, P.L., Mueller, R.A., & Breese, G.R.) pp. 1232-1239. Chapman & Hall, New York 1992.

Atkinson, L.L., Merchant, R.E., Ghatak, N.R., & Young, H.F. (1989). Sterile abscesses in glioma patients treated by intraparenchymal injection of lymphokine-activated killer cells and recombinant interleukin-2: case reports. Neurosurgery 25(5), 805-810.

Barba, D., Hardin, J., Sadelain, M., & Gage, F.H. (1994). Developement of anti-tumor immunity following thymidine kinase-mediated killing of experimental brain tumors. Proc. Natl. Acad. Sci. USA 91, 4348-4352.

Barba, D., Oldfield, E.H., Saris, S.C., Rosenberg, S.A., & Hamilton, J.M. (1988). Phase II immunotherapy of cystic primary brain tumours with interleukin-2/LAK cells. Proceedings Am. Soc. Clin. Oncology 7, 82.

Barbone, F., Delzell, E., Austin, H., & Cole, P. (1994). Exposure to epichlorohydrin and central nervous system neoplasms at a resin and dye manufacturing plant. Arch. Environ Health 49, 355-358.

Bates, S.E., Mickley, L.A., Chen, Y.N., Richert, N., Rudick, J., Biedler, J.L., & Fojo, A.T. (1989). Expression of a drug resistance gene in human neuroblastoma cell lines: modulation by retinoic acid-induced differentiation. Mol. Cell Biol. 9(10), 4337-4344.

Bello, M.J., de Campos, J.M., Kusak, M.E., Vaquero, J., Sarasa, J.L., Pestana, A., & Rey, J.A. (1994). Molecular analysis of genomic abnormalities in human gliomas. Cancer Genet. Cytogenet. 73, 122-129.

Benjamin, L.A., McGarry, R.C., & Hart, D.A. (1989). Effect of retinoic acid on human neuroblastoma: correlation between morphological differentiation and changes in plasminogen activator and inhibitor activity. Cancer Chemother. Pharmacol. 25, 25-31.

Biesalski, H.K., & Weiser, H. (1990). Microdetermination of retinyl esters in guinea pig tissues under different vitamin-A status conditions. J. Micronutrient Anal. 7, 97-116.

Biesalski, H.K. (1989). Comparative assessment of the toxicology of vitamin A and retinoids in man. Toxicology 57, 117-161.

Blaszek, I., Triana, K., Misset, J.L., & Mathe, G. (1990). Treatment of non blastic myelodysplastic syndromes by differentiation inducers. J. Cancer Res. Clin. Oncol. 116, 7.

Bollag, W. (1983). Vitamine A and retinoids: from inhibition to pharmacotherapy in dermatology and oncology. Lancet i, 860-865.

Bollag, W., Peck, R. (1991). Modulation of human immune functions *in vitro* by temarotene and its metabolite. Skin Pharmacol. 4 (3), 142-149.

Bourre, J.M. (1989). Nature, origin and role of fatty acids of the nervous system: An essential fatty acid, an alpha-linolenic acid, changing the structure and the cerebral function, Bull. Acad. Natl. Med. 173 (9), 1137-1148.

Boyd, A.S. (1989). An overview of the retinoids. Am. J. Med. 86, 568-574.

Brady, L.W., Miyamoto, C., Woo, D.V., Rackover, M., Emrich, J., Bender, H., Dadparvar, S., Steplewski, Z., Koprowski, H., Black, P., Lazzarom, B., Nairm, S., McCormack, T., Nieves, J., Morabito, M., & Eshleman, J. (1992). Malignant astrocytomas treated with Iodine-1235 labeled monoclonal antibody 425 against epidermal growth factor receptor: A phase II trial, Int. J. Radiat. Oncol. Biol. Phys. 22(1), 225-230.

Buck, J., Derguini, F., Levi, E., Nakanishi, K, & Hammerling, U. (1991). Intracellular signaling by 14-hydroxy-4,14,-retro-retinol. Science 254 (5038), 1654-56.

Burger, P.C., & Green, S.B. (1987). Patient age, histologic features, and length of survival in patients with glioblastoma multiforme. Cancer 59, 1617-1625.

Busa, W.B. (1988). Roles for phosphatidylinositol cycle in early development. Phil. Trans. R. Soc. London B 320, 415-426.

Carman, J.A., Smith, S.M., & Hayes, C.E. (1989). Characterization of a helper T lymphocyte defect in vitamin A-deficient mice. J. Immunol. 142, 388-393.

Chambers, R., Gillespie, G.Y., Soroceanu, L., Andreansky, S., Chatterjee, S., Chou, J., Roizman, B., & Whitley, R.J. (1995). Comparison of genetically engineered herpes simplex viruses for the treatment of brain tumors in a scid mouse model of human malignant glioma. Proc. Natl. Acad. Sci. USA 92, 1411-1415.

Chen, S.H., Shine, H.D., Goodman, J.C., Grossman, R.G., & Woo, S.L. (1994). Gene therapy for brain tumors: Regression of experimental gliomas by adenovirus-mediated gene transfer *in vivo*. Proc. Natl. Acad. Sci. USA 91, 3054-3057.

Chien, D.S., Sandri, R.B., & Tang-Liu, D.D.S. (1992). Systemic pharmacokinetics of acitretin, etretinate, isotretinoin, and acetylenic retinoids in guinea pigs and obese rats. Drug Metab. Disposit. 20, 211-217.

Cleary, S.F., Liu, L.M., & Merchant, R.E. (1990). Glioma proliferation modulated *in vitro* by isothermal radiofrequency radiation exposure. Radiat. Res. 121, 38-45.

Clerici, E., de Palo, G., Ferrario, E., Trabattoni, D., & Villa, M.L. (1992). Retinoids, breast cancer and NK cells. FASEB J. 6, A1433.

Cooke, N.E., & Haddad, J.G. (1989). Vitamin D binding protein (Gc-Globulin). Endocrine Reviews 10, 294-307.

Crawford, M.A., Doyle, W., Drury, P., Lennon, A., Costeloe, K., & Leighfieldm, M. (1989). n-6 and n-3 fatty acids during early human development, J. Intern. Med. 225(731), S159-59.

Creech-Kraft, J., & Juchau, M.R. (1992). Conceptual biotransformation of 4-oxo-all-trans-retinoic acid, 4-oxo-13-cis-retinoic acid and all-trans-retinoyl-beta-glucuronide in the rat whole embryo culture. Biochem. Pharmacol. 43, 2289-2292.

Creek, K.E., Silverman-Jones, C.S., & DeLuca, L.M. (1989). Comparison of the uptake and metabolism of retinol delivered to primary mouse keratinocytes either free or bound to rat serum retinol-binding protein. J. Invest. Dermatol. 92, 283-289.

Culver, K.W., Ram, Z., Wallbridge, S., Ishii, H., Oldfield, E.H., & Blaese, M. (1992). *In vivo* gene transfer with retroviral vector-producer cells for treatment of experimental brain tumors. Science 256, 1550-1552.

D'Amato, R.J., Loughnan, M.S., Flynn, E., & Folkman, J. (1994) Thalidomide is an inhibitor of angiogenesis. Proc. Nat. Acad. Sci. (Wash) 91, 4082-4085.

De Coster, R., Van Ginckel, R., van Moorselaar, R.J.A., Schulken, J.A., Debruyne, F.M.J., Bruynseels, J., & Denis, L. (1991). Antitumoral effect of R75251 in prostatic carcinoma experimental and clinical studies. Proc. Am. Ass. Canc. Res. 32, 213.

de Thé, H., Marchio, A., Tiollais, P., & Dejean, A. (1989). Differential expression and ligand regulation of the retinoic acid receptor alpha and beta genes. EMBO J. 8, 429-433.

de Thé, H., Vivanco-Ruiz, M.M., Tiollais, P., Stunnenberg, H., & Dejean, A. (1990). Identification of a retinoic acid responsive element in the retinoic acid receptor beta gene, Nature 343 (6254), 177-180.

Dean, B.L., Flom, R.A., Wallace, R.C., Khayata, M.H., Obuchowski, N.A., Hodak, J.A., Zabramski, J.M., & Spetzler, R.F. (1994). Efficacy of endovascular treatment of meningiomas: evaluation with matched samples. Am. J. Neuroradiol. 1, 1675-1680.

Dennert, G., & Lotan, R. (1978). Effects of retinoic acid on the immune system: stimulation of T killer cell induction. Eur. J. Immunol. 8, 23-29.

Diedrich, U. (1992). Function of oncogenes in neuroepithelial tumors. Fortschr. Neurol. Psychiatr. 60 (4), 146-153.

Diedrich, U., Lucius, J., Bittermann, H.J., Schlösser, M., Eckert, B., Behnke, J., & Pabst, B. (1995). Loss of alleles in brain tumor: Distribution and correlations with clinical course, J. Neurol. 242, 707-711.

Drott, P., Meurling, S., &. Gebre-Medhin, M. (1993). Interactions of vitamins A and E and retinol-binding protein in healthy Swedish children—evidence of thresholds of essentiality and toxicity. Scand. J. Clin. Lab. Invest. 53, 275-280.

Eckhoff, C., & Nau, H. (1990). Vitamin A supplementation increases levels of retinoic acid coumpounds in human plasma: Possible implications for teratogenesis. Arch. Toxicol. 64, 502.

End, D.W., Wouters, W., Van Ginckel, R., Garrabant, T.A., Van Wauwe, J., Coene, M.C., Bruynseels, J., De Coster, R., & Bowden, C. (1991). R75251, an antitumoral agent which inhibits retinoic acid (RA) metabolism. Proc. Am. Assoc. Canc. Res. 32, 213.

Ernestus, R.I., Röhn, G., Schröder, R., Els, T., & Paschen, W. (1992). Polyamine metabolism in gliomas, J. Neurol. 239(S2), S50.

Favennec, L. (1988). The biological effects of retinoids on cell differentiation and proliferation. J. Clin. Chem. Clin. Biochem. 2, 479-489.

Fetell, M.R., Houseplan, E.M., Oster, M.W., Fisher, P.B., Cote, D.N., Moser, F.G., & Sisti, M.B. (1987). Intratumoral administration of β-interferon in glioblastoma: Phase I study, Neurology 37: 334.

Fischer, I., Nolan, C.E., & Shea, T.B. (1987). Effects of retinoic acid on expression of the transformed phenotype in C6 glioma cells. Life Sci. 41, 463-470.

Forbes, A.R., & Goldberg, I.D. (1984). Radiation therapy in the treatment of meningeoma the Joint Center for Radiation Therapy experience 1970 to 1982. J. Clin. Oncol. 2, 1139-1143.

Formelli, F., Clerici, M., Campa, T., Di Mauro, M.G., Magni, A., Mascotti, G., Moglia, D., De Palo, G., Costa, A., & Veronesi, U. (1993). Five-year administration of fenretinide: Pharmacokinetics and effects on plasma retinol concentrations. J. Clin. Oncol. 11, 2036-2042.

Frey, J.R., Peck, R., & Bollag, W. (1991). Antiproliferative activity of retinoids, Interferon-a and their combination in five human transformed cell lines, Cancer Lett. 57, 223-227.

Fürstenberger, G., Rogers, M., Schnapke, R., Bauer, G., Höfler, P., & Marks, F. (1989). Stimulatory role of transforming growth factors in multistage skin carcinogenesis: Possible explanation for the tumor-inducing effect of wound healing in initiated NMRI mouse skin. Int. J. Cancer 43, 915-921.

Giguere, V., Ong, E.S., Segui, P., & Evans, R.M. (1987). Identification of a receptor for the morphogen retinoic acid. Nature 330, 624.

Glantz, M.J., Burger, P.C., Friedman, A.H., Radtke, R.A., Massey, E.W., & Schold, S.C., Jr. (1994). Treatment of radiation-induced nervous system injury with heparin and warfarin. Neurology 44, 2020-2027.

Gold, E.J., Mertelsmann, R.H., Itri, L.M., Gee, T., Arlin, Z., Kempin, S., & Clarkson, B. (1983). Phase I clinical trial of 13-*cis* retinoic acid in myelodysplastic syndromes. Cancer Treat. Rep. 67, 981-986.

Gulaya, N.M., Volkov, G.L, Klimashevsky, V.M., Govseeva, N.M., Artemenko, I.P.,& Gavriliuk, E.S. (1990). Modification of lipid composition of neuroblastoma C13oo N18 cells with liposomes alters the cholesterol content. Neuroscience 34, 785-792.

Hamada, H., Asakura, Y., Maeda, Y., Yokoyama, S., & Niiro, M. (1986). A study of direct antitumor activity of interferon-α against human glioma. Jpn. J. Cancer Chemother. 13, 464-471.

Harant, H., Lindley, I., Uthmann, A., Ballaun, C., Krupitza, G., Grunt, H., Huber, H., & Dittrich, C. (1995). Regulation of interleukin-8 gene expression by all-trans retinoic acid. Biophys. Biochem. Res. Comm. 210, 898-906.

Higuchi, M., Arita, N., Ohnishi, T., Hiraga, S., Kusakabe, M., Sakakura, T., & Hayakawa, T. (1995). Leptomeningeal dissemination of malignant glioma: immunohistochemical analysis of tenascin and glial fibrillary acidic protein. Clin. Neuropathol. 14, 223-227.

Hildebrand, J. (1994). The EORTC Brain Tumor Group: Age effect on the survival of patients with glioblastoma and anaplastic astrocytoma. In: *Cancer Treatment—An Update* (Banzet P, Holland, J.F, Khayat, D., & Weil, M., eds.) pp. 612-615. Springer, Paris Berlin.

Hong, W.K., & Itri, L.M. (1994). Retinoids and Human Cancer. In: *The Retinoids, Biology, Chemistry and Medicine* (Sporn, M.B., Roberts, A.B., & Goodman, D.S., eds.) p. 2. Raven Press, New York.

Hong, W.K., & Lippman, S.M. (1991): Chemoprevention of aerodigestive epithelial cancers. In: *Chemoimmuno Prevention of Cancer* (Pastorino, U., & Hong, W.K. eds.), pp. 141-146. Thieme Stuttgart, New York.

Horrobin, D.F. (1990). Essential fatty acids, lipid peroxidation and cancer. In: *Omega-6 Essential Fatty Acids: Pathophysiology and Roles in Clinical Medicine* (Horrobin D.F., ed.) pp. 351-377. Alan R Liss, New York.

Kim, J.H., Kim, S.H., Brown, S.L., & Freytag, S.O. (1994). Selective enhancement by an antiviral agent of the radiation-induced cell killing of human glioma cells transduced with HSV-tk gene. Cancer Res. 54, 6053-66.

Lanciotti, M., & Ponzoni, M. (1988). Rapido decremento del ciclo del fosfatidilinositolo durante la differenziazione di cellule de neuroblastoma indotta da acido retinoico. Boll. Ist Sieroter. Milan 67 (4), 309-315.

Lippman, S., Parkinson, D., Itri, L., Weber, R., Schantz, S., Ota, D., Schusterman, M., Krakoff, I., Gutterman, J., & Hong, W. (1992). 13-*cis*-retinoic-acid and interferon alpha 2a: effective combination therapy for advanced squamous cell carcinoma of the skin. J. Natl. Cancer Inst. 84, 235-241.

Lippmann, S.M., Batsakis, J.G., Toth, B.B., Weber, R.S., Lee, J.J., Martin, J.W., Hays, G.L., Goepfert, H., & Kong, W.K. (1993). Comparison of low-dose isotretinoin with beta carotene to prevent oral carcinogenesis. New Engl. J. Med. 328, 15-20.

Liu, L., Shack, S., Stetler-Stevenson, W.G., Hudgins, W.R., & Samid, D.(1994). Differentiation of cultured human melanoma cells induced by the aromatic fatty acids phenylacetate and phenylbutyrate. J. Invest. Dermatol. 103, 335-340.

Lo, C.F., Avvisati, G., Diverio, D., Biondi, A., Pandolfi, P.P., Alcalay, M., de Rossi, G., Petti, M.C., Cantu-Rajnoldi, A., Pasqualetti, D., Nanni, M., Fenu, S., Frontani, S., Frontani, M., & Mandelli, F. (1991). Rearrangements of the RAR-alpha gene in acute promyelocytic leukaemia: Correlations with morphology and immunophenotype. Brit. J. Haematol. 78, 494-499.

Lokeshwar, B.L., Selzer, M.G., & Block, N.L. (1992). Gunja-Smith Z: In vitro secretion of metalloproteinases and their inhibitors (TIMP) by normal, hypertrophic and malignant human prostate in tissue explant cultures, FASEB J. 6(4), A1359.

MacDonald, P.N., Bok, D., & Ong, D.E. (1990). Localization of cellular retinol-binding protein and retinol-binding protein in cells comprising the blood-brain barrier of rat and human. Proc. Natl. Acad. Sci. USA 87, 4265-4269.

Mahaley, M.S., Mettlin, C., Naturayah, N., Laws, E.R., & Peace, B.B. (1989). National survey of patterns of care for brain-tumor patients. J. Neurosurg. 71, 826-836.

Mahler, C., Verhelst, J., & Denis, L. (1993). Ketoconazole and liarozole in the treatment of advanced prostatic cancer. Cancer 71 (Suppl), 1068-1073.

Mangelsdorf, D.J., Ong, E.S., Dyck., J.A., & Evans, R.M. (1990). Nuclear receptor that identifies a novel retinoic acid response pathway. Nature 345, 224-229.

McElwee, N.E., Schumacher, M.C., Johnson, S.C., Weir, T.W., Greene, S.L., Scotvold, M.J., Hunter, J.R., Dinan, B.J., & Jick, H.(1991). An observational study of isotretinoin recipients treated for acne in a health maintenance organization. Arch. Dermatol. 127, 341-346.

McGarry, R.C., Bowen, T., Hayes, A., & Hammersley-Straw, D., (1989). Recognition of an *in vivo* response to human neuroblastoma modulation of antigen expression by retinoic acid. Medical Pediatr. Oncol. 17(6): 459-465.

Mehta, K., McQueen, T., Tucker, S., Pandita, R., & Aggarwal, B.B. (1994). Inhibition by all-trans-retinoic acid of tumor necrosis factor and nitric oxide production by peritoneal macrophages. J. Leukoc. Biol. 55, 336-342.

Melino, G., Farrace, M.G., Ceru, M.P., & Piacentini, M., (1988). Correlation between transglutaminase acitivity and polyamine levels in human neuroblastoma cells. Effects of retinoic acid and alpha-difluoromethylornithine. Exp. Cell. Res. 179(2), 429-45.

Meyskens, F.L. Jr., Goodman, G.E., & Alberts, D.S. (1985). 13-cis-retinoic acid pharmacology, toxicology, and clinical applications for the prevention and treatment of human cancer. CRC Critical Rev. Oncol. Hematol. 3, 75-101.

Meyskens, F.L., Gilmartin, G., Alberts, D.S., Levine, N.S., Brooks, R., Salmon, S.E., & Surwit, E.A. (1982). Activity of 13-*cis* retinoic acid against squamous epithelial premalignancies and malignancies. Cancer Treat. Rep. 66, 1315-1319.

Micksche, M., Cerni, C., Kokron, O., Titscher, R., & Wrba, H. (1977). Stimulation of immune response in lung cancer patients by vitamin A therapy. Oncology 34, 234-239.

Miller, V.A., Rigas, J.R., Muindi, J.R., Tong, W.P., Venkatraman, E., Kris, M.G., & Warrell, R.P., Jr. (1994). Modulation of all-*trans* retinoic acid pharmacokinetics by liarozole. Cancer Chemother. Pharmacol. 34, 522-526.

Mordan, L.J. (1989). Inhibition by retinoids of platelet growth factor-dependent stimulation of DNA synthesis and cell division in density-arrested C3H 10T1/2 fibroblasts. Cancer Res. 59 (4), 906-909.

Moriguchi, S., Jackson, J.C, & Watson, R.R. (1985). Effects of retinoids on human lymphocyte function. Human Toxicol. 4, 365-378.

Morimura, T., Takeuchi, J., Maeda, Y., & Tani, E. (1994). Preoperative embolization of meningiomas: its efficacy and histopathological findings. Noshuyo Byori 11, 123-129.

Muindi, J., Frankel, S.R., Miller, W.H., Jakubowski, A., Scheinberg, D.A., Young, C.W., Dmitrovsky, E., & Warrell, R.P. (1992). Continuous treatment with all-*trans* retinoic acid causes a progressive reduction in plasma drug concentrations: implications for relapse and retinoid "resistance" in patients with acute promyelocytic leukemia. Blood 79, 299-303.

Naganuma, H., Kimurat, R., Sasaki, A., Fukamachi, A., Nukui, H., & Tasaka, K. (1989). Complete remission of recurrent glioblastoma multiforme following local infusions of lymphokine activated killer cells. Case report. Acta Neurochir. (Wien) 99(3-4), 157-160.

Nakamura, K.D., & Hart, R.W. (1989). Proto-oncogne expression during retinoic acid-induced neural differentiation of embryonal carcinoma cells. Mech. Aging Dev. 48 (1), 53-62.

Nakamura, O., Maruo, K., Ueyama, Y., Nomura, K., & Takakura, K. (1986). Interactions of human fibroblast interferon with chemotherapeutic agents and radiation against human gliomas in nude mice. Neurol. Res. 8(3), 152-156.

Okuno, M., Chalala, T., Caraveo, V., Sasaki, A., Goodman, D.S., & Blaner, W.S. (1992). Regulation of cellular retinol-binding protein (CRBP) and retinol-binding protein (RBP) expression in adipocytes: Effects of retinoic acid (RA), dexamethasone (DEX) and triiodothyronine (T3). FASEB J. 6, A1782.

Orian, J.M., Vasilopoulos, K., Yoshida, S., Kaye, A.H., Chow, C.W., & Gonzales, M.F. (1992). Overexpression of multiple oncogenes related to histological grade of astrocytic glioma. Br J Cancer 66 (1), 106-112.

Ostertag, C.B., & Kreth, F.W. (1992). Iodine-125 interstitial irradiation for cerebral gliomas. Acta Neurochir. 119, 53-61.

Peiffer, J. (1983). Gehirn- und Rückenmarkstumoren. In: *Lehrbuch der Pathologie: Spezielle Pathologie für den zweiten Abschnitt der ärztlichen Prüfung.* Vol. 3. (Rotter, W., ed.) p. 72. Schattauer, Stuttgart.

Perez-Cruet, M.J., Trask, T.W., Chen, S.H., Goodman, J.C., Woo, S.L., Grossman, R.G., & Shine, H.D. (1994). Adenovirus-mediated gene therapy of experimental gliomas. J. Neurosci. Res. 39, 506-511.

Periquet, B., Lambert, W., Garcia, J., Lecomte, G., De Leenheer, A.P., Mazieres, B., Thouvenot, J.P., & Arlet, J. (1991). Increased concentrations of endogenous 13-*cis*- and all-*trans*-retinoic acids in diffuse idiopathic skeletal hyperostosis, as demonstrated by HPLC. Clin. Chim. Acta. 203, 57-65.

Pfeilschifter, J. (1990). Transforming growth factor-β. In: *Growth Factors, Differentiation Factors and Cytokines.* (Habenicht, A., ed.). pp. 56-64. Springer, Berlin Heidelberg.

Polednak, A.P., & Flannery, J.T. (1995). Brain, other central nervous system, and eye cancer. Cancer 75(1 suppl.), 330-337.

Pommerenke, E.W., Mattern, J., & Volm, M. (1992). Modulation of the multifactorial multidrug-resistance in tumors and normal tissue. J. Canc. Res. Clin. Oncol. 118, R51.

Rademaker, M., Wallace, M., Cunliffe, W., & Simpson, N.B. (1991). Isotretinoin treatment alters steroid metabolism in women with acne. Br. J. Dermatol. 124, 361-364.

Ram, Z., Culver, K.W., Walbridge, S., Frank, J.A., Blaese, R.M., & Oldfield, E.H. (1993). Toxicicty studies of retroviral-mediated gene transfer for the treatment of brain tumors. J. Neurosurg. 79, 400-407.

Reboul, P., Broquet, P., George, P., & Louisot, P. (1990). Effect of retinoic acid on two glycosyltransferase activities in C6 cultured glioma cells. Int. J. Biochem. 22, 889-893.

Reichel, H. (1990). Interactions of 1_,25-Dihydroxyvitamin D_3 and hematopoietic cells. In: *Growth factors, differentiation factors and cytokines.* (Habenicht, A., ed.) pp. 254-266. Springer, Berlin Heidelberg.

Riordan, J.R., Ling, V. (1985). Genetic and biochemical characterization of multidrug resistance. Pharmacol. Ther. 28, 51-75.

Roberts, A.B., Roche, N.S., & Sporn, M.B. (1985). Selective inhibition of the anchorage-independent growth of myc-transfected fibroblasts by retinoic acid. Nature 315(6016), 237-239.

Ron, E., Modan, B., & Boice, J.D. Jr. Alfandary, E., Stovall, M., Chetrit, A., & Katz, L. (1988). Tumors of the brain and nervous system after radiotherapy in childhood. N. Eng. J. Med. 319, 1033-1039.

Rosales, F., Kjolhede, C., & Chipaila, P. (1992). Lack of skin delayed hypersensitivity response (SKR) enhancement by vitamin A supplementation among children in Ndola, Zambia. FASEB J 6, A1661.

Ross, A.C. (1993). Cellular metabolism and activation of retinoids: roles of cellular retinoid-binding proteins. FASEB J. 7, 317-327.

Runde, V., & Aul, C. (1992). Treatment of acute promyelocytic leukemia. Onkologie 15, 356-363.

Rutka, J.T., De Armond, S.J., Giblin, J., McCulloch, J.R., Wilson, C.B.,& Rosenblum, M.L. (1988). Effect of retinoids on the proliferation, morphology and expression of glial fibrillary acidic protein of an anaplastic astrocytoma cell line. Int. J. Cancer 42, 419-427.

Rutz, H.P., & Little, J.B. (1989). Modification of radiosensitivity and recovery from X ray damage in vitro by retinoic acid. Int. J. Radiat. Oncol. Biol. Phys. 16, 1285-88.

Salcman, M., Scholtz, H., Kaplan, R.S., & Kulik, S. (1994). Long-term survival in patients with malignant astrocytoma. Neurosurgery 34, 213-219.

Schalinske, K.L., & Steele, R.D. (1991). 13-*cis*-retinoic acid alters methionine metabolism in rats. J. Nutr.121, 1714-1719.

Schluesener, H.J., Krone, A., & Meyermann, R. (1991). Entwicklung der Vielfach-Resistenz (multidrug resistance) in neuroglialen Tumoren, Verh. Dt Ges Neurol. 6, 432-433.

Schule, R., Umesono, K., Mangelsdorf, D.J., Bolado, J., Pike, J.W., & Evans, R.M. (1990). *Jun-Fos* and receptors for vitamins A and D recognize a common response element in the human osteocalcin gene, Cell 61(3), 497-504.

Serrou, B., & Cupissol, D. (1980). A retinoic acid derivative can restore depressed immune function of patients with advanced solid tumors. Allergol. Clin. Immunol. 8, 373.

Shimizu, K., Miyao, Y., Tamura, M., Yamada, M., Tamura, K., Mikoshiba, K., Hayakawa, T., & Ikenaka, K. (1995). Selective gene therapy of malignant gliomas using brain-specific promoters—ist efficacy and basic investigations. Biotherapy 9, 248-254.

Siegenthaler, G., Saurat, J.H. Retinoid binding proteins and human skin. Pharmacol. Ther. 40 (1), 45-54.

Steck, P.A., Hadi, A., Lotan, R., & Yung, W.K. (1990). Inhibition of epidermal growth factor receptor activity by retinoic acid in glioma cells. J. Cell. Biochem. 42, 83-94.

Strommer, K., Hamou, M.F., Diggelmann, H., & de Tribolet, N. (1990). Cellular and tumoral heterogeneity of EGFR gene amplification in human malignant gliomas. Acta Neurochir. Wien. 107(3-4), 82-87.

Summerbell, D., & Maden, M. (1990). Retinoic acid, a developmental signalling molecule. Trends in Neurosci. 13 (4), 142-147.

Szuwart, U., Bennefeld, H., & Behr, H. (1989). Classification of gliomas and medulloblastomas using the new TNM system for cerebral tumors. Neurosurg. Rev. 12, 233-238.

Thiele, C.J., Reynolds, C.P., & Israel, M.A. (1985). Decreased expression of N-myc precedes retinoic acid-induced morphological differentiation of human neuroblastoma. Nature (London) 313, 404-406.

Thomas, D.G.T., Lannigan, C.B., & Behan, P.O. (1975). Impaired cell-mediated immunity in human brain tumors. Lancet i, 1389-1390.

Trojan, J., Johnson, T.R., Rudin, S.D., Ilan, J., & Tykocinski, I. (1993). Treatment and prevention of rat glioblastoma by immunogenic C6 cells expressing antisense insulin like growth factor 1 RNA. Science 259, 94-97.

Tymianski, M., Willinsky, R.A., Tator, C.H., Mikulis, D., TerBrugge, K.G., & Markson, L. (1994). Embolization with temporary balloon occlusion of the internal carotid artery and *in vivo* proton spectroscopy improves radical removal of petrous-tentorial meningioma. Neurosurgery 35, 74-77.

Van der Meulen, D.L., Prasad, V.V., & Moskal, J.R. (1994). The identification of glioblastoma-associated, fucose-containing glycoproteins induced by retinoic acid. Mol Chem. Neuropathol. 21, 311-327.

Van Meir, E.G., Roemer, K., Diserens, A.C., Kikuchi, T., Rempel, S.A., Haas, M., Huang, H.J., Friedmann, T., de Tribolet, N., & Cavenee, W.K. (1995). Single cell monitoring of growth arrest and morphological changes induced by transfer of wild-type p53 alleles to glioblastoma cells. Proc. Natl. Acad. Sci. USA 92, 1008-1012.

Van Wauwe, J., Coene, M.C., Cools, W., Goossens, J., Lauwers, W., Le Jeune, L., Van Hove, C., & Van Nygen, G. (1994). Liarozole fumarate inhibits the metabolism of 4-keto-all-trans-retinoic acid Biochem. Pharmacol. 47, 737-741.

Vecht, Ch.J., Avezaat, C.J.J., van Putten, W.L.J., Eijkenboom, W.M.H., & Stefanko, S.Z. (1990). The influence of the extent of surgery on the neurological function and survival in malignant glioma. A retrospective analysis in 243 patients. J. Neurol. Neurosurg. Psychiatry 53, 466-471.

Venter, D.J., & Thomas, D.G. (1991). Multiple sequential molecular abnormalities in the evolution of human gliomas. Br. J. Cancer 63 (5), 753-757.

Wahl, M., & Carpenter, G. (1988). Regulation of epidermal growth factor-stimulated formation of inositol phosphates in A-431 cells by calcium and protein kinase C. J. Biol. Chem. 263, 7581-7590.

Walter, W., & Müller, W. (1959). Histologische Veränderungen an röntgenbestrahlten Gliomen. Zentralbl. Neurochir. 19, 79-90.

Wei, L.N., Blaner W.S., Goodman, D.S., & Nguyen-Huu, M.C. (1989). Regulation of the cellular retinoid-binding proteins and their messenger ribonucleic acids during P19 embryonal carcinoma cell differentiation induced by retinoic acid. Mol. Endocrinol. 3(3), 454-463.

Westarp, M.E., Westarp, M.P., Bollag, W., Bruynseels, J., Biesalski, H., Grossmann, N., & Kornhuber, H.H. (1994). Effect of six retinoids and retinoic acid catabolic inhibitor liarozole on two glioblastoma cell lines, and *in-vivo* experience in malignant brain tumor patients. In: *Cancer Treatment—An Update* (Banzet, P., Holland, J.F., Khayat, D., Weil, M., eds.), pp. 590-598. Springer, Paris.

Westarp, M.E., Westarp, M.P., Bollag, W., & Kornhuber, H.H. (1992). Hohe Retinoid-Serumdauerspiegel bei primären ZNS-Tumoren durch orale Kombination von 13cRA und Inhibitor R75251 (Permanently high serum retinoids in primary CNS tumors after oral 13cRA plus inhibitor R75251). Canc. Res. Clin. Oncol. 118 (Suppl), R122.

Westarp, M.E., Westarp, M.P., Bruynseels, J., Bollag, W., & Kornhuber, H.H. (1993a). Oral liarozole as a catabolic inhibitor potently increases retinoic acid *in vivo*: First experience from an ongoing therapeutic trial in highly malignant primary brain tumors. Onkologie 16, 22-25.

Westarp, M.E., Westarp, M.P., Grundl, W., Biesalski, H., & Kornhuber, H.H. (1993b). Improving medical approaches to primary CNS malignancies—retinoid therapy and more. Medical Hypotheses 41, 267-276 .

Westarp, M.E., Westarp, M.P., Welcker, B., Grossmann, N., Grundl, W., & Kornhuber, H.H. (1991). Retinoic acid and its catabolic inhibitor, R75251, in human malignant glioma cells. Europ. J. Cancer 27 (Suppl.3), 58.

Wolf, G. (1984). Multiple functions of vitamin A. Physiol. Rev. 64, 873-937.

Wouters, W., van Dun, J., Dillen, A., Coene, M.C., Cools, W., & DeCoster, R. (1992). Effects of liarozole, a new antitumoral compound, on retinoic acid-induced inhibition of cell growth and on retinoic acid metabolism in MCF-7 human breast cancer cells. Cancer Res. 52, 2841-2846.

Wu, J.K., Cano, W.G., Meylaerts, S.A., Qi, P., Vrionis, F., & Cherington, V. (1994). Bystander tumoricidal effect in the treatment of experimental brain tumors. Neurosurgery 35, 1094-1102.

Yu, V.C., Hochhaus, G., Chang, F.H., Richards, M.L., Bourne, H.R., & Sadee, W. (1988). Differentiation of human neuroblastoma cells: marked potentiation of prostaglandin E-stimulated accumulation of cyclic AMP by retinoic acid. J. Neurochem. 51(6), 1892-1899.

Yung, W.K.A. (1989). In vitro chemosensitivity testing and its clinical application in human gliomas. Neurosurg. Rev. 12, 197-203.

Yung, W.K.A., Lotan, R., Lee, P., Lotan, D., & Steck, P.A. (1989). Modulation of growth and epidermal growth factor receptor activity by retinoic acid in human glioma cells. Cancer Res. 49(4), 1014-1019.

Yung, W.K.A., Simaga, M., & Levin, V.A. (1993). 13-*cis*-retinoic acid: a new and potentially effective agent for recurrent malignant astrocytomas. 4th Internat. Congr. on Anti-Cancer Chemother. Paris, Abstract 238.

Zülch, K.J. (1960). Über die Strahlensensibilität der Hirngeschwülste und die sogenannte Strahlen-Spätnekrose des Hirns, Dtsch Med. Wochenschr. 85, 293-298, 309-310.

RETINOIDS AND LUNG CANCER

Andrew M. Arnold and Richard G. Tozer

Abstract 262
I. Introduction 262
II. Epidemiology 262
A. Limitations of Epidemiological Studies 262
B. Studies of Dietary Intake 263
C. Studies Using Serum Levels 265
D. Summary of Epidermiological Data 266
II. Laboratory Studies 266
A. Studies of Retinoid Activity *In Vivo* 266
B. Studies *In Vitro* 267
C. The Molecular Biology of Retinoids in Lung Cancer 269
D. Summary of Laboratory Studies 272
IV. Clinical Studies 272
A. Prevention of Lung Cancer with Retinoids 273
B. Retinoids as Adjuvant Therapy 277
C. Retinoids in the Treatment of Advanced Disease 278
V. Conclusion 281
References 281

Advances in Organ Biology
Volume 3, pages 261-287.

ISBN: 0-7623-0285-2

ABSTRACT

Data from epidemiological and preclinical studies support the exploration of retinoids as preventative or therapeutic agents to combat lung cancer. To date though, retinoids or carotenoids have failed to reverse bronchial premalignant lesions or to reduce the incidence of lung cancer. Two large trials have actually demonstrated that beta carotene administration produces an increase in lung cancer incidence. Studies in patients with established disease have also been negative. These results have left researchers looking for new insights into retinoid action. Recent discoveries have provided new information concerning the molecular mechanisms by which retinoids are involved in the development of lung cancer and the role they play in maintaining a healthy bronchial epithelium. The search for more active retinoids continues; however it is unlikely that retinoids will have a significant impact effect when used as single agents. It is expected that more scientifically based trials of retinoids in combination will eventually produce advances.

I. INTRODUCTION

Despite a reduction in the prevalence of smoking in most western countries, lung cancer remains a major World health problem (Resnicow et al., 1991; Brown and Kessler, 1988). Unfortunately, secondary preventive efforts, using sputum cytology and chest X-rays, in high risk populations have proven ineffective. For this reason, better screening techniques and therapeutic strategies for lung cancer control require continual evaluation (Wagner et al., 1995).

The majority of lung cancers are of squamous origin, therefore, considerable attention has focused on a possible therapeutic role for retinoids used alone or in combination with other active agents or treatment modalities (Lippman et al., 1987a; 1987b). Theoretically, retinoids have the potential to (1) prevent lung cancer, (2) act as adjuvants for the management of early stage disease, and (3) have an impact in the management of patients with more advanced disease.

Thus, throughout the 1980s, there was considerable hope that the epidemiological and preclinical studies that suggested beneficial effects of retinoids would translate into therapeutic advances in the clinical setting. While there has been some progress, a number of recently completed studies have raised many new questions for researchers interested in the use of retinoids to combat the lung cancer epidemic.

This chapter systematically examines many aspects of the relationship between retinoids and lung cancer, reviews interesting new laboratory findings, and critically appraises some of the methodological pitfalls in the field of retinoid research.

II. EPIDEMIOLOGY

A. Limitations of Epidemiological Studies

Two authors have comprehensively reviewed the epidemiological data concerning the association between retinoids and carotenoids and lung cancer (Fontham,

1990; Willett, 1990). Many of the methodological issues are discussed in depth in these analyses. When reviewing epidemiological data, concerning a causal effect or association, it must be remembered that a prospective cohort study carries considerably more weight than does a retrospective case-control study. Unfortunately cohort studies are more expensive and thus are less frequently performed. In general, the epidemiological studies have taken one of two approaches: (1) the study of dietary intake and (2) the study of serum levels of carotenoids and retinol. Many of the studies suffer from inaccuracies intrinsic to such surveys.

Studies of dietary intake range from those categorizing intake using very general food groups (e.g., by fruit, vegetable, or other food type) to those attempting to specifically quantify carotenoid and retinoid intake. The more recent studies have used more sophisticated questionnaires such as the food frequency method. This seeks to measure intake over a prolonged period rather than relying on quantifying current intake or on recalled dietary data (which could lead to bias). Studies of serum levels are also subject to methodological flaws as well as those related to sampling and limitations of laboratory techniques.

While some of the studies specifically set out to examine lung cancer, others extracted the data concerning lung cancer from broader surveys examining all types of cancer. Some studies failed to provide evidence for the adequacy of controls and failed to distinguish between the true effects of retinol versus beta carotene. Finally, many studies have not accounted for the effects of smoking or gender and have not provided adequate documentation of histology or of specific lung cancer type.

B. Studies of Dietary Intake

A large number of studies have examined the relationship between dietary intake of fruits, vegetables, carotenoids, and retinol and the incidence of lung cancer. Generally, results have been expressed in terms of relative risk when comparing subjects with a high versus low intake. The results of these studies are summarized in table 1.

Table 1. Summary of Epidemiological Studies Examining the Association of Certain Food Types and Lung Cancer*

	Case Control Studies			*Cohort Studies*		
Intake	*+ ve*	*- ve*	*Relative Risk (High vs. Low)*	*+ ve*	*- ve*	*Relative Risk (High vs. Low)*
Vegetables and Fruits	9	1	0.30 - 0.80	3	0	0.55 - 0.71
Dietary Carotenoid Intake	16	0	0.40 - 0.80	1	0	0.14
Retinol	4	5	0.38 - 1.25	0	1	2.0

Source: * Adapted from Fontham (1990).

Case Control Studies of Dietary Intake

Ten case control studies have examined the relationship between dietary intake of fruits and vegetables and lung cancer risk (Maclennan et al., 1977; Metlin et al., 1979; Pisani et al.,1986; Ziegler et al., 1986; Koo, 1988; Fontham et al.,1988; Le Marchand et al., 1989; 1993; Jain et al., 1990; Mayne et al., 1994; Candelora et al., 1992). As the reported studies come from different areas of the world, with widely varying mixes of fruit and vegetable in the local diets, it is not surprising that a range of results are reported. Despite this, virtually all the studies showed an inverse relationship between the relative risk of lung cancer and a higher intake including recent studies conducted entirely in nonsmokers (Mayne et al., 1994; Dorgan et al., 1993). Relative risks, reported in these studies varied from 0.3-0.8. In general the relationship between a higher intake and protection from lung cancer was stronger for males as well as for squamous cell type. In a subsequent further analysis of one of these studies, Dorgan and colleagues (Dorgan et al., 1993) noted that the protection from consumption of yellow/green vegetables appeared to be predominantly in white rather than black smokers.

Vegetables and fruit often contain high quantities of other potentially protective substances, such as vitamin C and selenium. Thus such results must be interpreted with caution and cannot stand alone as evidence supporting a retinoid or carotenoid effect.

Sixteen case control studies have examined for the association of dietary carotenoid intake and lung cancer risk (Ziegler et al., 1986; Koo, 1988; Fontham et al.,1988; Le Marchand et al., 1989; 1993; Jain et al., 1990; Mayne et al., 1994; Kolonel et al., 1983; Hinds et al., 1984; Ziegler et al., 1984; Samet et al.,1985; Wu et al., 1985; Middleton et al., 1986; Bond et al., 1987; Byers et al., 1987; Pastorino et al., 1987; Dartiques et al., 1990). In all studies there was an inverse association between intake and lung cancer risk (relative risk reduction, ranging from 0.4 to 0.8 when comparing high intake with low). By contrast, in the nine studies specifically examining retinol intake and lung cancer risk, only four showed a significant trend and, in two of these studies, the association noted was only for adenocarcinoma (Ziegler et al., 1986; 1984; Koo, 1988; Fontham et al., 1988; Le Marchand et al., 1989; Samet et al.,1985; Byers et al., 1987; Pastorino et al., 1987; Legardeur et al., 1990).

Cohort Studies of Dietary Intake

Three prospective cohort studies support the data from the case control studies and suggest a protective effect from green and yellow vegetables (Hirayama, 1979; 1984; Shibata et al., 1992). One study is noteworthy in that it was confined only to female nonsmokers (Pastorino et al., 1987). Just one cohort study has looked specifically at dietary carotenoids and this was strongly positive with a relative risk reduction 0.14 (Shekelle et al.,1981). The same study showed no risk reduction when examining dietary retinol intake.

C. Studies Using Serum Levels

As blood samples are often stored for nonspecific purposes at the outset of many longitudinal health surveys, there are a reasonable number of prospective studies examining the relationship between serum levels of retinoids and carotenoids and lung cancer rates. These are summarized in Table 2.

Case Control Studies using serum levels

The four case control (retrospective) studies reported have demonstrated significantly lower mean levels of carotenoids (Hirayama, 1984; Kune and Watson, 1989; Legardeur et al., 1990; Smith and Walker, 1991) in patients who developed lung cancer. Two studies (Hirayama, 1984; Kune and Watson, 1989) showed significantly lower levels of retinol. However, because of the diverse effects that a new diagnosis of cancer has on metabolism and dietary intake these results can be largely dismissed.

Cohort Studies using serum levels

Eleven prospective studies have looked at either mean serum levels of carotenoid and retinoid or both. The mean levels reported vary considerably from study to study and are probably attributable to analytical technique. The studies reported variable times between blood collection and cancer diagnosis. In some studies, with a short interval between collection and diagnosis, occult cancer could already be present at the time of sampling. With respect to serum carotenoid, five of seven studies (Stahelin et al., 1984; 1991; Willett et al., 1984; Nomura, 1985; Menkes et al., 1986; Wald et al.,1988; Connett et al., 1989) have shown lower levels in lung cancer cases compared to controls. These findings are thus consistent with the studies of dietary carotenoid intake. In some studies the effect appears greater in smokers than nonsmokers and one study reported a greater protective effect against squamous and small cell lung cancers than adenocarcinomas (Hirayama. 1979).

Of ten prospective studies examining retinoid levels (Stahelin et al., 1984; 1991; Willett et al., 1984; Nomura, 1985; Menkes et al., 1986; Wald et al.,1988; Peleg et al., 1984; Friedmann and Blaner, 1986; Wald et al., 1980; Kark et al. 1981; Salonen

Table 2. Summary of the Epidemiolgical Studies Examining the Association Between Serum Levels of Carotenoids and Retinoids and Lung Cancer Rates*

	Case Control Studies		*Cohort Studies*	
Serum	+ve	-ve	+ve	-ve
Carotenoids	4	0	5	2
Retinoids	2	0	3	7

Source: * Adapted from Fontham (1990).

et al., 1985), only three were positive and each of these had considerable methodological problems, including a short interval from initial sampling to diagnosis and small numbers of cases of lung cancer.

D. Summary of Epidemiological Data

Despite the deficiencies of many epidemiological studies, the findings fairly consistently suggest a protective effect of carotenoids but are much less consistent concerning the role of retinol. The possibility remains that carotenoid or retinoid intake is associated with some unrelated but truly protective compound(s) as yet undiscovered. Conversely, classification errors that arise from questionnaire-based dietary studies may minimize the true protective role of retinoids and carotenoids and an important effect may be underestimated. There is a suggestion that carotenoids may provide more effective protection against squamous and small cell lung cancers and may provide the greatest benefit in subjects with longer smoking histories. Finally, as carotenoids and retinol may prevent cancer at different points in the process of carcinogenesis, it is important to continue to consider and investigate their roles separately.

III. LABORATORY STUDIES

The bronchial epithelium is normally lined by pseudostratified, ciliated, columnar cells. The observation that deficiency or excess of retinoid can profoundly affect such epithelium has been known for 70 years. The early studies of Wolbach and Howe showed that deficiency of vitamin A causes disappearance of normal mucociliary epithelium with replacement by keratinizing cells (keratinizing squamous metaplasia). (Wolbach and Howe, 1925). The powerful effects of retinoids were again demonstrated by Fell and Mellanby when they added retinyl acetate to embryonic chick ectoderm and observed differentiation from a keratinized epithelium to that of a columnar, ciliated type similar to that found in the respiratory tract (Fell and Mellanby, 1953).

Since these early studies a large amount of data has emerged from the laboratory concerning retinoids and lung cancer. Recently molecular biological studies have added to the more traditional studies conducted on cell lines and in animal models.

A. Studies of Retinoid Activity *In Vitro*

Many studies of retinoid activity have been carried out using organ culture techniques. The initial techniques were crude, but a number of systems indicated the ability of retinoids to reverse changes induced by carcinogens in tissue originating from the trachea or lungs (Mossman et al., 1980; Laznitski and Bollag, 1982). To provide a technique that gives more consistent results, the hamster tracheal organ culture has been used as an initial screen for retinoids which might have potential as agents against lung cancer. The technique is sensitive and relies on measuring retinoid-induced differentiation in a serum-free medium. This allows comparison of the effects of retinoid analogues (Table 3) and hence an examination of their structure-activity relationships (Sporn and Roberts, 1984). In this system, the retinoids all-*trans* retinoic acid and 13-*cis* retinoic acid have persistently emerged as the most active retinoids. Retinol, retinyl acetate, and etretinate are less active. Synthetic retinoids in the arotinoid class have not yet been tested as extensively. However, Eliason and coworkers (Eliason et al., 1993) studied the effects Ro 40-8757 (mofarotene), which has a morpholine structure, on two lung adenocarcinoma lines (a549 and HLC-1) and one large cell lung cancer line (LXFL 529). Only the large cell lung cancer line was more effectively inhibited by Ro 40-8757 than all-*trans* retinoic acid or 13-*cis* retinoic acid.

Studies in cell lines also allow exploration of retinoid pharmacology. Addition of 13-*cis* retinoic acid to small cell lung cancer lines, growing in a serum-free, hormonally defined, medium increased expression of retinoic acid receptor beta (RARB) mRNA and caused significant, dose-dependent growth inhibition as demonstrated by an *in vitro* proliferation assay. This effect was attenuated by the addition of serum and the responsible component was shown to be serum albumin. Preincubation of the serum with triglycerides blocked this effect (Davis et al., 1995).

Reviewing the *in vitro* comparisons between retinoids, it is tempting to speculate that interventions prescribing dietary (physiological) levels of retinoids alone are unlikely to be effective against lung cancer. To date, the most likely candidate retinoids for clinical use in lung cancer intervention trials are all-*trans* retinoic acid and 13-*cis* retinoic acid.

B. Studies *In Vivo*

Most animal experiments have been carried out using carcinogen-induced respiratory changes in rodents. The changes include the development of metaplasia

Table 3. Comparison of the *In Vitro* Activity of Some Common Retinoids as Measured by Suppression of Keratinization in Tracheal Organ Culture.*

Retinoid	*Activity*	ED_{50} *M*
all-*trans* retinoic acid	+++	3×10^{-11}
13-*cis* retinoic acid	+++	3×10^{-11}
Retinol	++	7×10^{-10}
Retinaldehyde	++	3×10^{-10}
Retinyl Acetate	++	1×10^{-9}
Etretinate	++	2×10^{-8}
4-HPR	++	3×10^{-10}

Source: *Adapted from Sporn and Roberts (1984).

and frank tumors. Studies conducted *in vivo* have generally examined one of two aspects: (1) the effects of dietary retinoid deficiency on tumor growth and (2) the effects of increased intake of either dietary or pharmacological doses on tumor growth. In addition, two distinct endpoints have been used: (1) prevention or delay of tumor formation or (2) tumor regression.

The studies reported have been contradictory. Some have suggested a protective role for retinoids, while others have demonstrated a negative effect. This has been interpreted as reflecting interspecies variation as well as differences in the exposure to carcinogen and dose or class of retinoid. Thus some studies have used very high doses of carcinogen unlikely to be overcome by any dose of retinoid. In addition, earlier studies maintained animals under suboptimal conditions and insufficient attention was paid to dietary levels of retinoids.

In 1967, Saffiotti and colleagues (1967) reported that vitamin A palmitate markedly reduced squamous metaplasia and squamous cell cancers in an experimental lung cancer model. (Saffiotti et al., 1967) Syrian golden hamsters received intratracheal instillations of the carcinogen benzo[a]pyrene (BP). After one week, one group received intragastric vitamin A palmitate. This retinoid is now regarded as only moderately active, and it is likely that more active retinoids might achieve even better results.

Using the same model, Smith and colleagues (smith et al., 1975) demonstrated that higher doses of retinyl acetate were actually detrimental. When these studies were repeated under slightly different circumstances, the detrimental effects of the higher dose of retinyl acetate disappeared.

Port and colleagues used low doses of intratracheal benzo[a]pyrene:ferric oxide to induce lung cancer in hamsters at a rate that approximated that of lung cancer in heavy smokers. During their lifetime, 10 percent of the control animals developed tumors, compared to 1.5 percent of animals receiving 1.5 mg of 13-*cis* retinoic acid and 1 percent of animals receiving 4.5 mg of 13-*cis* retinoic acid (Port et al., 1975).

Cone and Nettesheim (1973) used intratracheal instillation of 3-methylcholanthrene (MCA) to induce metaplastic nodules and lung tumors in rats. The animals were given intragastric retinyl acetate in either low or high concentration prior to the carcinogen. There was an unequivocal reduction in the mean number of tumors and tumor volumes in the rats treated at high dosage. (Cone and Nettesheim, 1973) In a subsequent study, Nettesheim and colleagues found that retinyl acetate retained its protective effect even when given late after the administration of carcinogen (Nettesheim et al., 1976). In a third experiment, Nettesheim and Williams demonstrated that the addition of 13-*cis* retinoic acid or all-*trans* retinoic acid provided no further protective benefit when compared to moderated dietary levels of retinyl acetate (Nettesheim and Williams, 1976).

Stinson and colleagues used the carcinogen N-Methyl-N-nitrosourea to induce tracheal tumors in Syrian golden hamsters. They tested the effect of 13-*cis* retinoic acid, ethyl retinamide, and N-(2-hydroxyethyl) retinamide given for up to six months. In each case there was an increase in tumors in treated animals versus controls. (Stinson et al., 1981) Yun and coworkers (Yun et al., 1995) induced pulmonary adenomas within nine weeks in the inbred mouse line A/J and the non inbred line N:GP(S) with benzo[a]pyrene. Retinoic acid did not reduce the incidence of malignancy in this situation.

A few studies have utilized transplanted cell lines in mice. Hubert and colleagues (Hubert et al., 1983) have studied the effect of 13-*cis* retinoic acid or all-*trans* retinoic acid on inhibition of squamous cell and large cell tumors transplanted in BALB/c mice. Both retinoids significantly inhibited tumor growth. The increased lifespan, seen in mice with the squamous cell tumour, was similar to or greater than that obtained using a number of cytotoxic agents, although these studies were not done concurrently.

C. The Molecular Biology of Retinoids in Lung Cancer

The effects of retinoids in lung cancer appear to be mediated through two distinct classes of nuclear receptors: the RARs and the RXRs (Rusch and Dmitrovsky , 1995). Each class is encoded by three distinct genes (A, B, and C). These receptors belong to the steroid/thyroid superfamily of nuclear receptors and are capable of forming stable homo- and heterodimers. The dimerized receptors then function as ligand-activated nuclear transcription factors. Target genes are regulated by the binding of the receptor to retinoic acid-response elements (RARE's). Normal human tracheobronchial epithelial cells will show induction of RAR-beta (RARB) mRNA and growth inhibition following treatment with retinoic acid.

Several themes have arisen regarding the relationships of the RAR's and RXR's and lung cancer. A deletion of the short arm of chromosome three, a common finding in lung cancer, maps closely to the RARB gene and has initiated several studies examining the role of this receptor in the development of lung cancer. The RARB gene contains an RARE in its promoter and is the best studied retinoic acid-responsive gene.

Houle and colleagues (Houle et al., 1993) compared the ability of cell lines derived from lung tumors (cell lines CALU-1 and H157) to establish colonies in nude mice following transfection with RARB. Cells were 50 percent less tumorigenic than untransfected or mock-transfected cells. Tumors that did develop from RARB transfected cells had a reduced growth rate and increased latency. Tumors from these cells demonstrated low levels of RARB mRNA. It was concluded that RARB functions as a tumor suppressor gene in epidermoid lung tumorigenesis. Expression of RARB is markedly increased with the application of retinoic acid to epithelial tissues. It was noted that constitutive RARB gene expression in both human lung cancer cell lines and human primary lung cancers is abnormal.

Zhang and coworkers (Zhang et al., 1994) noted that RARB expression cannot be stimulated by retinoic acid in several human lung cancer cell lines. It appeared that in some cell lines, the RARB response element became unresponsive in its natural promoter context but became active when linked to the heterologous tk promoter. Similar observations were made by Geradts and colleagues.(Geradts et al., 1993) who studied 22 small cell and 15 non-small cell lung cancer cell lines. The cell lines varied greatly in their constitutive expression of RARB mRNA and degree of growth inhibition after treatment with retinoic acid. Furthermore, several of the cell lines failed to show RARB induction following retinoic acid treatment. They were able to demonstrate possible alterations within the coding region of the RARB gene using cDNA-PCR, but these were not confirmed through single-strand conformation polymorphism analysis studying the entire open reading frame of RARB. No alterations were found in the structure or expression of the RARA gene through Northern and Southern blot analysis.

Kim and colleagues (Kim et al., 1995) studied a lung carcinogenesis model consisting of normal, premalignant, and malignant human bronchial cells. Treatment with all-*trans* retinoic acid inhibited the growth of normal cells but was less effective with premalignant and malignant cells. This retinoid refractoriness correlated with increased expression of RARB. Receptor function appeared to be normal in these cells as assayed by DNA-binding and transcriptional transactivation properties. They subsequently studied RARB function in biopsies of primary human non-small cell lung cancers, and also observed normal RARB function. They concluded that defects in retinoid signaling may not be intrinsic to the retinoid receptors but may represent effects from other factors.

Preliminary results from Ross and Canada (1994) demonstrated that treatment of the human lung adenocarcinoma cell line Lu-CSF-1 with retinoic acid suppressed proliferation by 75 percent. RARB mRNA and IL-1 beta mRNA were both dramatically upregulated. Exogenous IL-1 beta could also inhibit cellular proliferation, suggesting that the antiproliferative effect of retinoic acid may be mediated through induction of IL-1 beta. Jakowlew and colleagues studied the effects of retinoic acid on NCI-H727 lung cancer cells. (Jackowlew et al., 1994). Preliminary results showed that the retinoid increased TGF-beta 1 and TGF-beta 3 mRNAs in exponentially growing cells and also increased expression of RARB and RXRA mRNAs. There was no effect on expression of TGF-beta receptor mRNA and colony formation of these cells was inhibited. Of the RARs and RXRs, RARB appears to be a consistent theme in the modulation of the malignant phenotype by retinoids. Defects may appear at several levels, from the promoter region of the RARB gene itself to downstream components in its signaling pathway.

Abnormalities in factors regulated by the retinoid receptors have also been studied. Anderson and colleagues (Anderson et al., 1995) demonstrated that Stromolysin-3, a matrix metalloproteinase that cleaves various serine protease inhibitors, such as the alpha 1 proteinase inhibitors, is upregulated in a panel of primary non-small cell lung cancer cell lines. The induction of Stromolysin-3 was inhibited by all-*trans* retinoic acid. Retinoic acid treatment of a highly metastatic lung carcinoma cell line (C87) markedly reduced the expression of the beta 4 integrin subunit as demonstrated by immunofluorescence, immunoprecipitation, and Northern blot analysis. This was correlated with inhibition of the clonogenic potential of the cell line, in soft agar, as well as chemotactic and chemoinvasive capacity. There was also a reduction in the ability of cells injected into mice to form colonies in the lungs, but the growth potential of the cell line was only moderately reduced. The investigators concluded that reduction of the malignant phenotype in this cell line by retinoic acid was due to reduction in the expression of the beta 4 integrin subunit (Gaetano et al, 1994).

Chen and coworkers (Chen et al., 1994) demonstrated that all-*trans* retinoic acid decreased protein kinase C activity in the metastatic Lewis lung carcinoma cell line LL/2. There was decreased attachment to reconstituted basement membrane and endothelial cells, decreased invasion through matrigel, and decreased number of metastasis following injection into mice. Consequently there is evidence that retinoids can modulate a cell's invasive potential through regulation of proteases, kinases, and components of the extracellular matrix.

Zhang and colleagues (Zheng et al., 1995) studied induction of tissue transglutaminase (TGase II) in the rat tracheobronchial epithelial cell line SPOC-1. Increased levels of TGase II were associated with induction of apoptosis as demonstrated by studies of DNA fragmentation and histology. This cell line was shown to express RARA, RARB, RARC, and RXRB. The use of receptor specific agonist/antagonist retinoids demonstrated that TGase II was induced by RARA. The agonist retinoids to RARA also induced apoptosis. Use of a dominant negative mutant of RARA confirmed these results suggesting that apoptosis in SPOC-1 cells is regulated by an RARA-dependent signaling pathway. Vollberg and colleagues (Volllberg et al., 1992) also showed high levels of TGase II in several lung cancer cell lines and demonstrated that retinoic acid was effective in inducing TGase II in a time- and dose-dependent manner.

Other studies have also demonstrated that retinoids can alter components of signaling pathways that affect cell cycling and apoptosis. Maxwell and Mukhopadhyay (1994) showed that 5 micromolar retinoic acid increased levels of p53 after 16 to 48 hours of treatment of the non-small cell lung cancer cell lines H460a and H226b. Levels of p53 mRNA did not increase, suggesting that the retinoid affected post-translational modifications of p53 decreasing protein turnover. Pulse-chase experiments confirmed this result. Cell lines that expressed a mutant p53 did not show these changes.

Maxwell (1994) studied expression of retinoblastoma protein (RB) in non-small cell lung carcinoma cell lines treated with retinoic acid. With one exception, growth was uniformly inhibited after 16 hours of treatment and the cells became refractory to the retinoid after 48 hours. After 16 hours the hyperphosphorylated form of RB (p110) increased five to ten-fold. In the H460a and H226b cells, p110 was converted to the underphosphorylated p105 form after 24 hours of treatment. The p105 form, and a 60 kDa species, became the predominant form after 48 hours of treatment. After 72 hours, retinoblastoma protein became undetectable. In H322j and H358 cells, both the p110 and p105 forms were induced by eight hours of treatment and levels remained unchanged for up to 72 hours. Maxwell concluded that retinoic acid inhibited cell growth by maintaining high levels of RB through post-translational mechanisms that either stabilized the molecule or down-regulated an RB-specific protease. He also concluded that other molecular events must be involved as the RB-deficient cell line H596b has similar growth properties as other cell lines following treatment with retinoic acid.

Insertion of v-Ha-ras into the c-myc overexpressing small cell lung cancer line NCI-H82 leads to morphologic changes resulting in cells that resemble non-small cell lung cancer (Kalemkerian et al., 1994). Treatment of the cells with all-*trans* retinoic acid prior to insertion of v-Ha-ras blocked the phenotypic changes and decreased c-myc mRNA and increased L-myc mRNA.

Finally, retinoids may also have effects independent of their receptors. Anees and Steven (1994) showed reversible inhibition of the cell surface protease GB in lung cancer cell lines. The inhibition was due to reversible binding at a site other than the catalytic site.

D. Summary of Laboratory Studies

In summary, the data from screening experiments have shown marked differences in activity for different retinoids. Studies in whole animal models have indicated potential for both chemoprevention and therapy of established disease. Molecular biology demonstrates pleiotropic effects of retinoids at the cellular level ranging from modulation of molecules, central to control of cellular proliferation, to those associated with metastasis. Abnormalities in RARB appear central in many studies though other receptors, notably RARA, are also involved. Abnormalities in these receptors may appear at the level of their expression to alterations in the components of their signaling pathways. That the addition of retinoids can affect molecules as diverse as RB, p53, cytokines, proteases, and markers of apoptosis is not surprising taken in the context of the diverse functions of the retinoic acid receptors. The heterogeneity of the molecular responses to retinoids mirrors the heterogeneity of lung cancer as a disease and argues that retinoids can be either useful or detrimental in various subsets of lung cancer. Heterogenity will also complicate screening of individuals for whom retinoids may be a useful modality.

IV. CLINICAL STUDIES

As discussed earlier, preclinical studies unequivocally indicate that retinoids can prevent, delay and effect tumor regression in experimental lung cancer. This is supported by epidemiological evidence that reduced retinoid or carotenoid intake or serum levels leads to a higher incidence of the disease. However, true evidence for the usefulness of retinoids must come from controlled trials performed in human subjects. Clinical studies, in a number of settings, have now been completed. As discussed ahead, a number of the trials have produced unexpected results. In addition to the usual endpoints of cancer incidence or mortality, these studies must address the side-effects of retinoids, long-term toxicity, and the potential elevating effects of retinoids on lipids. In prevention studies, particularly, the chosen designs must pay attention to issues such as recruitment and compliance (Zelen, 1988).

A. Prevention of Lung Cancer with Retinoids

Smoking cessation is the most effective way to prevent lung cancer; however, there remains a group of subjects who are unable to quit yet appear to be motivated to reduce their lung cancer risk. There is a further large group of subjects that has quit smoking but remains at significant risk for many years (Halpern et al., 1993). Identification of these individuals and the administration of effective chemoprevention has the potential to save thousands of lives. This section reviews the principles of lung cancer chemoprevention strategies with retinoids and discusses some recently reported trials.

Two, complementary experimental approaches to lung cancer chemoprevention are being taken: (1) trials examining intermediate endpoints as surrogates for the development of lung cancer and (2) trials using lung cancer incidence as the primary endpoint. These approaches are not exclusive and studies of intermediate endpoints can be incorporated into larger studies whose long-term goal is to study incidence. The potential advantages and disadvantages of each approach are summarized in Table 4

Intermediate Endpoints

The bronchial epithelium is normally lined by pseudostratified, ciliated, columnar cells. Smokers generally show squamous metaplasia with or without atypical cells. Using well defined criteria, the atypia can be graded as mild, moderate or severe (Auerbach et al., 1956; Saccommanno et al., 1974). Higher grades of atypia are generally regarded as pre-malignant (Auerbach et al., 1957). Bronchial atypia can be studied directly by bronchoscopic biopsy or washings and, indirectly, by sputum sampling. The sensitivity of these techniques can be improved by immunostaining (Mulshine et al., 1994). It is not surprising that early studies employing intermediate endpoints as surrogates for lung cancer looked for microscopic improvements in either bronchial biopsies or sputum samples.

In the future, it is likely that molecular or genetic markers will be included, in such studies, and may lead to improved diagnostic precision. Thus recently Kelly et al. (1994) have demonstrated *ras* mutations in cells obtained by sputum cytology in patients known to have lung cancer. Similarly Mills and colleagues (Mills et al., 1995) have detected *ras* oncogene mutations from bronchoalveolar lavage specimens of some patients diagnosed with lung cancer. It is also likely in the future that such techniques can identify those at risk who have not yet developed frank malignancy (Mao et al., 1994; Sorenson et al., 1994).

Table 4. The Advantages and Disadvantages of Incidence Versus Intermediate Endpoints in Lung Cancer Prevention Trials Using Retinoids.

Factor	*Incidence*	*Intermediate Endpoint*
Number of subjects required	Large	Relatively small
Cost	Expensive	Relatively inexpensive
Time	Long time required	Relatively short
End-point characteristics	(i) Gold standard (ii) Requires meticulous follow-up and linkages with tumour registries	(i) Requires validation (ii) May be subject to observer variation (iii) May require invasive techniques
Subject compliance	Major Issue	Less of a problem
Co-intervention/contamination	Major issues	Less of a problem
Provides biological information	Only if incorporated into study	Main purpose of study
Value for testing new agents	Less valuable	(i) Useful for phase I, II and III studies (ii) Useful for testing combinations (iii) Useful for establishing starting doses
Clinical relevance of positive result	Could be immediate	Require confirmation in long-term studies
Potential for immediate obsolescence of test agent	Large potential	Less potential

Saccommanno and colleagues (Saccomanno et al., 1982) identified a group of 16 subjects with either moderate or severe atypia on sputum sampling. The subjects were given 13-*cis* retinoic acid in a dose of 0.5 - 3.0 mg./kg. for up to six months. On completion of this uncontrolled study, no significant improvement (or deterioration) was noted in the level or degree of atypia. Degenerative alterations were seen in many cells but the significance of this finding is unclear. The small numbers studied, the lack of a control group, and widely varying drug dosages make any further interpretation of these findings difficult.

The first study, using direct sampling, was reported by Mathé and coworkers (Mathé et al., 1982). In an uncontrolled trial subjects with at least a 15 pack-year smoking history were given the synthetic retinoid etretinate 25 mg. orally daily for six months. Based upon pretreatment samples obtained by fibre-optic bronchoscopy, an index of metaplasia was developed that correlated with number of pack-years of smoking. Fifty-two subjects with an index of metaplasia greater than 15 percent, received etretinate and 30 were evaluable after six months. Of 27 subjects who continued to smoke, 20 developed a lower index, two were unchanged, and five developed a higher index. Three subjects, who stopped smoking and previously had indices of 19 percent, 40 percent, and 74 percent had no evidence of metaplasia on re-examination. The authors concluded that etretinate may benefit some subjects with a high degree of metaplasia and may have greatest impact on those who stop smoking while taking the drug. However this study can be criticized on several counts including: (1) lack of a control group, (2) failure to provide information on compliance and on the large number of unevaluable subjects, (3) failure to indicate if the biopsy specimens were read blind, and (4) no validation of the reliability of the index of metaplasia.

To address the deficiencies of uncontrolled studies and to use a noninvasive technique, our group at McMaster University has used changes in sputum cytology as a study endpoint (Browman et al., 1989). We identified subjects with bronchial atypia, on sputum sampling, and assessed the effect of a six-month course of etretinate at the same dose used by Mathé and colleagues (Mathé et al., 1982). Attention was paid to recruitment of a highly motivated group of subjects (Arnold et al., 1989) and compliance was monitored by pill counts and serum sampling throughout the trial (Browman et al., 1989; Arnold et al., 1990). The number of subjects dropping out during the six months study period was low and subjects requiring dose reductions, due to toxicity, were equally distributed in both treatment arms indicating a strong placebo effect. One hundred and fifty subjects were randomized into the study and 138 subjects were able to provide satisfactory sputum samples for the final analysis. Comparing the pre and post-treatment distributions, an overall reduction in atypia was seen but the final distribution was identical for both groups, indicating that the etretinate had no overall effect (Arnold et al., 1992).

The lack of major toxicity may indicate that this dose was too low to effect a change on sputum atypia. (Band et al., 1989) and it is feasible that higher doses of etretinate may have an impact on sputum atypia. In experimental systems, relatively high doses of retinoids have been needed to inhibit tumor development (Crocker et al., 1970). If a strategy using chemoprevention is to ultimately impact on lung cancer incidence it is very unlikely that any but minor side-effects would be acceptable to a population of smokers. Finally, it is now clear from laboratory studies that etretinate may be less active against lung tumors than other retinoids.

In another approach, van Poppel and collelagues have studied the effect of the administration of beta carotene on the frequency of micronuclei seen in sputum samples (van Poppel et al., 1992). One hundred and fourteen heavy smokers were administered beta carotene, 20 mg. daily, or placebo for 14 weeks. A 27 percent reduction in micronuclei count was demonstrated in the active treatment arm while plasma beta carotene levels increased 13 percent. In a complementary study performed in the same population, the investigators were not able to demonstrate that beta carotene supplementation prevented oxidative DNA damage and hence cancer risk (Van Poppel et al., 1992).

At the MD Anderson Hospital, Lee and coworkers have completed a rigorously controlled trial designed to assess the activity of 13-*cis* retinoic acid in reversing squamous metaplasia found in the bronchial epithelium of chronic smokers (Lee et al., 1994). Eighty-six smokers with biopsies obtained from six sites and with dysplasia and/or a metaplasia index of greater than 15 percent were randomly assigned to receive either 13-*cis* retinoic acid 1 mg/kg or placebo daily for six months. Sixty-nine were reevaluated at the completion of treatment. In the group as a whole, the metaplasia index decreased over time. However, the magnitudes of the final mean metaplasia index changes did not differ significantly in the two treatment groups. It was noted that the change was solely in those who reduced their smoking during the study.

Thus results of recent studies in chemoprevention, using intermediate endpoints, emphasize the utmost importance of controls. The most likely explanation for improvements seen in uncontrolled studies is the phenomenon of regression towards the mean. It is also logical to assume that during the study periods under scrutiny subjects might reduce their smoking and hence would demonstrate a degree of natural improvement.

In summary, a number of chemoprevention trials using intermediate endpoints have been completed. One uncontrolled study was positive; however, the experimental design used leaves the results open to question. Controlled studies have indicated that the retinoids used have been ineffective.

Cancer Incidence as the Endpoint

The true test of the hypothesis that retinoids or carotenoids can prevent lung cancer will only come from the results of large randomized trials using lung cancer incidence as the primary endpoint. These studies are major undertakings requiring large resources of time and money. The sample sizes required for successful completion run to many thousands of subjects making issues such as recruitment, long-term compliance, and redundancy of the basic study question significant problems for investigators. Nevertheless, three recently completed studies have yielded considerable information on many of the administrative, methodological, pharmacological, and toxicity issues and are the foreruners of future large chemoprevention trials.

In 1994 a joint NCI(US)-Finnish group published results of the Alpha-Tocopherol, Beta Carotene Cancer Prevention Study. The trial was set up in 1985, to study the effect of vitamin E and beta carotene on the incidence of lung and other cancers in male smokers (Alpha-Tocopherol, Beta Carotene Cancer Prevention Study, 1994). Using a factorial design, a total of 29,133 male smokers, 50 to 69 years of age, from southwestern Finland, were randomly assigned to one of four regimens: alpha-tocopherol (50 mg per day) alone, beta carotene (20 mg per day) alone, both alpha-tocopherol and beta carotene, or double placebo. Eight hundred and seventy-six new cases of lung cancer were diagnosed during the duration of the trial. Unexpectedly, a higher incidence of lung cancer was seen among the men who received beta carotene compared to controls (18% change in incidence 95% confidence interval, 3% to 36%). Total mortality was eight percent higher (95% confidence interval, 1% to 16%) among the participants who received beta carotene than among those who did not, primarily due to lung cancer and ischemic heart disease.

Just as this chapter was completed (in March 1996), a second major study conducted in Seattle, Washington, by Omenn and colleagues was unexpectedly terminated 21 months early. The CARET study differs from the Finnish study in that a combination of a retinoid plus beta carotene was tested. The treatment arm comprised beta carotene (30 mg./day) plus retinyl palmitate (25,000 IU/day) (Omenn et al., 1994). The resultls have not yet been formally published, however, preliminary information was given at a press release and reported in the Journal of the National Cancer Institute News section (Smigel, 1996).

14,254 current or former smokers and an additional 4,060 asbestos-exposed subjects, most of whom are smokers, were randomized. The sample size had been calculated based upon: (1) expected lung cancer rates from age matched data from the SEER databank; (2) an expected maximum potential reduction of lung cancer incidence of 33 percent, and (3) an assumption of 70 percent compliance at three years. After an average of four years on study, 28 percent more participants taking the combination developed lung cancer with a 17 percent higher overall mortality. The excess cardiovascular mortality was 27 percent. By contrast, in the Finnish study, those taking alpha-tocopherol appeared to be slightly protected against heart disease (Alpha-Tocopherol, Beta Carotene Cancer Prevention Study, 1994).

Goodman and colleagues (Goodman et al., 1994) have taken advantage of the CARET study to test the hypothesis that, in the Finnish study, there might have been a negative effect of beta carotene on alpha-tocopherol levels. Thus they have studied alpha tocopherol levels in subjects receiving beta carotene in the CARET study. After six years of beta carotene supplementation there was no suggestion that alpha-tocopherol levels are diminished.

The Finnish and CARET studies were set up to specifically address the issue of lung cancer. A third study, the Physicians Health Study (Hennekens et al., 1996) examined the effect of beta carotene supplementation (50 mg on alternate days) in 22,071 males and examined the incidence of all cancers. Eleven percent of the population were current smokers and there was no difference in the incidence of lung cancer after 13 years of follow-up.

The results of these three large chemoprevention studies represent a setback to those who believe that beta carotene supplementation may have a protective effect in those at high risk for lung cancer. They go in the face of the epidemiological and animal data reviewed earlier. The studies have generated much comment and speculation (Greenberg R and Sporn M, 1996) but to date no good explanation has been put forward.

B. Retinoids as Adjuvant Therapy

While chemoprevention studies have been disappointing, there may be a role for the use of retinoids as adjuvants to conventional treatment modalities, particularly surgery. Indirect support for the use of retinoids in this setting comes from a study of 103 patients with treated squamous cell cancer of the head and neck. Hong et al. (1990) demonstrated a marked reduction in the occurrence of second primary malignancies in patients treated with the retinoid 13-*cis* retinoic acid.

Pastorino and collealgues (Pastorino et al., 1993) provided more direct evidence in a study of 307 patients with stage I non-small cell lung cancer. Following curative resection, the patients were randomized to receive retinol palmitate administration (orally 300,000 IU daily for 12 months) or no treatment. After a median follow-up of 46 months, the number of patients with either recurrence or new primary tumors was 56 (37%) in the treated arm and 75 (48%) in the control arm ($p = .045$). This encouraging result requires confirmation from other studies; however, it is the first direct evidence of possible benefit of a retinoid given in the adjuvant setting. Longer follow-up will be required to assess the effect upon survival. It is possible that retinoid treatment of a longer duration might produce further benefit.

In Europe, the EORTC is conducting a multicenter study examining the use of retinyl palmitate and N-acetylcysteine as an adjuvant to conventional therapy in patients with early head and neck cancer and non-small cell lung cancers. The biological rationale for use of the combination is somewhat unclear. Nevertheless the study uses a factorial design that will examine the effect of each drug separately as well as in combination.

Thus, to date, the use of retinoids in the adjuvant setting against smoking-related cancers has been somewhat encouraging and merits confirmation in larger trials and exploration of the use of newer retinoids. For new studies being designed, it is appropriate to examine nuclear retinoic acid receptor levels to obtain further biological information. In the future it may be feasible to treat in a manner similar to hormonotherapy in breast cancer.

C. Retinoids in the Treatment of Advanced Disease

Single Agent Studies

Based upon laboratory evidence of antitumor activity, early investigations using retinoids attempted to treat advanced lung cancers. As part of a phase II study, Goodman (1986) treated 17 patients with non-small cell lung cancer with oral retinol (200,000 units) but observed no responses . Mishke and coworkers (Mishke et al., 1977) treated nine patients with squamous cell lung cancer with either retinyl palmitate or 13-*cis* retinoic acid and reported minor responses. Meyskens reported two minor responses in eight patients treated with 13-*cis* retinoic acid alone (Meyskens e tal., 1982). Using the same retinoid Grunberg and Itri (1987) reported one partial response out of 25 patients.

All-*trans* retinoic acid has shown very high activity in patients with acute promyelocytic leukemia essentially acting as a differentiating agent (Castaigne et al., 1990). Treat and colleagues (Treat et al., 1994) have evaluated this retinoid, as a single agent, in 28 patients with metastatic, measurable, non-small cell lung cancer at a dose of 175 mg/m2 orally on a daily basis. Two partial responses occurred in patients with adenocarcinoma.

Despite these isolated reports of clinical activity, it is highly unlikely that retinoids alone could ever significantly impact upon patients with advanced squamous cell carcinoma of the lung. For this reason, exploration of the use of retinoids in combination with other active agents should continue.

Retinoids in Combination With Other Agents

A number of agents have been shown to act additively or synergistically with retinoids. One combination that has received considerable recent attention is 13-*cis* retinoic acid given together with interferon alfa-2a (Eisenhauer et al., 1994). The proposed rationale for combining such agents is that: (1) both agents produce their effects on differentiation and proliferation by distinct molecular mechanisms and (2) experimental data demonstrates additive as well as synergistic effects of their combination use.

Tomida (1982) reported that naturally occurring interferon alfa-2a enhanced the differentiating effect of retinoic acid in cultured HL-60 cells (Tomida et al., 1982). Langer and Pestka demonstrated that retinoic acid could produce an increase in the receptors for interferon alfa-2a. (Langer and Pestka, 1985; Lin and Sartorelli, 1987) The extent of differentiation appears to be highly dependent on the actual combination of retinoid and interferon chosen (Peck and Bollag, 1991). Cancers of squamous cell origin would seem to be ideal candidates for trials of such therapy.

In two provocative reports the combination of 13-*cis* retinoic acid and interferon alfa-2a demonstrated a response activity of 50 percent or more in patients with both heavily pretreated advanced squamous cell skin cancers and in previously untreated squamous cell carcinoma of the cervix. The dose of 13-*cis* retinoic acid was 1 mg/kg daily in both studies and the dose of interferon alfa-2a was three million and six million respectively (Lippman et al., 1992a; 1992b).

Based upon the preliminary experimental and encouraging clinical findings in squamous cell cancers of other sites, the National Cancer Institute of Canada Clinical Trials Group conducted a phase II study of the combination of 13-*cis* retinoic acid (1 mg./kg.) and interferon alfa-2a (three million units) in 34 lung cancer patients who had received no prior chemotherapy (Arnold et al., 1994). Two cohorts were studied—sixteen patients with inoperable squamous cell lung cancer and a further 16 with either large cell or adenocarcinoma. Thirty-two patients were evaluable for response. Of 16 evaluable patients with squamous cell lung cancer only one short-lived partial response was observed. No responses were observed in 16 further patients with other non-small cell histologies (predominantly adenocarcinomas). In another study, Rinaldi and colleagues (Rinaldi et al., 1993) administered the identical regimen to 17 patients with squamous cell histology. Again, only one partial response was observed (Table 5).

Table 5. The Combination of 13-CRA and alpha interferon in Lung Cancer.

Total	*Prior XRT*	*Prior Chemo*	*CR(%)*	*PR*
Squamous Lung	17	11	0	1(6%)
Squamous Lung	17	4	0	1(6%)
Non-Squamous Lung	17	4	0	0

Source: * Modified from Eisenhauer (1994).

The toxicity observed was typical for the combination of interferon plus a retinoid and in general the regimens were well tolerated. However, in view of the poor response rate, this approach is not recommended. It is interesting, however, to speculate why cancers from different sites, but of similar histology, should show such discrepant response rates. It is unlikely that previous treatment had a significant impact. None of the lung cancer patients had received previous systemic therapy and only the minority had received radiation. Cervical cancer is virally induced whereas lung cancer clearly is not. There is some data to show that both 13-*cis* retinoic acid and interferon alfa-2a inhibit HPV-transformed cell growth and *in vivo* viral carcinogenesis (Repucci et al., 1991; Androphy, 1986). However, it is unlikely that viral inhibition is the sole explanation of the difference. Future studies of 13-*cis* retinoic acid plus interferon alfa should attempt to correlate tumor response with tumor biology and in particular the known retinoid receptors.

Methodological Issues

With the exception of small cell lung cancer, lung cancer does not usually demonstrate dramatic tumor shrinkage when exposed to single-agent systemic therapy. It is highly improbable, therefore, that conventional phase II studies of new retinoids will demonstrate response rates near the conventional 20 percent rate usually used to sclcct agents for further investigation. Thus, to date, responses to retinoids have been observed only infrequently and have been of a minor nature. Based upon preclinical observations, it is likely that any future role of retinoids, in the setting of established lung cancer, will be most likely in the adjuvant setting or maintenance phase after a remission has been obtained with other modalities. Alternatively, based upon sound principles developed in laboratory models, retinoids may be used in combination with chemotherapy or biological agents.

It is therefore suggested that, once phase I studies of new retinoids have established an appropriate maximally tolerated dose (MTD), a modified phase II program be established using randomization to retinoid or placebo in addition to the active agent. Due to their unique mechanism of action, endpoints for lung cancer, such as disease-free survival, reduction in second primary tumors, or prolonged periods of disease stabilization should be the primary focus of such trials.

These points are illustrated by the recent fate of the experimental arotinoid mofarotene (Arnold et al., 1994). This retinoid appears to have a unique mechanism of action and acts additively or synergistically with a number of active chemotherapeutic agents and biological response modifiers. In the laboratory it shows activity in a number of lung cancer models. Mofarotene entered Phase I clinical trials in solid tumors, but (as might be expected) no major responses were seen. The arotinoid was not developed further, following a company management decision based upon these preliminary findings. Due to overoptimistic and inappropriate early expectations, the opportunity was lost to test the arotinoid in some potentially novel phase II trials.

V. CONCLUSION

Despite the squamous cell origin of many lung cancers, the use of retinoids to prevent or treat this disease remains conjectural. Nevertheless, there is accumulating evidence that this class of compounds may eventually play a role in lung cancer control. To date, observations from epidemiological, laboratory, and small, uncontrolled clinical studies have not held up to the rigor of well-designed controlled trials. Beta carotene appears to be harmful in a chemoprevention setting.

The most useful areas for future exploration would appear to be in an adjuvant setting and in combination with cytokines or combination chemotherapy. Retinoids alone are very unlikely to produce responses similar to those obtained by combination chemotherapy or radiation. Newer retinoids such as the arotinoids or 9-*cis* retinoic acid have yet to be tested against lung cancer.

There is important new information emerging concerning the molecular biology of retinoids and lung cancer. It is hoped that such information may lead to a more rational approach retinoid use in the future. It will be important to include biological studies as part of clinical trial protocols.

REFERENCES

Anderson, I.C., Sugarbaker, D., Ganju, R., Tsarwhas, D., Richards, W., Sunday, M., & Shipp, M.A. (1995) Stromelysin-3 is overexpressed by stromal elements in primary non-small cell lung cancers and regulated by retinoic acid in pulmonary fibroblasts. Cancer Res. 55, 4120-4126.

Androphy, E.J. (1986). Papilloma viruses and interferon. Ciba Found Symp. 120, 221-234.

Anees, M., & Steven, F.S. (1994). Retinoic acid inhibition of a tumour protease immobilized on cell surfaces and in free solution. J. Enzyme Inhib. 8, 51-59.

Arnold, A., Ayoub, J., Douglas, L., Hoogendoorn, P., Skingley, L. Gelmon, K., Hirsh, V., & Eisenhauer, E. (1994) Phase II trial of 13-*cis*-retinoic acid and interferon αin non-small cell lung cancer. J. Natl. Cancer Inst. 86, 306-309.

Arnold, A., Johnstone, B., Stoskopf, B., Skingley, P., Browman, G., Levine, M., & Hryniuk, W. (1989). Recruitment for an efficacy study in chemoprevention- the concerned smoker study. Prev. Med. 18, 700-710.

Arnold, A., Kowaleski, B., Tozer, R., & Hirte, H. (1994). The arotinoids: early clinical experience and discussion of future development. Leukemia 11, 1817-24.

Arnold, A.M., Browman, G.P., Johnstone, B., Skingley, P., Booker, L., & Levine, M.N. (1990). Chemoprevention for lung cancer—evidence for a high degree of compliance. Cancer Detect Prev. 14, 521-525.

Auerbach, O., Gere, B., & Forman, J.B. (1957). Changes in the bronchial epithelium in relation to smoking and cancer of the lung. N. Engl. J. Med. 256, 97-104.

Auerbach, O., Petrick, G., & Stout, A.P. (1956). The anatomical approach to the study of smoking and bronchogenic carcinoma. Cancer. 9, 76-83.

Band, P., Deschamps, M., & Israel, L. (1989). Retinoid chemoprevention timing and dose intensity. Cancer Invest. 7, 205-201.

Bond, G., Thompson, F., & Cook, R. (1987). Dietary vitamin A and lung cancer: results of a case control study among chemical workers. Nutr. Cancer. 9, 109-121.

Browman, G., Arnold, A., Levine, M.N., D'Souza, T., Turner-Smith, L., Cayco, R., Johnstone, B., & Birse, R. (1990). Use of screening phase data to evaluate observer variation of sputum cytodiagnosis as an outcome measure in a chemoprevention trial. Cancer Res. 50, 1216-1219.

Browman, G.P., Arnold, A., Booker, L., Johnstone, B., Skingley, P., & Levine, M.N. (1989). Etretinate blood levels in monitoring of compliance and contamination in a chemoprevention trial. J. Natl. Cancer Inst. 81, 795-798.

Brown, C., & Kessler, L. (1988). Projections of lung cancer mortality in the United States: 1985-2025. J. Natl. Cancer Inst. 80, 43-51.

Byers, T., Graham, S., Haughey, B., Marshall, J., & Swanson, M. (1987). Diet and lung cancer risk: findings from the Western New York Diet Study. Am. J. Epidemiol 127, 351-363.

Candelora, E.C., Stockwell, H.G., Armstrong, A.W., & Pinkham, P.A. (1992). Dietary intake and risk of lung cancer in women who have never smoked. Nutr. Cancer. 17, 263-270.

Castaigne, S., Chomienne, C., & Daniel, M.T. (1990). All-*trans* retinoic acid as a differentiation therapy for acute promyelocytic leukemia 1. Clinical results. Blood. 76, 1704-1709.

Chen, Z., Gundimeda, U., & Gopalakrishna, R. (1994). Retinoids induce a decrease in protein kinase C activity in lung carcinoma cells and suppress invasion and metastasis (Meeting abstract). Proc. Ann. Meet. Am. Assoc. Cancer Res. 35, A335.

Cone, W., & Nettesheim, P. (1973). Effects of vitamin A on 3-methyl-cholanthrene induced squamous metaplasia and early tumours in the respiratory tract of rats. J. Natl. Cancer Inst. 50, 1599-1608.

Connett, J., Kuller, L., & Kjelsberg, M. (1989). Relationship between carotenoids and cancer. The Multiple Risk Factor Intervention Trial (MRFIT). Cancer 64, 126-134.

Crocker, J., & Sanders, L. (1970). Influence of vitamin A and 3,7-dimethyl-2,6-octadienal (citral) on the effect of benzo(a)pyrene on the hamster trachea in organ culture. Cancer Res. 30, 1312-1318.

Dartiques, J.F., Dabis, F., Gros, N., Moise, A., Bois, G., Salomon, R, Dilhuydy, J.M., & Courty, G. (1990). Dietary vitamin A, beta carotene and risk of epidermoid lung cancer in South-western France. Eur. J. Epidemiol. 6, 261-265.

Davis, I., Mathias, A., Unsworth, E., Miller, M., Cuttitta, F., Mulshine, J., & Jakowlew, S. (1995). Analysis of small cell lung cancer cell growth inhibition by 13-*cis* retinoic acid: importance of bioavailability. Cell Growth Differ. 65, 485-492.

Dorgan, J.F., Ziegler, R.G., Schoenberg, J.B., Hartge, P., McAdams, M.J., Falk, R.T., Wilcox, H.B., & Shaw, G.L. (1993). Race and sex difference in associations of vegetables, fruits and carotenoids with lung cancer risk in new Jersey. Cancer Causes Control 4, 273-281.

Eisenhauer, E.A., Lippman, S.M., Kavanagh, J.J., Paredes-Espinoza, Arnold, A., Hong, W.K., Masssimini, G., Schleuniger, U., Bollag, W., Holdener, E.E., & Krakoff, I. (1994). Combination 13-*cis*-Retinoic acid and interferon α-2a in therapy of solid tumours. Leukemia 8, 1622-1625.

Eliason, J.F., Kaufmann, F., Tanaka, T., & Tsukaguchi, T. (1993). Anti-proliferative effects of the arotinoid Ro 40-8757 on human cancer cell lines in vitro. Br. J. Cancer 67, 1293-1298.

Fell, H., & Mellanby, E. (1953). Metaplasia produced in cultures of chick ectoderm by high vitamin A. J. Physiol. 119, 470-488.

Fontham, E., Pickle, L., Haenszel, W., Correa, P., Lin, Y., & Falk, R. (1988). Dietary vitamin A and C and lung cancer risk in Louisiana. Cancer 62, 2267-2273.

Fontham, E.T. (1990). Protective dietary factors and lung cancer. Int. J/ Epidemiol. 19, 832-842.

Friedmann, G., & Blaner, W.D.S.G. (1986). Serum retinol and retinol binding protein levels do not predict subsequent lung cancer. Am. J. Epidemiol. 123, 781-789.

Gaetano, C., Melchiori, A., Albini, A., Benelli, R., Falcioni, R., Modesti, A., Modica, A., Scarpa, S., & Sacchi, A. (1994). Retinoic acid negatively regulates beta 4 integrin expression and suppresses the malignant phenotype in a Lewis lung carcinoma cell line. Clin. Exp. Metastasis 12, 63-72.

Geradts, J., Chen, J.Y., Russell, E.K., Yankaskas, J.R., Nieves, L., & Minna, J.D. (1993). Human lung cancer cell lines exhibit resistance to retinoic acid treatment. Cell Growth Differ. 410, 799-809.

Goodman, G. (1986). Phase II trial of retinol in patients with advanced cancer. Cancer Treat. Rep. 70, 1023-1024.

Goodman, G.E., Metch, B.J., & Omenn, G.S. (1994). The effect of long-term beta-carotene and vitamin A administration on serum concentrations of alpha-tocopherol. Cancer Epidemiol. Biomarkers Prev. 3, 429-32.

Grunberg, S., & Itri, L. (1987). Phase II study of isotretinoin in the treatment of advanced non-small cell lung cancer. Cancer Treat. Rep. 71, 1097-1098.

Halpern, M.T., Gillespie, B.W., & Warner, K.E. (1993). Patterns of absolute risk of lung cancer moertality in former smokers. J. Natl. Cancer Inst. 85, 457-464.

Hinds, M., Kolonel, L., Hankin, J., & Lee, J. (1984). Dietary vitamin A, carotene, vitamin C and risk of lung cancer incidence in Hawaii. Am. J. Epidemiol. 119, 227-236.

Hirayama, T. (1979). Diet and cancer. Nutr. Cancer. 1, 67-81.

Hirayama, T. (1984). Lung cancer in Japan: effects of nutrition and passive smoking. In: *Lung Cancer: Causes and Prevention.* (Mizell, M., Pelayo, C. eds). pp. 175-195. Verlag Chemie International. Deerfield Beach.

Houle, B., Rochette-Egly, C., & Bradley, W.E. (1993). Tumor-suppressive effect of the retinoic acid receptor beta in human epidermoid lung cancer cells. Proc. Natl. Acad. Sci. USA 90, 985-989.

Hubert, D., Holiat, S., Smith, W., & Baylouny, R. (1983). Inhibition of transplanted carcinomas in mice by retinoids but not by vitamin C. Cancer Treat. Rep. 67, 1061-1065.

Jain, M., Burch, J., Howe, G., Risch, H., & Miller, A. (1990). Dietary factors and risk of lung cancer: results from a case-control study, Toronto 1981-1985. Int. J. Cancer. 45, 287-293.

Jakowlew, S.B., Mathias, A., Mulshine, J.L. & Moody, T.W. (1994). Retinoic acid stimulates expression of transforming growth factor-beta, retinoic acid receptor and retinoid x receptor mRNAs in NCI-H727 lung cancer cells (Meeting abstract). Proc. Ann. Meet Am. Assoc. Cancer Res. 35, A546.

Kalemkerian, G.P., Jasti, R.K., Celano, P., Nelkin, B.D., & Mabry, M. (1994). All-*trans*-retinoic acid alters myc gene expression and inhibits in vitro progression in small cell lung cancer. Cell Growth Differ. 5, 55-60.

Kark, J., Smith, A., Switzer, B., & Hames, C. (1981). Serum vitamin A (retinol) and cancer incidence in Evans County, Georgia. J. Natl. Cancer Inst. 66, 7-16.

Kelly, K. (1994). Evalutaion of k-ras mutations in sputum samples by single stranded conformation polymorphism. Lung Cancer. 11(suppl), 225. Abstract.

Kim, Y.H., Dohi, D.F., Han, G.R., Zou, C.P., Oridate, N., Walsh, G.L., Nesbitt, J.C., Xu, X.C., Hong, W.K., & Lotan, R. (1995). Retinoid refractoriness occurs during lung carcinogenesis despite functional retinoid receptors. Cancer Res. 55: 5603-5610.

Kolonel, L., Nomura, A., Hinds, M., Hirohata, T., Hankin, J., & Lee, J. (1983). Role of diet in cancer incidence in Hawaii. Am. J. Epidemiol. 119, 227-236.

Koo, L. (1988). Dietary habits and lung cancer risk among Chinese females in Hong Kong who have never smoked. Nutr. Cancer. 11, 155-172.

Kune, G., Kune, G., & Watson, L. (1989). Serum levels of beta-carotene, vitamins A, and zinc in male lung cancer cases and controls. Nutr. Cancer. 12, 169-176.

Langer, J., & Pestka, S. (1985). Changes in binding of interferon IFN-aA to HL-60 cells during myeloid differentiation. J. Interferon Res. 5, 637-649.

Laznitski, I., & Bollag, W. (1982). Prevention and reversal by a retinoid of 3,4-benzpyrene-and cigarette smoke condensate-induced hyperplasia and metaplasia of rodent respiratory epithelia in organ culture. Cancer Treat. Rep. 66, 1375-1380.

Le Marchand, L., Hankin, J.H., Kolonel, L.N., Beecher, G.R., Wilkens, L.R., & Zhao, L.P. (1993). Intake of specific carotenoids and lung cancer risk. Cancer Epidemiol. Biomarkers Prev. 2, 183-187.

Le Marchand, L., Yoshizawa, C., Kolonel, L., Hankin, J., & Goodman, M. (1989). Vegetable consumption and lung cancer risk: a population based case-control study in Hawaii. J. Natl. Cancer Inst. 81, 1158-1164.

Lee, J.S., Lippman, S.M., Benner, S.E., Lee, J.J., Ro, J.Y., Lukeman, J.M., Morice, R.C., Peters, E.J., Pang, A.C., Fritsche, H.A., & Hong, W.K. (1994). Randomized placebo-controlled trial of isotretinoin in chemoprevention of bronchial squamous metaplasia. J. Clin. Oncol. 12, 937-945.

Legardeur, B.Y., Lopez, A., & Johnson, W.D. (1990). A case control study of vitamins A, E, and C in lung cancer patients. Nutr. Cancer. 14, 133-140.

Lin, J., & Sartorelli, A. (1987). Stimulation by interferon of the differentiation of human promyelocytic leukemia (HL-60) cells produced by retinoic acid and actinomycin D. J. Inter. Res. 7, 379-387.

Lippman, S., Kessler, J., & Meyskens, F. (1987a). Retinoids as preventive and therapeutic anticancer agents (part I). Cancer Treat. Rep. 71, 391-405.

Lippman, S., Kessler, J., & Meyskens, F. (1987b). Retinoids as preventive and therapeutic anticancer agents (part II). Cancer Treat. Rep. 71, 493-515.

Lippman, S.M., Kavanagh, J.J., Parades-Espinoza, M., Delgadillo-Madrueno, F., Paredes-Casillas, P., Hong, W.K., Holdener, E., & Krakoff, I.H. (1992a). 13-*cis*-retinoic acid and interferon α-2a: highly effective therapy for advanced squamous cell carcinoma of the cervix. J. Natl. Cancer Inst. 84, 241-245.

Lippman, S.M., Parkinson, D.R., Itri, L.M., Weber, R.S., Schantz, S.P., Ota, D.M., Schusterman, M.A., Krakoff, I.H., Gutterman, J.U., & Hong, W.K. (1992b). 13-*cis*-retinoic acid and interferon α-2a: effective combination therapy for advanced squamous cell carcinoma of the skin. J. Natl. Cancer Inst. 84, 235-241.

MacLennan, R., DaCosta, J., Day, N., Law, C., Ng, Y., & Shanmugaratnum, K. (1977). Risk factors for lung cancer in Singapore Chinese, a population with high female incidence rates. Int. J. Cancer 20, 854-860.

Mao, L., Hruban, R.L., & Boyle, J.O. (1994). Detection of oncogene mutations in sputum precedes diagnosis of lung cancer. Cancer Res. 54, 1643-1637.

Mathé, G., Gouveia, J., & Hercend, T. (1982). Correlation between precancerous bronchial metaplasia and cigarette consumption, and preliminary results of retinoid treatment. Cancer Detect. Prevent. 5, 461-466.

Maxwell, S.A., & Mukhopadhyay, T. (1994). Transient stabilization of p53 in non-small cell lung carcinoma cultures arrested for growth by retinoic acid. Exp. Cell Res. 214, 67-74.

Maxwell, S.A. (1994). Retinoblastoma protein in non-small cell lung carcinoma cells arrested for growth by retinoic acid. Anticancer Res. 14, 1535-1540.

Mayne, S.T., Janench, D.T., Greenwald, P., Chorost, S., Tucci, C., Zaman, M.B., Melamed, M.R., Kiely, M., & McKneally, M.F. (1994). Dietary beta carotene and lung cancer risk in US nonsmokers. J. Natl. Cancer Inst. 5, 33-38.

Menkes, M., Comstock, G., Vuilleunier, J., Helsing, K., Rider, A., & Brookmeyer, R. (1986). Serum beta-carotene, vitamin A and E, selenium and the risk of lung cancer. N. Engl. J. Med. 315, 1250-1254.

Metlin, C., Graham, S., & Swanson, M. (1979). Vitamin A and lung cancer. J. Natl. Cancer Inst. 62, 1435-1438.

Meyskens, F., Gilmartin, E., & Alberts, D. (1982). Activity of isotretinoin against squamous cell cancers and neoplastic lesions. Cancer Treat. Rep. 66, 1315-1319.

Micksche, M., Cerni, M., & Kokron, O. (1977). Stimulation of immune response in lung cancer patients by vitamin A therapy. Oncology. 34, 234-2388.

Middleton, B., Byers, T., Marshall, I., & Graham, S. (1986). Dietary vitamin A and cancer- a multisite case-control study. Nutr. Cancer 8, 107-116.

Mills, N.E., Fishman, C.L., Scholes, J., Anderson, S.E., Rom, W.N., & Jacobsson, D.R. (1995). Detection of K-ras oncogene mutations in bronchoalveolar lavage fluid for lung cancer diagnosis. J. Natl. Cancer Inst. 87, 1056-1060.

Mossman, B., Craighead, J., & Macpherson, B. (1980). Asbestos induced epithelial changes in organ cultures of hamster trachea: Inhibition by retinyl methyl ether. Science. 207, 311-313.

Mulshine, J., Tockman, M., & Scott, F. (1994). Global strategies for the early detection of epithelial cancers. In: *Early Detection of Cancer: Molecular Markers* (Srivastava, S., ed.). Futura Publishing Co., Armonk, NY.

Nettesheim, P., & Williams, M. (1976). The influence of vitamin A on the susceptibility of the rat lung to 3-methylcholanthrene. Int. J. Cancer. 17, 351-357.

Nettesheim, P., Cone, V., & Synder, C. (1976). The influence of retinyl acetate on the post initiation phase of preneoplastic lung nodules in rats. Cancer Res. 36, 996-1002.

Nomura, A., Stemmermann, G., Heilbrun, L., Salkeld, R., & Vuilleumier, J. (1985). Serum vitamin levels and risk of cancer of specific sites in men of Japanese ancestry in Hawaii. Cancer Res. 45, 2369-2372.

Omenn, G.S., Goodman, G., Thornquist, M., Grizzle, J., Rosenstock, L., Barnhart, S., Balmes, J., Cherniack, M.G., Cullen, M.R., & Glass, A. (1994). The beta-carotene and retinol efficacy trial (CARET) for chemoprevention of lung cancer in high risk populations: smokers and asbestos-exposed workers. Cancer Res. 54 (Suppl), 2038s-2043s

Pastorino, U., Infante, M., Maioli, M., Chiesa, G., Buyse, M., Firket, P., Rosmentz, N., Clerici, M., Soresi, E., & Valente, M. (1993). Adjuvant treatment of stage I lung cancer with high-dose vitamin A. J. Clin. Oncol 11, 1216-22

Pastorino, U., Pisani, P., & Berrino, F. (1987). Vitamin A and female lung cancer: a case control study on plasma and diet. Nutr. Cancer 10, 171-179.

Peck, R., & Bollag, W. (1991). Potentiation of retinoid induced differentiation of HL-60 and U937 lines by cytokines. Europ. J. Cancer 27, 53-57.

Peleg, I., Heyden, S., Knowles, M., & Hames, C. (1984). Serum retinol and risk of subsequent cancer: extension of the Evans County Georgia, study. J. Natl. Cancer Inst. 73, 1455-1458.

Pisani, P., Berrino, F., Macaluso, M., Pastorino, U., Crosignani, P., & Baldasseroni, A. (1986). Carrots, green vegetables and lung cancer: a case-control study. Int. J. Epidemiol. 15, 463-468.

Port, C., Sporn, M., & Kaufman, D. (1975). Prevention of lung cancer in hamsters by 13-cis retinoic acid. Proc. Am. Assoc. Cancer Res. 16, 21.

Repucci, A.D., Di Lorenzo, T.P., & Abramsom, A.L. (1991). *In vitro* modulation of human laryngeal papilloma cell differentiation by retinoic acid. Otolaryngol Head Neck Surg. 105, 528-532.

Resnicow, K., Kabat, G., & Wynder, E. (1991). Progress in decreasing cigarette smoking. Important Adv. Oncol. 205-213.

Rinaldi, D.A., Lippman, S.M., Burris, H.A., Cirou, C., Von Hoff, D.D., & Hong, W.K. (1993) Phase II study of 13-*cis*-retinoic acid and interferon α-2a in patients with advanced squamous cell lung cancer. Anticancer Drugs 4, 33-36.

Ross, H.J., & Canada, A.L. (1994). Antiproliferative response of human lung cancer cells to trans-retinoic acid may involve endogenous IL-1 beta (Meeting abstract). Proc. Ann. Meet. Am. Assoc. Cancer Res. 35, A1690.

Rusch ,V.W., & Dmitrovsky, E. (1995). Molecular biologic features of non-small lung cancer. Clinical implications. Chest. Surg. Clin. N. Am. 51, 39-55.

Saccomanno, G., Archer, V.E., Auerbach, O., Saunders, R.P., & Brennan, L.M. (1974). Development of carcinoma of the lung as reflected in exfoliated cells. Cancer 33, 256-270.

Saccomanno, G., Moran, P., & Schmidt, R. (1982). Effects of 13-*cis* retinoids on premalignant and malignant cells of lung origin. Acta. Cytologica. 26, 79-85.

Saffiotti, U., Montesano, R., Sellakumar, A., & Borg, S. (1967). Experimental cancer of the lung. Inhibition by vitamin A of the induction of tracheobronchial squamous metaplasia and squamous cell tumours. Cancer 20, 857-864.

Salonen, J., Salonen, R.R.L., Maenpaa, P., Alfthan, G., & Puska, P. (1985)Risk of cancer in relation to serum concentrations selenium, and vitamin A and E: matched case-control analysis of prospective data. Br. Med. J. 290, 417-420.

Samet, J., Skipper, B., Humble, C., & Pathak, D. (1985). Lung cancer risk and vitamin A consumption in New Mexico. Am. Rev. Resp. Dis. 131, 198-202.

Shekelle, R., Lepper, M., & Liu, S. (1981). Dietary vitamin A and the risk of cancer in the Western electric study. Lancet 2, 1185-1190.

Shibata, A., Paganini-Hill, A., Ross, R.K., Yu, M.C., & Henderson, B.E. (1992) Dietary beta-carotene, cigarette smoking and lung cancer in men. Cancer Causes Control 3, 207-214.

Smigel, K. (1996). J. Natl. Cancer Inst. 88, 145.

Smith, A.H., & Walker, K.D. (1991). Serum beta-carotene in persons with cancer and their immediate families. Am. J. Epidemiol. 133, 661-671.

Smith, P., Rodgers, A., Herndon, B., & Newberne, P.M. (1975). Vitamin A (retinyl acetate) and benzo -a. pyrene-induced respiratory tract carcinogenesis in hamsters fed a commercial diet. Cancer Res. 35, 11-16.

Sorenson, G., Pribish, D., & Maurer, L. (1994). Detection of mutated k-ras sequences in blood from patient with a pulmonary carcinoma. Lung Cancer 11, (suppl 1), 57 abstract.

Sporn, M., & Roberts, A. (1984). Biological methods for analysis and assay of retinoids-relationships between structure and activity. In: *The Retinoids*. (Sporn, M., Roberts, A., Goodman, D.S. eds.). pp. 236-279. Academic Press, Orlando.

Stahelin, H., Rosel, F., Buess, E., & Brubacher, G. (1984). Cancer, vitamins and plasma lipids: prospective Basel study. J. Natl. Cancer Inst. 73, 1463-1468.

Stahelin, H.B., Gey, K.F., Eichholzer, M., & Ludin, E. (1991). Beta-carotene and cancer prevention: the Basel study. Am. J. Clin. Nutr. 53 (Suppl), 265-269.

Stinson, H., Reznik, G., & Donshoe, R. (1981). The effect of three retinoids on tracheal carcinogenesis induced by *N*-methyl-*N*-nitrosurea in hamsters. J. Natl. Cancer Inst. 66, 947-951.

Alpha-Tocopherol, Beta Carotene Cancer Prevention Study. (1994) The effect of vitamin E and beta carotene on the incidence of lung cancer and other cancers in male smokers. N. Engl. J. Med. 330, 1029-1035.

Tomida, M., Yamamoto, Y., & Hozumi, M. (1982). Stimulation by interferon of induction of differentiation of human promyelocytic leukaemia cells. Biochem Biophys. Res. Commun. 104, 30.

Treat, J., Greenberg, R., Bratschi, J., Gorman, G., Meehan, L., & Friedland, D. (1994) First hints in non-small cell lung cancer (NSCLC) Leukemia 8 Suppl 3, 55-58.

van Poppel, G., Poulsen, H., Loft, S., & Verhagen, H. (1995). No influence of beta carotene on oxidative DNA damage in male smokers. J. Natl. Cancer Inst. 87, 310-311.

van Poppel, G., Kok, F.J., & Hermus, R.J. (1992). Beta-carotene supplementation in smokers reduces the frequency of micronuclei in sputum. Br. J. Cancer 66, 1164-1168.

Vollberg, T.M., George, M.D., Nervi, C., & Jetten, A.M. (1992) Regulation of type I and type II tranglutaminase in normal human bronchial epithelial and lung carcinoma cells. Am. J. Respir. Cell Mol. Biol. 7, 10-18.

Wagner , H., & Ruckdeschel, J.C. (1995). Screening, early detection and early intervention strategies for lung cancer. Cancer Control 2, 493-502.

Wald, N., Idle, M., Boreham, J., & Baily, A. (1980) Low serum vitamin A and subsequent risk of cancer. Lancet 2, 813-815.

Wald N, Thompson S, Densem J, Boreham J, & Bailey A. (1988). Serum beta-carotene and subsequent risk of cancer: results fro the BUPA study. Br. J. Cancer 57, 428-433.

Willett, W.C., Polk, B.F., & Underwood, B.A. (1984). Relation of serum vitamins a and e and carotenoids to the risk of cancer. N. Engl. J. Med. 310, 430-434.

Willett, W.C. (1990). Vitamin A and lung cancer. Nutr. Rev. 48, 201-211.

Wolbach, S., & Howe, P. (1925). Tissue changes following deprivation of fat-soluble A-vitamin. J. Exp. Med. 42, 735-769.

Wu, A., Henderson, B., Pike, M., & Yu, M. (1985). Smoking and other risk factors for lung cancer in women. J. Natl. Cancer Inst. 74, 747-751.

Yun, T.K., Kim, S.H., & Lee,Y.S. (1995). Trial of a new medium-term model using benzo(a)pyrene induced lung tumor in newborn mice. Anticancer Res. 15, 839-845.

Zelen, M. (1988). Are Primary cancer prevention trials feasible? J Natl Cancer Inst. 80, 1442-1444.

Zhang, L.X., Mills, K.J., Dawson, M.I., Collins, S.J., & Jetten, A.M. (1995) Evidence for the involvement of retinoic acid receptor RAR alpha-dependent signaling pathway in the induction of tissue transglutaminase and apoptosis by retinoids. J. Biol. Chem. 270, 6022-6029.

Zhang, X.K., Liu, Y., Lee, M.O., & Pfahl, M. (1994). A specific defect in the retinoic acid response associated with human lung cancer cell lines. Cancer Res. 54, 5663-5669.

Ziegler, R., Mason, T., & Stemhagen, A. (1986). Carotenoid intake vegetables and the risk of lung cancer among white men in New Jersey. Am. J. Epidemiol. 123, 1080-1093.

Ziegler, R., Mason, T., & Stemhagen, A. (1984). Dietary carotene and vitamin A and the risk of lung cancer among white men in New Jersey. J. Natl. Cancer Inst. 73, 1429-1435.

INDEX

A-box helix, 53
A-domain, 55
Actin, 145
Activator-1 site (AP-1), 86, 204
  anti-AP1 activity, 171
Acyl-CoA-independent enzymes, 38
Agonists, RXR-specific, 63
AHPN, 6-[3-(1-adamantyl)4-hydroxyphenyl]-2-napthalene carboxylic acid, 162, 166, 171
AMP-dependent protein kinase, 60
Amphibian appendages, 110-115
Angiogenesis, 198, 211
Anopthalmia, 87
Antagonists, RAR-specific, 58
Antiproliferative effects, 59
Apoptosis, 161-186
  angiogenesis, 202-203
  characteristics, 163-4
  defined, 162
  dependent mechanisms, 171-180
  effect of retinoids on, 162
  genes involved in, 164-168
  induction, 177-179, 180-182
  inhibition, 179-180
  receptor independent mechanisms, 180-184
  regulation of, 170-186
  tumor, 202
Arginine-glycin-aspartic acid (RGD), 208
Astrocytoma, human, 235
ATRA, 237
Axolot, 94

B-cell proliferation, 41
Basement membranes, 205
bFGF, 212
Boswellia acids, 232
5'-Bromdeoxyuridine, 144
Bronectin, 206

Ca^{2+} signalling pathway, 142, 150-151
CAAT box, 49
Cadherin, 208
Calcium, intracellular, 142
Calmodulin, 152
cAMP, 147
Cancer, 18, 161, 162, 201, 219-226
  breast, 149
  hematopoietic cells, 221-223
  lung, 262-281
  mammary, 224-226
  risk, 40
  therapy, 209
  vitamin A and, 198
Carbonic anhydrase, 201
Carcinogens, 266
  N-methyl-N-nitrosourea, 268

Carcinoma
  F19 embryonal, 144
  F9 embryonal, 149, 201
  mammary, 224-226
  orpharyngea, 235
  skin, 241
  squamous cell, 198
Cardiovaccular system , 87
Caretenoids, 38, 40
  lung cancer and, 262, 275
  9-*cis* and, 62
β-Carotene, 5, 81, 82, 198
Cell (*see also* Tumor cell)
  adhesion, 145
  cycle, 147
  death of, 163-170 (*see also* Apoptosis)
  endometrial stromal, 61
  endothelial, 203
  glanndular differentiation, 61
  hematopoietic, 212-223
  heptoma, 170
  HL-60, 147
  homeostasis, 44
  leukemia, 170, 171
  lung carcinoma, 170
  lysis, 209
  melanoma, 60
  mucosecretory differentiation, 61
  neroblastoma, 143, 213
  proliferation, 58, 152
  shape regulation, 145
  surface receptors, 45
  teratocarcinoma, 212
Cellular retinol binding proteins (*see* CRPB)
Cellular retinoic acid binding proteins (*see* CRABP)
cFos, 86
Chemoprevention, 198
Chick embryo (*see also* Embryo)
  early embryos, 117-125
  limb, 94, 115, 125-129
Chromatin, 47, 54, 164
C4-hydroxylation, 46
Chylomicrons, 3, 38
CNS central nervous system, 94, 95
  developement, 102
  embryos, 95
  neoplasms, 249-250
Coactivators, 57, 59
Collagenase, 86, 213
Congenital abnormalities, 80, 88
Connexins, 210
CRABP Cellular retinoic acid binding proteins, 6, 2-4, 38-47, 61, 84, 87, 201
Cranial ganglia, 95
Craniofacial region, 84
CRPB, Cellular retinol binding proteins, 41, 83 (*see also* CRABP)
Cyclic AMP response elcment binding protein, 56
Cytokines, 144, 198
Cytomegalovirus, 55
Cytoskeletal
  depolymerization, 147
  dynamics, 145
  elements, 147

Dehydrogenase
  alcohol, 84
  aldehyde, 84
  short chain alcohol, 45
Dependent mechanisms, retinoid receptor, 171
Developmental defects, retinoid-related, 80
Diacylglycerol (DAG), 146
Differentiation, 164, 165
Dimerization, 52-54
Dimethylsulphoxide (DMSO), 147
DNA ladder, 164
DNA-binding domain (DBD), 53

Dominant-negative receptors, 58
DR-2 RARE, 42

E-cadherin, 209
E-RAPB, epididymal retinoic acid binding protein, 40, 50
E1a, 56
EGF receptor activation, 147, 148
Electrophoresis, 164
Embryogenesis, mouse, 80
Embryo
  CNS and, 96, 118-119
  development, 80-89, 94-128, 142
  differentiation, 142, 201
  eye in, 87, 88
  face in, 84, 87, 120
  heart in, 119
  neural crest in, 85, 119
  proliferation, 201
  RBP and, 81
  retinoic acid synthesis in, 82-84
  signalling properties, 101
  skin patterning, 120-121
  tail bud in, 121-123
  uptake of retinol, 81
  vitamin A deficiency (*see* Vitamin A)
  yolk sac, 81, 82, 84
Endogenous retinoids
  and distribution, 105
  and retinoic acid synthesis, 98-100
Endometrial adenocarcinoma, 38
Engrailed protein, 95
Enterocytes, 38
Enzymes
  cross-linking of proteins, 166
  cytochrome P450, 43
Epididymal retinoic acid binding protein, E-RAPB, 40
Equilibrium dissociation constants, 50
Erythropoiesis, 103
Ethyl TTNPB, 21-24, 55, 59
  modification, 21-26
Excenephaly, 88
Extracellular matrix (ECM), 143

Fangliosides, 147
Fastrulation, 109
Fatty-acid binding proteins, 41
Felatinases, 204, 205
Fenretinide, 246
Fetus, 80
Fibroblast, 60
Fibronectin, 207
Filamin, 145
Flow cytometry, 164
Fluconazole, 46
Focal adhesion kinases, 146
Frog, 111, 235

G1/S transition, 169
Gametogenesis, 80
Gap junctional communication (GJC), 210-213
GDP, 148
Gel shift, 58
  growth factors (FGF), 212
Gene
  $\alpha_5\beta_1$, 143
  apoptosis and, 164-168
  18A2/*mts* 1, 147, 152
  AP-1 directed, 150
  Bcl-2 family, 164-166
  c-fos, *c-jun*, 150, 222
  c-myc,169-170
  c-src, *c-yes*, 148
  cell cycle-related, 168-170
  collagen type IV, 86
  differential induction, 59
  Dlx1, *Emx*, *En*, 86
  environment interactions, 89
  F-crystaline, 52
  Fg8f-8, 128
  fos, 205
  fyn, *Grb7*, 148
  homeobox, 101, 103

Hox, 85, 86, 99
ICE family, 166
iyn, 148
jun, 205
knockout, 51
Krox-20, 86, 95, 104
laminin B1, 86
luciferase reporter, 224
*p*53, 168-9, 233
pax2, 86, 95
proto-oncogene, 236
RAR family, 171
S-100 family, 151
Shh, 82, 98
sonic hedgehog, 127
TGase, 166-167
TGF, 12, 21, 82, 86, 270
Wnt, 86
Glioma (*see also* Tumor)
anaplastic astrocytoma, 232
glioblastoma multiforme, 232
therapy, 232-235
Globin levels, 103
Glucocorticoid receptor (GR), 58
Glycosidases, 207
Granulocyte macrophage colony stimulating factor (GM-CSF), 169
Growth factor, production, 200
GTP, 148

Heart, 51, 87, 88, 119
Heat shock proteins, 142
Henson's node, 118
Heparanase, 207
Heparin, 212
Heptocytes, 39
Herpes simplex virus (HSV-TK), 250
Heterodimers, 52, 54, 57, 62
RAR-RXR, 52, 56, 59
RXR, 53, 58
Hindbrain, 85, 99
Homeostasis, embryogenesis, 161
Homeostatic regulation, 163
Homodimers, 172
RXR, 52-55, 59, 62
Hormone receptors
dimers, 51
steroid and thyroid, 52
Hormone response elements, 54, 172
HPLC analysis, 106, 115, 116
4-HPR, N-(4-hydroxylphenyl)amide, 14, 19, 162, 165, 169-171
Hyaluronate, 209, 211
Hydrophobic residues, 23,13

ICAM-1, 209, 210
Immune response, 211
surveilance, 210
Inositol triphosphate, IP3, 146
Integrins, signal transduction, 144-146
alpha3 beta1, 143
alpha$_{IIb}$ *beta3,* 146
beta2, 209
receptors, 142
role of, 143-146
Intercellular adhesion molecule-1 (ICAM-1), 151
Interferon, 203, 211, 241, 251, 279
therapy, 250
Interleukin-2, 211
Intestinal mucosa, 42
Intracranial pressure, 253
Ionizable groups, 12
IP3, inositol triphosphate, 146, 150
Isoforms, 49
Isomerization, 5, 55

Ketoconazole, 46
Kinases, 148, 236
Krox-20, 86, 95, 104
enhancer, 49
Krox-24, 42

Laminin, 203, 206, 207
LDG1069, 24
LDL receptors, 40

Lecithin-retinol acetyl transferase (LRAT), 45
Lens agenesis, 88
Lesions, 201
Leukemia cells, 21, 151, 221
  acute promylocytic, 56, 38, 88, 222
  myelocytic, 58
  promyelocytic, 198, 199, 201
LFA-1, 209, 210
Limb
  supernumeracy, 86
  bud developement, 84, 94, 125-129
  regeneration, 94, 110-117, 235
Lipoproteins, low density, 40
Lipoxgenase activity, 47
Liver, 39, 51, 81, 87
LRAT, lecithin-retinol acetyl transferase, 45
Lung cancer, 223-224, 261-281
  biology of retinoids and, 269-272
  c-myc insertion, 272
  carotenoids, 262
  clinical studies, 264-266, 272-281
  endpoint, primary, 276-277
  epidemiological data, 266
  etretinate, 274
  intermediate endpoints 273-276
  laboratory studies, 272
  large cell, 267
  Lewis cell line, 271
  non-small cell, 271
  prevention, 262, 273-277
  serum levels, 265
  treatment, 278-281
  v-Ha-ras insertion, 272
Lungs, vitamin A defficiency, 87
Lymphocytes, 170
Lymphokine killer, 211

Mammalian development, 81
Matrilysin (MMP-7), 205
Matrix metallaproteinases (MMPs), 204
Melanin, differentiation marker, 202
Melanoma, 60, 201, 207
Membrane permiability, 60
  fluidity, 60
  plasma, 60
Metabolites (*see* Retinoid metabolism)
Metalloproteinase, 203
  inhibitors, 234
Metastasis, 211
Microcephaly, 87
Micropthalmia, 87
Microtubule polymerizatin, 145
Mitogen-activated protein kinase (MAPK), 147
Monomers, RXR, 55
Morphogen, defined, 36
Morphogenesis, 51
Mouse
  curly tail, 89
  embryos, 86
  tumorigenesis, 44
  embryogenesis, 80
Myoblastin, 151

Napthalenyl benzoic acid, 21-26
N-CoR, 57
Neurite outgrowth, 144
Neuroblasoma, 55, 170, 201, 202
Newt, 94
NGFI-B, 53, 62
N-(4-hydroxylphenyl)amide, 14, 162 (*see also* 4-HPR)
Night blindness, 87
Nitric oxide, 235
Nonreceptor kinases, 142
Nuclear receptors, 9, 85-86, 108-110, 171-177
  activators, 57
  access of retinoic acid to, 84-85
  LXR, 53, 62
  NURR-1, 53, 62
  TR , 53
Null mutation, 87

Oncogenesis, 38
β-Oxidation, 46, 62

P450 isoenzymes, 46, 61, 238
  p53, 271
Palindromic RARE, 52
Papillomas, 198
PDGF, 147
Peroxisome proliferator-activated receptors (PPAR), 49
Phorbol ester, 144, 204, 210
Phospholipase C (PLC), 149
Phosphoprotein, p53, 151
Phosphotyrosine kinase, 236
Placenta, 80, 81, 85
Plasmin, 204-206
Platelet derived growth factor (PDGF), 204
PLC-gamma, 149
Polar groups, 12
Polarity-specific receptors, 57
Post-natal lethality, 50
Pregnancy, 80, 87
Preinitiation complex, 54, 56
Promoters, 150
Prostoglandin H synthase, 46, 47
Protease inhibitors, 204, 207, 213
Proteins (*see also* Retinol binding protein)
  $\alpha_5\beta_1$, 143
  CBP, 51, 58
  CD44, 209
  E1a-like, 51
  fusion protein, PML-RARα, 58
  G, 146
  *hsp*28, 151,152
  retinoylation of, 6

Raf activation, 148
RA synthesis, endogenous retinoids, 98-100
RA (*see* Retinoic acid)
RAR (*see* Retinoic acid receptor)
RARE (*see* Retanoic acid, response element)
Ras, activation, 148
RBP (*see* Retinol binding proteins)
Regeneration, endogenous retinoid, 115-116
  tail, 112
Respiratory system, 84
Response elements, 52, 57
Retina, 84, 88
Retinal, 2
  9-*cis,* 46
  11-*cis*, 9, 37
  dehydrogenase, 40, 45, 46, 83
  oxidation, 46
  pigment epithelium (RPE), 39
  reductase, 40
Retinoblastoma protein, 271
Retinoic acid (RA)
  all-*trans*, 36, 46, 59, 61, 104, 198
  analogues, 13-26
  catabolism, 238-241
  cellular binding proteins, 42
  9-*cis*, 24, 36, 37, 40, 96
  13-*cis*, 8, 38, 43
  clinical application, 241-245
  conjugation reactions, 6
  derivatives, 235
  developmental function, 60-63
  3,4-didehydro, 10
  13,14-dihydro-9-*cis*, 46
  excentric cleavage, 45
  4-hydroxy, 46
  isomers, 46, 60
  lizarole and, 241-246
  mechanisms of action, 59-60
  metabolism, 44-47
  modification (*see* Retinoid analogues)
  oil drop injection, 104
  oral derivatives, 237-238
  oxidation, 5, 9-11
  oxidizing retinol to, 4

4-oxo, 43, 46, 49, 50, 53, 54, 56, 61, 62, 198
response element (RARE), 42, 48, 50-51, 59, 85, 99
signaling, 84-89, 105
structure and formation, 2-5
syndrome, 245-246, 253
synthesis, 45, 82-84
transcription factors, 149
treatment, duration, 242
vitamin A, 4
Retinoic acid receptor (RAR), 2, 9, 13, 37, 40, 47-59, 100 (*see also* Nuclear receptors)
antagonists, 24-26
dependent mechanisms, 171-184
development function, 50-52
dimerization, 52-54
EGF, 149
expression, 50
families of, 85-86,
functional redundancy, 51
half-site response element, 52
interacting proteins, 57
isoforms, 48
specific ligands, 58-59
steroid superfamily, 47
subtypes, 23, 149
transcriptional regulation, 54-58
Retinoid analogues, 13-26
amide derivatives, 14
Retinoid metabolism, 2, 5-6, 44, 46
conjugation, 5-13
5,6-epoxy, 46
4-hydroxy-RA, 10
isomerization, 6, 8-9
oxidation, 6, 9-11
4-oxo-RA, 10
Retinoids (*see also* RXR)
active derivatives, 1-26
activity *in vitro*, 266-267
activity *in vivo*, 267-269
additives and, 279-280
adjuvant therapy, 277-278
advanced disease in, 278
antiinvasive/ antimetastatic effects, 201
apoptosis and (*see* Apoptosis)
cancer, 219 (*see also* Cancer)
cellular binding proteins, 41
cellular effects, 113-115
defects, 87-88
embryonic/mammilian development, 80-89
extracellular matrix synthesis, 211
formation, 2-5
growth factor activities, 142-152, 211-213
in patterning, 93-139
isomerization, 5
molecular mechanisms of, 35-63
radiotherapy, 246-249
receptors (*see* Retinoic acid receptors)
signal transduction, heat shock proteins, 151-152
storage, 4
structure, 1-26
synergistic activity, 279-280
synthetic 2, 21-26, 182
theraputic effects, 201
Retinol (*see also* Retinol binding protein, Vitamin A)
dehydrogenase, 40, 45, 83
cystolic NADP dependent, 45
microsomal NADP-dependent (RoDH), 45
metabolism, 41, 44-47
micro-encapsulated, 251
reduction, 2
retro derivatives, 11, 41
vitamin A, 6
Retinol binding proteins (RPB) 4, 42-44
metabolism and, 44-47
embryonic synthesis, 81-83
Retinotaurine, 12

Retinoyl-β-glucuronide, 12
Retinoyl-Co-A, 12
Retinoylation, 12, 60
Retinyl esters, 3, 4, 38, 45, 81, 82
Retinyl palminate, 245
  CoA, 6
Rhobomere-specific, 43
Riboprobe, 109
RNA polymerase II, 54
RoDH I, 45
RPB (*see* Retinol, binding protein)
RPE, Retinal pigment epithelium, 39
RXR, retinoid X receptors, 23- 24, 37, 40, 42-59, 221-226 (*see* also Retinoic acid receptors)
  apoptosis and, 162 (*see also* Apoptosis)
  developmental function of, 50-52
  isotypes, 85

Serine proteinase, 151
Signal transduction
  cytoskeletal transduction, 146-148
  growth factor, 148
  integrin mediated, 143-146
Skin, 164
  patterning, 120
SMRT, Silencing mediator for retinoic acid and thyroid hormone receptors, 57
Son of sevenless (Sos), 148
Sp1 response elements, 42
  transcription factor, 49
Sputum cytology, 275
Squamous metaplasia, 51
Src homology domains (SH), 146
Sterility, male, 51
Stromelysin 3, 204, 205, 270
SUGI, 57
Syndactyly, 88

T3 receptor-asscociating cofactors (TRACs), 57
Tail bud, 121
TATA box, 42, 48-49
  binding protein, 54
Taurine conjugation, 62
TBP, 57
  associated factors (TAFs), 54
T-cell
  activation, 41
  hybridomas, 169
  receptor, 168
Tensin, 149
Teratogens, 17, 52
  effects, 51
  potential, 24,26
Terminal deoxnucleotidyl transferase (TdT), 164
Testis generation ,50
Tetramers, RXR, 55
Tetrogenicity, 36
TFII, 54, 57
Thrombospondins (TSP), 144
Thyroid hormone receptors (TR), 49
TIFI-1, 57
Tissue inhibitors of metallaproteinases, 204
T lymphocites, 209, 211
TPA tumor promoter, 143, 147, 205
TRACs, T3 receptor-asscociating cofactors, 57
Transcription factors, 54, 58, 103
  retinoic acid, 149
Transcriptional activation, 55, 57
  induction, 59
  repression, 57- 59
Transgenic mice, 44, 50
Transglutaminase, 166
Transient transfection, 56, 58
Transthyretin, 39
Trunk, 84
TTNPB (*see* Ethyl TTNPB)
Tubulin cytoskeleton, 145
  polymerisn, 147

Tumor cell
  adhesion, 197, 207-210
  angiogenesis, 202, 203
  apoptosis, 202
  CNS and, 232-233, 252-253
  cycloreductive therapy, 246
  gap junctional communication, 210
  growth factor 200-202
  host immune response to, 210
  induction, 268
  integrins and, 207
  invasion and metastasis, 142, 197, 200, 203-207
  medication, 250-251
  necrosis factor (TNF), 204, 211
  neuro-ectodermal, 236
  pancreatic tumourigenesis, 44
  patient characteristics, 240
  retinoid mediation and, 198-200
  stroma, 211
  suppressor activity, 51
Type IV collagen, 203- 204, 207, 208

Ultraspiracle, 85
Urogenital tract, 87

Vascular cell adhesion molecule-1 (VCAM-1), 144
Ventral retina anomaly, 88
Vertebrae, 88
Vinculin, 145
Vitamin A, 36-38 (*see also* Retinol)
  defficiency, 51, 84, 86-88, 201
  palmitate, 268
  cancer, 198
  maternal deficiency, 80-82
  main sources, 2-4
  synthtic analogues, 161-162
Vitamin D receptors (VDR), 49, 172
Vitamin D_3, 251
  1,25-dihydroxy, 203
Vitronectin, 208

Xenopus, 62, 94, 101-110
  all-*trans*-RA detection in, 105-108

Zebrafish, 25, 50, 57, 94, 100, 104, 184
Zinc finger, 52, 172
ZPA, Zone of polarizing activity, 36